The IACUC Administrator's Guide to Animal Program Management

The IACUC Administrator's Guide to Animal Program Management

William G. Greer
Ron E. Banks

CRC Press
Taylor & Francis Group
Boca Raton London New York

CRC Press is an imprint of the
Taylor & Francis Group, an **informa** business

CRC Press
Taylor & Francis Group
6000 Broken Sound Parkway NW, Suite 300
Boca Raton, FL 33487-2742

First issued in paperback 2020

© 2016 by Taylor & Francis Group, LLC
CRC Press is an imprint of Taylor & Francis Group, an Informa business

No claim to original U.S. Government works

Version Date: 20151005

ISBN 13: 978-0-367-57502-1 (pbk)
ISBN 13: 978-1-4398-4905-7 (hbk)

Library of Congress Cataloging-in-Publication Data

Greer, William G., 1961- , author.
 The IACUC administrator's guide to animal program management / William G. Greer and Ron E. Banks.
 p. ; cm.
 Includes bibliographical references and index.
 ISBN 978-1-4398-4905-7 (alk. paper)
 I. Banks, Ron E., author. II. Title.
 [DNLM: 1. Animal Care Committees--organization & administration--United States. 2. Animals, Laboratory--United States. QY 54]

 SF407
 636.088'50973--dc23   2015033140

Visit the Taylor & Francis Web site at
http://www.taylorandfrancis.com

and the CRC Press Web site at
http://www.crcpress.com

Contents

Preface

The purpose of the *IACUC Administrator's Handbook* is to assist the IACUC administrator charged with developing, managing, and overseeing a program of animal care and animal use. There are more than sufficient regulations and policies governing animal care and use and for very good reasons. These federal standards must be applicable to all circumstances and situations. They tend to be a collection of expectations and goals that provide organizations the flexibility to develop practices for achievement. The regulatory expectations may be clear, but the manner of developing the operations and practices to assure compliance while ensuring animal welfare is often not prescribed. In this reference, the authors provide many options and possibilities for specific operational practices (e.g., how to build a well-functioning IACUC, what a functional protocol template looks like) to satisfy the regulatory requirements.

The material provided in this handbook is a compilation of several years of best practices meetings among IACUC administrators across the country. Best practices meeting attendees from private, public, governmental, and academic organizations have helped to shape and develop the information offered in the *IACUC Administrator's Handbook*. It is through the insight of several hundred colleagues— their successes as well as their failures—that the authors have distilled suggestions and considerations for your local animal care and use program.

There is no desire to replace or modify any existing regulatory document. It is reasonable and appropriate for regulations and policies to evolve. In concert with the maturing framework of animal care and use programs, the operational practices must also transform to best support the organization, guide the research community, and assure animal well-being in all activities involving sentient beings.

This work is not intended to replace other useful references or manuals regarding programmatic function but rather to complement them. The primary difference the reader may find in this handbook is the transparent and open nature of describing processes that have been time tested and proven to satisfy the regulatory requirements. There is no intent that organizations model their programs according to this book. The desire is that IACUC administrators can refer to and assimilate the wisdom and experience distilled from hundreds of colleagues into the foundations of their own programs.

The authors had specific goals in preparing this handbook:

1. Assuring animal well-being and welfare by building flexible and self-correcting programs of animal care and use
2. Combating the ignorance of necessary functional activities such as assuring regulatory compliance while ensuring animal welfare
3. Providing answers to common questions of individuals assigned animal program duties
4. Creating a sense of transparency of what has worked and what has not worked among organizations

5. Developing a community of IACUC administrator professionals who become recognized as core and critical members of the research efforts of the organization, and who take a seat at the program leadership table along with the institutional official, the attending veterinarian, the IACUC, and the researcher

The examples and suggestions in this handbook may be used to help guide foundational development of a comprehensive animal care and use program. IACUC administrators must always be cognizant that—as noted in the National Institutes of Health's *Guide for the Care and Use of Laboratory Animals*, eighth edition (referred to as the *Guide*)—the concept and application of performance standards, in accordance with goals, outcomes, and considerations defined in the *Guide*, is essential to this process.

It is with deep gratitude that the authors recognize the hundreds of IACUC administrators who selflessly serve their organizations and researchers. It is this same community that has provided the bulk of this handbook's good ideas, although all situations are attributed to that fanciful organization, "Great Eastern University," where everything that can go wrong has and anything that can work right has not.

Let those eight last words of a dysfunctional or necrotic program of animal care and use never be said of your home organization: "We have never done it that way before." It is our hope that every reader finds one to several new or novel ideas that will be applicable to his or her program, support strong research, improve suboptimal conditions, and correct noncompliant behavior. May each of us review our practices and, from the reading of this book, utter the eight exciting words of new beginnings: "Our practices are not sufficient for today's environment."

Best wishes for the benefit of the animals, the researchers, the organizations.

Bill Greer
Penn State University

Ron Banks
Duke University

Introduction

The National Institutes of Health (NIH) Office of Laboratory Animal Welfare (OLAW) is pleased to endorse this volume of best practices for operation of animal care and use programs (ACUP) at Public Health Service (PHS)-assured institutions. But more than this static collection of operational practices, we support your practice of coming together to share your methods of operating programs at your institutions—large and small, academic, government, nonprofit, and for-profit organizations.

The PHS system of animal activities oversight is based on the concept of self-monitoring and self-reporting by a local institutional committee, the Institutional Animal Care and Use Committee (IACUC). IACUC administrators have a significant impact on efficient, compliant IACUC operation.

ACUPs are ever-evolving. Research and technology change as investigators use new and increasingly sophisticated animal models. Policy and regulation interpretations evolve, too, as OLAW, the United States Department of Agriculture (USDA), Animal and Plant Health Inspection Service (APHIS), Animal Care (AC), and the Association for Assessment and Accreditation of Laboratory Animal Care International (AAALAC) refine their understanding of the needs of both animals and biomedical researchers.

ACUPs are complex operations. Changes to one component influence other aspects of the program. All of the practices described in this document are fully compliant with OLAW's expectations, provided the rest of the program supports the practice.

The take-away message from this comprehensive volume is that there are many ways to operate an ACUP that provides support for the highest quality biomedical research in a context of humane animal care. When you, as IACUC administrators, come together to share your institutional practices and learn from one another, you are moving animal welfare and biomedical research forward in a solid, responsible, practical way.

I hope that readers of this book will consider attending Best Practice Meetings to add their voice and expertise to the collective knowledge that this forward-thinking group continues to develop.

Susan Brust Silk, MS
Director, Division of Policy and Education,
NIH Office of Laboratory Animal Welfare

Acknowledgments

We would like to express our sincere thanks to volunteer research administrators who devoted their time and knowledge to pre-reviewing some of the materials contained in this document:

Carolyn A. Berger, University of Michigan
Astrid Haakonstad, University of Michigan
Julie Laundree, University of Michigan
Joanna Lyons, The Pennsylvania State University
Alison D. Pohl, University of Connecticut Health Center
Susan Brust Silk, the Office for Laboratory Animal Welfare
Eileen Morgan, the Office of Laboratory Animal Welfare

Authors

William G. Greer. Bill received his BS degree in microbiology from The Pennsylvania State University and is currently earning an MS degree in adult education. After receiving his BS degree, he worked as a research technician for an animal vaccine production company, Intervet Inc. (formerly Tri Bio Laboratories), served as the production manager, and was responsible for satisfying the regulatory requirements associated with the development and production of animal vaccines. In 2002, he was hired by Penn State University to oversee the Institutional Animal Care and Use Committee (IACUC), Institutional Biosafety Committee (IBC), and the University Isotope Committee (UIC). He currently serves as the associate director in the Office for Research Protections at Penn State. Bill is a certified IACUC Professional Administrator, Laboratory Animal Technician, and hazardous materials shipping professional. He established the IACUC Administrators Best Practices Meetings and led the team that formalized the IACUC Administrator Association (IAA). He also established the Committee on Institutional Cooperation IACUC Administrators' working group (Big Ten Universities). In 2011, Bill was awarded the Penn State Vice President for Research Outstanding Leadership Award, and he currently serves as the institution's Biosafety Committee Chairman. Bill continues to serve as an ad hoc consultant to the AAALAC Council and on the Council for Certified Professional IACUC Administrators and as the IAA President and Chairman of the Board of Directors. He has over 25 years of experience in research compliance management and oversight.

Ron E. Banks, DVM. Receiving his veterinary degree from Auburn University's School of Veterinary Medicine, Dr. Banks served 34 years in the United States Army Veterinary Corps and is currently the director of the Office of Animal Welfare Assurance at Duke University, Durham, North Carolina. Along his professional journey, he has contributed to such bodies as AAALAC's Council on Accreditation (retiring from the council as a Council Member Emeritus in 2011); he was a Charter Diplomate of the American College of Animal Welfare and a Charter Member of the IACUC Administrator's Association. He is a Fellow of the National Academies of Practice, and board-certified with the American College of Laboratory Animal Medicine, the American College of Veterinary Preventive Medicine, and the American College of Animal Welfare. Along with Bill Greer, he has facilitated Best Practice meetings since its inception in 2005 and continues to serve as a member of the Board of Directors of the IACUC Administrators Association.

Acronyms and Common Terms

AAALAC	Association for the Assessment and Accreditation of Laboratory Animal Care International
AALAS	American Association for Laboratory Animal Science
Administrator	Animal Care and Use Program Administrator
Ag Guide	The *Guide for Care and Use of Agricultural Animals in Research and Teaching*
ARP	Animal Resource Program
AV	Attending veterinarian
AWAR	Animal Welfare Act Regulations
BP	Best Practice
BP Meeting	IACUC Administrators Best Practice Meeting
CEO	Chief executive officer
CPIA	Certified Professional IACUC Administrator
CRO	Contract research organization
DAR	Department of Animal Resources
DMR	Designated member review
EHS	Environmental Health and Safety
FCR	Full committee review
FDA	Food and Drug Administration
GEU	Great Eastern University
Guide	The *Guide for Care and Use of Laboratory Animals*
GWU	Great Western University
IAA	IACUC Administrators Association
IACUC	Institutional Animal Care and Use Committee
IO	Institutional Official
NIH	National Institutes of Health
OHSP	Occupational Health and Safety Program
OSHA	Occupational Health and Safety Administration
PAM	Post approval monitoring
PHS	Public Health Service
PHS Policy	Public Health Service Policy on Humane Care and Use of Laboratory Animals
PI	Principal investigator
POC	Point of contact
PPE	Personal protection equipment
PRIM&R	Public Responsibility in Medicine and Research
SOP	Standard operating procedure
USDA	United States Department of Agriculture
VA	Veterans Administration

CHAPTER 1

An Introduction to Best Practices Meetings

INTRODUCTION

National regulations, policy requirements, and guidelines apply to animal care and use programs at a broad spectrum of organizations across the nation, including the academic, private for-profit and non-profit sectors, and the public sector. Although the requirements are common to all, organizations employ diverse practices in their animal care and use programs to enable them to meet the requirements while also fulfilling the goals of their specific programs. This wealth of practices is based on many program needs including the species used, procedures performed, or the research goals of the organization. However, this diversity can create challenges to those responsible for conducting (administering, managing) the programs—Institutional Animal Care and Use Committee (IACUC) administrators, facility managers, and attending veterinarians (AVs). These individuals are all charged with the successful operation of the organizations' animal care and use programs and must determine: What is the right way to perform a practice? What is the best practice? What is a common practice?

For example, regulations and policies require humane end points for animals that, when reached, necessitate an action. The animal(s) must be removed from the study or a procedure must be undertaken to alleviate the pain and distress of the animal(s). Many organizations have established programs with well-defined and effective policies on humane procedural end points, while others have not considered the issue and will await a troublesome situation before establishing a process or effecting a solution. In some cases, there are variations in selected end points among organizations, which can lead to variable outcomes. Some organizations may choose a conservative approach and prematurely remove animals from a study, thus assuring that no animal reaches a humane end point. This approach may compromise the research by preventing the accumulation of data that would occur just prior to reaching an end point. Conversely, other programs using a less stringent approach may allow animals to achieve or exceed the humane end point, and in doing so may inadvertently allow some animals to experience pain or distress beyond that approved by the IACUC. Assuming the IACUC has appropriately selected end points, animals should not exceed defined end points and thus not experience unnecessary unrelieved pain and distress. If an IACUC humane end point is exceeded, this operation is not compliant with the committee-approved animal activity. If the IACUC has approved an end point that exceeds the humane end point, without scientific justification, then the IACUC is noncompliant. This diversity in procedures will lead to varying outcomes in identical or similar studies.

Consequently, what options are available to the IACUC administrator to influence the institutional leaders and IACUC members to select appropriate practices that facilitate good science while providing humane animal care and use that is compliant with regulatory guidance?

HISTORY OF BEST PRACTICES MEETINGS

In 2005, approximately fifty IACUC administrative staff members held a meeting to discuss operations practices of their animal care and use programs. The goal of this meeting was to explore the similarities and differences of programs developed under a single regulatory climate.

Attendees discovered that there were as many programmatic differences as there were attendees, and while most differences were subtle, there were also significant variations in process and function. Attendees appreciated and benefited from the creativity and innovation used by their colleagues to solve operation problems at the represented organizations.

The IACUC administrators had begun the process of peer education—a process of using the experiences of others to enhance their own understanding and knowledge of tools and processes that could assist their organizations in achieving "best practices" (BP) for given situations. After that initial meeting, many of the IACUC administrators continued networking. They soon concluded that a means to share their knowledge and a method to educate their colleagues could result in an enhanced consistency among the programs they managed. The IACUC administrators also expressed a strong desire to develop mutually supportive and professionally beneficial relationships with their peers.

This concept of education and networking remains the driving force for IACUC administrators who have developed a process for sharing successful ideas and effective resolutions for institutional program procedures. The BP meetings are the result of those 2005 discussions. That BP meeting and subsequent BP meetings have provided IACUC administrators a secure forum to discover workable, practical solutions for overcoming challenges and a mechanism to share their successes. IACUC administrators who have struggled to find solutions have used the BP meetings to enlist assistance and seek options that would benefit their organization.

The ability to openly share concerns in a secure environment has benefitted many animal care and use programs. The BP meetings have provided the forum to creatively network, share resources, and facilitate peer-to-peer counseling. The BP meetings have encouraged IACUC administrators to meet challenges and have provided opportunities to engage in role-playing activities which assist individual assessment and professional development for the overall benefit of the programs in which these administrators serve.

IACUC administrators from the United States and Canada have continued to hold annual best practices meetings since 2005. Through 2010, more than three hundred IACUC administrators have regularly attended the annual BP meeting. Approximately 55 percent of BP meeting attendees have come from academia, 35 percent from private industry, and roughly 10 percent from US government agencies and Canadian organizations. All initial meetings were held in State College, Pennsylvania, near The Pennsylvania State University. There was a sense of community, which came with a single location removed from the regular circuit of national meetings. The meeting format was casual: a roundtable discussion with sixty administrative colleagues from about fifty organizations. Having a maximum of sixty attendees facilitated small group discussion and interaction between attendees. For 2 days each year, IACUC administrators discussed their issues and challenges with similarly minded colleagues. Each BP meeting agenda was determined by the attendees prior to meeting. This format ensured that the regulatory issues important to the attendees were discussed at that meeting. At each BP meeting, a rotating roster of attendees served as the presenters, trainers, facilitators, and trainees—the ultimate peer-instruction process.

Regulatory oversight and program accreditation representatives also participated in BP meetings. On day 1, representatives from the Office of Laboratory Animal Welfare (OLAW), the US Department of Agriculture (USDA), and the Association for Assessment and Accreditation of Laboratory Animal Care (AAALAC) attended. These agency representatives listened to administrators' concerns and contributed to the development of compliant best practices suggestions; they also provided regulatory or accreditation points of view when necessary. On day 2, only the AAALAC representative participated. This allowed attendees an additional opportunity to ask questions regarding the best practices for accredited organizations. Each BP meeting was filled with discussions of efficient methods for conducting everyday work practices. The discussions did not focus specifically on regulations, but rather on methodologies used to satisfy the requirements of the animal welfare laws and policies.

As the value of BP meetings became known to the community, individual registration requests increased. Organizers determined that the small size of the meeting (maximum of sixty attendees) was a necessary component of the successful format. Therefore, a second meeting was added in 2008; there were now two BP meetings each year. Each maintained the same structure and format. The agenda was driven by the attendees, so each meeting was unique and all were successful. The group continued to meet in State College, Pennsylvania, where the cost was reasonable, the mechanics of conducting the meeting were well developed, and the attendees were comfortable in the location.

The news continued to spread and more organizations desired to participate in the process of peer education and networking. In 2010, the BP meeting organizers decided to conduct a meeting at a location other than Penn State University. This expansion was intended to make the meeting accessible to additional organizations. The first meeting not held at Penn State was in Durham, North Carolina.

The Durham meeting was slightly different from the previous meetings. Most attendees were local, which meant there were fewer after-hours and supper conversations. As in all previous meetings, the agenda was determined by the attendees. As most attendees were from the Southeast, the agenda was focused on the interests of organizations in that location. However, the same small-group-with-discussion format was maintained and another successful outcome was achieved.

The BP Meeting Format

To facilitate discussions at BP meetings, program participants volunteered to lead a discussion on a topic of interest (e.g., occupational health and safety programs). They prepared 5–10 minutes of PowerPoint slides on a challenge they encountered. In some cases, the presenter also reported on how his or her organization resolved the issue. The brief background presentation was followed by an open discussion involving all participants; each contributed institutional insight, provided guidance, and on occasion suggested appropriate programmatic corrective actions.

Attendees have routinely noted that lessons learned from BP meetings have had positive impacts upon institutional processes. The brainstorming efforts of attendees have resulted in the development of common practices—for example, for methods of ensuring protocol and grant congruency, operating effective pre-review processes, more accurate procedures for tracking animals used in research, and options for assuring compliant institutional responsibilities while collaborating with peers.

Limitations of the Meeting Format

In 2011, approximately 50 percent of attendees were return participants—a strong indicator of the value and benefit of this process. However, it is also a limiting factor that prevents inclusion of new IACUC administrators in the meetings. Several dozen individuals were not able to attend the 2009 best practices meeting due to the space limitations.

During that meeting, BP meeting attendees discussed concerns with space limitations and the developing library of valuable resources, and they determined that there were two actions necessary for the good of the IACUC administrator community: (1) Develop a professional educational association to facilitate continued networking and resource sharing, and (2) provide more opportunities for IACUC administrators to attend BP meetings.

IACUC ADMINISTRATORS' ASSOCIATION

At the 2009 BP meeting, the attendees discussed the potential of an association. BP meeting attendees decided that the formation of an organization focused on IACUC administrators would provide a network to foster communication among them, offer sharing of resources such as templates and forms, and provide a sense of identity to a group of individuals who are not veterinarians (who are represented by the American College of Laboratory Animal Medicine [ACLAM]), animal care staff (who have the American Association for Laboratory Animal Science [AALAS] as their core educational entity), or scientists (who have scientific societies where they share their knowledge and skills). BP meeting attendees believed that forming a professional organization would generate new educational and training opportunities and provide a professional organization for this highly specialized group of professionals. Initiatives suggested by attendees included hosting a website where members could find resources for program management, successful standard operating procedure (SOP) templates, useful institutional policies, and various protocol templates. Attendees were excited about publishing an e-newsletter in which members could describe specific programmatic successes at their organizations. An association would serve as the repository for reports of prior BP meetings, as a clearinghouse for webinars specifically designed for IACUC administrators, and as a technological hub for connecting IACUC administrators.

The 2009 BP meeting attendees voted to establish an organizing committee to explore the logical and natural maturation of the concept of the IACUC administrators' association. A nationwide survey of over two hundred previous BP meeting attendees was performed. The survey provided a 40 percent return rate and, of those, over 90 percent indicated an interest in and desire to join such an organization. Several dozen respondents indicated a desire to assist in the development of an association. So, the organizing committee elected by the 2009 BP meeting attendees developed several advisory groups from those indicating an interest, and all began the process of exploring the options and opportunities for an IACUC administrators' organization.

In 2010, the organizing committee members reported to the BP meeting attendees that they recommended the formation of an IACUC administrators' association. Included in the recommendation were a set of bylaws, a mission statement, and general procedures. Subsequently, the IACUC administrators' association was established as a 501-C3 (not for profit and tax exempt) educational association to foster cost-effective opportunities for mentoring and educating IACUC administrators. The association's target community is individuals who are administering animal programs—those who are "in the trenches" of program management and program oversight. While all organizations are welcome, the association maintains a keen interest in those smaller organizations that do not have as many resources or as deep a knowledge pool as larger organizations generally enjoy. The association encourages accreditation with AAALAC, certified professional IACUC administrator (CPIA) certification through Public Responsibility in Medicine and Research (PRIM&R), and engagement with AALAS.

BP MEETINGS

One of the most important advantages of establishing a formal association was the means to develop an organizational foundation for hosting BP meetings three to four times a year and to provide the opportunity to attend BP meetings in different regions of the country. Such an endeavor would ensure the continuation of excellent peer mentoring and strengthen the sharing, teaching, and training efforts of professional IACUC administrators across the nation.

While the effort to expand BP meetings continues, this reference book was written as a means of collating the available information and knowledge from the BP meetings held to date. The authors have extracted the content from the pages and pages of notes captured from various situations and scenarios presented at past BP meetings. This information is provided to you, our colleagues, to enable you to enhance your program of animal care and use. The scenarios are all based upon mythological organizations where everything wrong commonly occurs—Great Eastern University (GEU); Great Western University (GWU); and Great Central University (GCU). You will find that aspects of many scenarios apply to your own organization! That's the beauty of the process. We can discuss the troubles of others and take the benefits of their experiences home to improve, enhance, or modify our own practices.

In all cases, the recommendations and conclusions are intended to be fully compliant with federal law, regulation, and policy. The reader is encouraged to adopt any of the conclusions or suggestions identified in this document, if such adoption would enhance programmatic outcomes, improve animal well-being, or augment the foundation for reliable and successful research or teaching outcomes.

CHAPTER 2

The IACUC Administrator's Office Structure

SCENARIO

Great Eastern University (GEU) has grown to a size at which the old processes no longer work as well as they used to. The organization wishes to develop an office structure that meets the needs of the research community, is fully compliant with regulatory expectation, and has no conflicts of interest regarding the various aspects of the program.

GEU's animal resource program (ARP) director, who is also the attending veterinarian (AV) and serves on the IACUC, has administratively overseen the IACUC for the past 15 years. During a recent meeting, the institutional official (IO) suggested establishing a separate office to manage research support, protocol compliance, and administrative support to the IACUC. The AV did not favor the change and adamantly justified the need to manage the program since he is responsible for ensuring the health and welfare of the research and teaching animals.

The IO remarked that continuing the current arrangement was like "the fox watching the henhouse." He noted the changes in the program over several years and expressed concern that public perception of the reporting structure would be that the veterinary staff members control the program, which was designed to ensure that the organization's veterinary care program satisfied the federal regulations and policies governing animal use. The IO's conclusion was that GEU would establish a research compliance office to guarantee positive public perception and to strengthen the organization's overall program of animal care and use.

Discussions at past best practices (BP) meetings identified two common office structures. In some cases, organizations have the administrative office reporting to the ARP director, and in other situations a compliance office is established that often reports directly to a senior level administrative leader—for example, the office for the vice president of research. But in most cases, organizations are shifting away from the IACUC administrator reporting to the AV, in large part believing there is an inherent conflict in such relationships.

OFFICE MODELS

Attendees at BP meetings have outlined at least two different office models being effectively used by organizations.

Research Compliance Office

A current trend in research administration is for organizations to establish an office charged with the specific task of ensuring the organization is compliant with the federal laws and policies governing the use of animals in research, teaching, and testing. The animal care team typically comprises experts in animal care and use (e.g., veterinarians, veterinary technicians, certified animal care technicians, and other individuals with expertise in animal care and use). A research compliance team is frequently composed of staff members with expertise in the federal mandates, policies, and, sometimes, accreditation requirements. A survey of IACUC administrators (approximately 135 respondents) conducted during BP meetings suggested that as the number of annual submissions to the IACUC increases at an organization, the likelihood that those organizations will establish a research compliance office increases. For example, the data suggested that 85 percent of the organizations processing more than four hundred submissions a year established a research compliance office to manage the organization's compliance issues. Frequently an organization's compliance office becomes a department under the office of the vice president for research.

Administration through the Director of the ARP

A second popular office structure is to have IACUC administration report to the director of the ARP. Under this model, those administering the daily activities of the IACUC program and those responsible for the animal care program report to the same individual. The primary advantage of this process is attributed to program efficiency. It allows a management unit to care for the animals and address compliance concerns directly without the need to communicate to another office, but rather only inform the IACUC.

This model has wide acceptance. However, the perception exists that when the AV oversees the program of animal care and use and also manages its compliance, there is the lack of a check-and-balance system, which can compromise program integrity. Survey data suggest the current trend is for the AV. to oversee the program of veterinary care, with an IACUC administrator managing the overall animal care and use program compliance. The IACUC is the connect between the two individuals.

In this model, the compliance and veterinary care portions of GEU's animal care and use program are overseen by the ARP director. As a result, the director not only supervises veterinarians caring for the animals but also provides regulatory guidance to the IACUC, which is ultimately responsible for confirming program compliance. In this case, the strength of validating program compliance by an independent group—a separate compliance office—is missing.

For example, consider the following scenario: During a semiannual facility inspection, an IACUC member identified a cat with bandages on a front paw. As a result the committee member asked to review the animal's health records. The records indicated that the animal's foot was bandaged because it was accidently caught in the cage door. In addition, the records indicated the animal was observed daily by veterinary staff members, but that the bandages had not been changed or the lesion hadn't been checked for the past 7 days. In the process, the veterinary technician discovered the injury had progressed into a severe infection, which resulted in the need to euthanize the animal.

The ARP director brought the matter to the attention of the IACUC. During discussions with the IACUC, the ARP director stated that from a regulatory standpoint an animal experiencing health issues should receive direct and regular veterinary care until it recovers from the problem. She noted that the cat in this case was observed daily by veterinary staff members. IACUC members expressed concern that the care the animal received was inadequate and that the issue should be identified as a deficiency in the veterinary care program and must be reported to the Office of Laboratory Animal Welfare (OLAW) and the Association for Assessment and Accreditation of Laboratory Animal Care (AAALAC) International. The ARP director stated that the animal was treated daily by veterinary staff and although the infection unfortunately led to euthanasia, it was not as a result of inadequate veterinary care.

The concern with this particular situation is that the ARP director was also the AV and may have been affected by negative repercussions if the issue was defined as inadequate veterinary care. The IACUC administrators were conflicted since they worked directly for the ARP director but knew the regulators had previously observed similar circumstances at other organizations and had issued a finding of "inadequate veterinary care."

In this particular situation, the ARP director should have provided the IACUC with information and redacted herself from committee deliberations. The committee should have addressed the situation in the context of the regulations and ensured that the veterinary care program was functioning according to the standards. The ARP director's interpretation of the regulations to the committee was not specific or well defined, but favorable to the AV's direct reports. The committee should have pursued the idea of inadequate veterinary care and further investigated what exactly was done during the daily check by veterinary staff.

ARP directors can effectively manage the compliance portion of the animal care and use program, but situations may arise that require immediate separation from the discussion to prevent either direct or indirect conflicts that may compromise program integrity.

HOW MANY FULL-TIME EMPLOYEES DOES IT TAKE TO EFFECTIVELY OPERATE AN IACUC ADMINISTRATIVE OFFICE?

BP meeting attendees have concluded that the number of staff members needed to effectively operate an IACUC administrative office is dependent upon specific variables; for example, the number of new proposal submissions, whether new submissions are all reviewed by the IACUC or if a defined subset is reviewed by the full committee, whether proposals are reviewed annually or triennially, whether annual progress reports are reviewed as an administrative action or as full committee, and other factors all determine how many full-time employees (FTEs) are required in the IACUC office. Many organizations find that AAALAC accreditation requirements, decentralized housing facilities, and working with US Department of Agriculture (USDA)-covered species are all factors that require an IACUC administrative office to have more FTEs than similar programs in which these features do not exist.

The Submission Variable

Perhaps the most obvious and measurable component of an animal care and use program that may influence an organization's number of full-time IACUC administrative staff is the number of submissions processed annually. The number of protocols is frequently used as a surrogate for the size of the program. In all situations, an organization must have an employee responsible for receiving IACUC submissions and distributing them to the IACUC for review. Regardless of whether a program is developed in house or a data/tracking system is purchased from a vendor, a significant number of person-hours will be spent on the review process, and the number of hours will increase with submissions. In addition, once a submission is approved by the IACUC, additional hours will be required for an administrative staff member to monitor the activity of the project (i.e., the project expiration date). Survey data collected during past BP meetings suggest that, as organizations' programs grow, the number of submissions handled by a single FTE increases. According to BP meeting survey data, organizations maintaining two hundred or less active protocols typically have one full-time IACUC administrative professional. Organizations maintaining hundreds of active protocols tend to have one FTE per every 200–250 active protocols. This roughly works out to between fifteen and thirty new protocols a month per IACUC administrative employee, but that figure must be adjusted according to species (primate studies require more effort than mouse breeding activities). Based on discussions with institutional representatives, personnel resource dedication is also governed by organizations' financial capabilities.

BP meeting attendees also reported that those organizations managing less than two hundred active protocols generally handle and process protocols physically, while those maintaining five hundred or more protocols tend to use some form of electronic system. For example, in smaller organizations, a paper copy of the application is submitted to the IACUC office and is physically reviewed by the IACUC administrator. In some instances it is returned to the principal investigator (PI) for revisions, but eventually paper copies are made by the IACUC administrator and forwarded to the IACUC members for their review. Once the review process is completed the documents are compiled into a master document and physically filed by the IACUC administrator.

Organizations having larger protocol bases and those without significant budgetary restraints generally purchased computer software that allowed them to automate parts of their process. In this case, the IACUC administrator is able to review an electronic copy (includes PDF files) of the submission and forward it via e-mail to the committee members for review. Once the submission has been approved by the IACUC, the IACUC administrator saves the document electronically to a network server.

The Variable for Number of Animal Users

In addition to the number of submissions an organization receives annually, the number of faculty or staff using animals could also have an impact on the total FTEs required to operate the office. In some situations, the IACUC administrator established office hours specifically to meet with PIs, students, or others interested in discussing issues relating to protocol preparation. Some IACUC administrators develop and present continuing education in the form of classroom presentations, online training, or e-mail notices. Most meet with PIs submitting to the IACUC for the first time to assist them with the unique activities related to protocol preparation and submission. During these meetings, the IACUC administrator often discusses the protocol development process and nuisances related to the IACUC submission process.

The "Other Factors" Variable

There are other factors that have a significant impact upon personnel resource needs—whether an organization is Public Health Service (PHS) assured, AAALAC accredited, uses USDA-covered species, or the amount of external funding received.

PHS Assured

If an organization receives PHS funds (e.g., National Institutes of Health [NIH], Food and Drug Administration, Centers for Disease Control and Prevention), then PHS assurance is required.

Maintaining an assurance requires a cyclic resubmission of the assurance document to OLAW as well as an annual report on the status of the animal care and use program. Each document is substantial and many hours may be required to create a complete, accurate submission.

PHS assurance also requires semiannual review of the program (the way we do business) and the facilities (the way we care for animals). There is little variation in the time and resources required to conduct a program review among programs of varying sizes, but the amount of time an IACUC administrator must dedicate to facility inspections can be significant. In some programs the IACUC administrator serves as scribe and/or participant during the inspection process; that is especially time consuming for the IACUC administrator. Most of the additional documentation required for having a PHS assurance is the same as that required for maintaining AAALAC accreditation and will be covered in the next section.

AAALAC International Accreditation

Organizations maintaining AAALAC accreditation must prepare and maintain a written document that describes their program for animal care and use. This process involves describing the activities of the program in an outline shell developed by AAALAC. The outline is exhaustive and includes all of the intricacies and details of the program. A complete program description will include, for example, a list of all animal facilities and relevant information about each—for example, room sizes, heating, ventilating, and air conditioning (HVAC) parameters, whether there are outside windows, the materials that make up the floor and walls (tile ceilings, block walls), whether the door is metal or wood, and whether it has a viewing window. In addition to describing the physical composition of each room, the document must also discuss IACUC work practices such as how designated member and full committee reviews are conducted, the committee composition, the credentials of those serving the program, the occupational health and safety program, the organization's emergency disaster plan and other relevant procedures. The organization must maintain and provide to AAALAC as part of the written program description a current list of approved projects being overseen by the committee. The list should include, for example, the PI's name and contact information, the project/protocol number, the species used in the study, and a brief summary of the activities. The IACUC administrator should also prepare and update regularly an average daily animal census sheet. Once the document is complete, the IACUC administrator will need to develop processes to monitor it and keep it current.

In addition to preparing and maintaining the organization's program description, the IACUC administrator should also develop plans to work with AAALAC. AAALAC accreditation must be renewed every 3 years. The renewal process involves providing an updated program description every three years and working with the AAALAC team to coordinate and perform the site visit. The IACUC administrator will also need to collect and maintain the information provided in annual reports to AAALAC. For example, the IACUC administrator will update the facilities information and provide records of annual animal use on the report. Incidents of noncompliance will also need to be reported to AAALAC by the IACUC administrator. When animal use activities occur outside the organization and those animals are owned by the accredited organization, the IACUC administrator will need to monitor the accreditation status of the collaborating organization. In the event the collaborating organization is not accredited, the IACUC administrator will need to alert the home organization grants office and IACUC and potentially make provisions to cover the research work under the organization's accreditation. Frequently, when organizations are AAALAC accredited, the IACUC administrator serves as a point person for AAALAC and handles all accreditation-related activities.

USDA Implications

An IACUC administrator managing a program that uses USDA-covered species will need to develop a process for gathering information needed for the year-end annual USDA report. The IACUC administrator will monitor protocols and animal use and record in a database or equivalent data collection system the total number of each USDA-covered species used in research, teaching, or testing activities. The database must also record the pain category of the intended animal use—that is, whether animals are being held for USDA-covered activities, used in procedures that involve only minimal pain and distress, used in procedures that involve painful or distressful procedures that were relieved using pain-relieving medications, or used in procedures that involve unrelieved pain and distress. In addition, the IACUC administrator will need to specifically identify those animals used in unrelieved painful or distressful procedures and document the reasoning of the PI as to why pain-relieving medications could not be used. The IACUC administrator must maintain a list of those procedures approved by the IACUC that deviate from the expectations provided in the Animal Welfare Act regulations (AWAR). At the end of each federal fiscal year (October 1 of the prior year through September 30 of the current year), each IACUC administrator will prepare a report that includes the information maintained throughout the year and provide it to the USDA not later than December 1 of the current year.

FUNDING IMPLICATIONS

Organizations conducting a significant amount of animal research that is supported by external funding (e.g., PHS agencies, National Science Foundation) generally require more administrative personnel than those organizations that use only internal funding streams (e.g., many pharmaceutical companies). For example, PHS agencies require an institutional IACUC review and approve all of the animal use procedures identified in a grant before funds can be released to the organization. Organizations that receive funds from the PHS may need to dedicate a significant portion of an IACUC administrator's time to ensuring the PIs have the animal use procedures in their grant approved by the IACUC in a timely manner. This process may include having the IACUC administrator conduct protocol and grant concordance reviews on behalf of the organization. In addition, many organizations ask the IACUC protocol and grant concordance reviews (addressed in a later chapter) on behalf of the administrator to develop and execute methods for ensuring that PHS-funded research is conducted as described in the grant and the approved IACUC proposal and that procedural modifications to the animal activities are brought to the attention not only of the granting agency, but also of the IACUC for review and approval. The process may involve acquiring copies of a PI's annual reports to funding agencies or regular interactions with the PI.

SUMMARY

Benchmarking is a challenging method for determining the necessary size of an IACUC administrative support function and the number of IACUC administrators needed. The size of the supporting office and the number of professional IACUC administrative staff members an organization hires to manage and maintain its program can depend on a number of activities conducted through the office. Although "active protocols maintained" is not the only parameter to consider, it does offer a starting point for determining office size, staff number, or oversight requirements within an animal care and use program. The data outlined in Table 2.1 were acquired as a result of surveying those that attended past best practices meetings and may provide additional staffing guidance.

Who does what? Once an organization identifies what activities the IACUC administrative office will conduct and oversee, it may want to consider organizing the responsibilities into specific positions. For example, an organization with approximately five hundred active IACUC protocols that receives a significant amount of PHS funding, uses USDA-covered species, has decentralized housing (e.g., animals housed at different locations across a state), and has a small biosafety program may decide to hire at least four compliance liaison professionals to assure an adequate level of internal assessment, training, and assistance.

Organizations may choose to have one compliance liaison manage protocol reviews, including biosafety committee reviews (assuming the biosafety program is small). They may decide to assign one staff member to conduct congruency reviews, facilitate official correspondence between the organization and regulating and accrediting organizations, and handle noncompliance activities. They may also decide that another person is needed to monitor the approved protocols (postapproval monitoring) and that the fourth staff member will manage the general daily activities associated with the program.

Once the responsibilities are identified, organizations frequently assign a descriptive title to each position based on the duties. Although institutional practices vary, on average organizations call the individual managing protocol reviews and coordinating IACUC meetings, the IACUC administrator. Those monitoring approved

Table 2.1   Raw Data to Determine the Size of the Animal Care and Use Program

Number of Managed Protocols	Sample Size	Average Full Time IACUC Professionals	Protocols per Full Time IACUC Professional	Average IACUC Members	Total Protocols per IACUC Member
Up to 100	73	3	33	8	13
100–200	19	6	36	11	19
200–300	11	2	188	13	23
300–400	8	2	174	15	27
400–500	7	2	250	16	32
500–1000	5	3	385	16	62
Over 1000	3	5	213	6	61

protocols are identified as postapproval monitors/liaisons, and those conducting concordance reviews and handling noncompliance are the compliance coordinators. The unit manager appears to be consistently identified as either the associate/assistant director or the director of research compliance.

In this particular scenario the duties are distributed among the staff as described in the following:

IACUC administrator. The IACUC administrator will receive and process all of the IACUC and biosafety submissions received in the office. He or she will prescreen the documents and forward them to the committee for consideration. He or she will also develop meeting agendas, facilitate meeting activities, and complete the meeting minutes. The IACUC administrator will also serve as the primary compliance advisor to the IACUC, ensuring that the committee is up to date on all relevant protocol issues.

Compliance liaison (postapproval monitor, or PAM). The compliance liaison will regularly visit with PIs conducting research and assure that they are working within the parameters of their protocols. He or she may answer specific questions or provide on-the-spot training. The compliance liaison may schedule and coordinate the semiannual facility inspections.

Compliance coordinator. The compliance coordinator typically coordinates all activities associated with noncompliance. He or she may facilitate the IACUC investigation, communicate as appropriate with the PI, ensure that any IACUC-imposed sanctions are satisfied according to the plan and schedule submitted to oversight agencies, and file reports with the appropriate organizations (i.e., OLAW, the USDA, and AAALAC). The compliance coordinator will also conduct protocol and grant concordance reviews to ensure that the PI and the organization are in compliance with the terms and conditions of the NIH Grants Policy Statement (http://grants.nih.gov/grants/policy/nihgps_2012/index.htm).

Director. The director will provide guidance and supervise the professional compliance staff members. He or she will be ultimately responsible to ensure that those activities conducted by the compliance office are done accurately and in accordance with the federal policies and mandates. The director may also investigate methods to improve the overall efficiency of the office practices.

HIRING IACUC ADMINISTRATIVE STAFF MEMBERS

Once the duties and titles of each position have been identified, the organization then identifies and hires the appropriate people for each position. Organizations may begin this process by developing announcements that will be used to advertise the position vacancies. The announcements not only discuss the position responsibilities, but also identify the necessary qualifications an individual must have to be considered for the position.

Credentials: What Is Important?

The required credentials identified by organizations vary, but some consistencies were identified during past IACUC administrators' BP meetings. For example, most organizations agreed that the director should have experience managing complex programs, such as an organization's research compliance program, and supervising people. He or she should have a detailed understanding of the standards governing research activities, as well as supervisory and leadership experience. Organizations frequently look for individuals with both a baccalaureate degree and over 5 years of experience or a master's degree in a relevant field. The IACUC administrator, compliance coordinator, and compliance liaison must have a detailed understanding of the standards governing the use of animals in research, teaching, and testing. He or she must be able to provide guidance to PIs and committee members. Frequent credentials identified for the IACUC administrator include a baccalaureate degree with 3–5 years' experience and having earned or being in the process of earning the CPIA certification. Critical interpersonal skills include the ability to remain calm in a stressful situation and to communicate effectively. Those serving an organization as an IACUC administrator or compliance liaison frequently interact with individuals who are emotionally attached to their work. It is best to have an employee who keeps a calm demeanor and cool head in these circumstances. Another critical skill is the ability to independently assume responsibility for tasks while working as part of a team. In other words, the most successful IACUC administrators are always looking at the situation to find contributing causes for the distress (Did the protocol have insufficient explanation? Were the cages serviceable?) and thinking of ways to engage systemic solutions to prevent the development of similar problems at a later time or in a similar situation.

Position Announcements

Organizations develop varying position announcements to advertise organization vacancies. The announcements typically include the position titles, duties, and qualifications. BP meeting attendees indicated that when their organization announced the positions specifically through multimedia supporting IACUC activities, such as the IACUC administrators' listserv, Comp Med, or using PRIM&R and Laboratory Animal Welfare Training Exchange (LAWTE) job listing boards, their chance of getting relevant and qualified candidates to apply for the position increased. Sample position announcements are provided in Appendices 1, 2, and 3.

Conducting Interviews

After the vacancy listing has been posted, and the prospective candidates have been identified, organizations begin to schedule the interviews. Those participating in past BP meetings identified a number of processes that can be used during the interview process to assist with the identification of the best candidate. For example, some organizations require the candidates to give a presentation to a select group (e.g., IACUC members, other office staff members, and PIs) on, for example, "the importance of complying with research regulations" or "the ethical use of animals in research." The presentation gives interviewers the opportunity to see how well the candidate organizes his or her thoughts and to determine how well he or she can speak in front of an audience.

Some organizations give candidates a case study to consider and discuss with a group of seasoned coordinators. For example, a candidate may be asked to consider reusing a dog in research activities after it has already endured two major multiple survival surgeries as part of another project. The scenario may indicate that it is the appropriate thing to do since the reuse of animals will conserve valuable animal resources. The candidate may be asked to discuss the dilemma of saving the resource versus the ethical standards of not putting the dog through additional research after it has endured the past survival surgeries.

Additional activities conducted during interviews by other organizations include (1) questioning by the IACUC and administrative staff members, (2) lunchtime interviews with perhaps the director and immediate supervisor to evaluate how effectively the candidate can communicate, and (3) asking the candidate to provide a preliminary screening of a proposal to determine his or her level of regulatory proficiency.

· Organizations frequently have each of those participating in the interview process complete an applicant evaluation form, which identifies whether the interviewers believe the applicant should be further considered for the position.

Conducting Reference Checks

A primary component of hiring IACUC administrators may be the reference check process. Organizations are often interested to know what role applicants have played in past animal care and use programs at other organizations. If a desired candidate has never worked in animal research, the organization is usually interested in the candidate's opinion of animals being used for research activities. For example, it may not be a good situation for the organization and the position might not be a good fit for the candidate that does not support the use of animals for research activities. Those opposing animal use may not be able to critically consider the risk/gain benefits of animal use and might provide a convoluted approach on whether the research animal regulations are being appropriately applied.

REPORTING LINES: WHO REPORTS TO WHOM?

Organizations frequently develop direct reporting lines based on responsibilities and indirect reporting lines based on critical and needed interactions. For example, an IACUC administrator may administratively (direct report) report to the director, but may operationally (indirect report) report to the IACUC and the IACUC chair.

Direct Reporting Lines

Organizations have identified multiple reporting lines and instituted various ways for staff management to organize them. By default the director is at the top of the organization chart and provides oversight to those working in the department. However, throughout the industry others within departments have different reports. A common method used is to have the IACUC administrator report directly to the director and have others report to the IACUC administrator as he or she is the most experienced in knowledge of regulatory requirements. In addition, since the IACUC administrator should provide overall administrative support to the animal care and use program, those activities of the compliance liaisons and the compliance coordinators, which have a direct impact upon a portion of the program, must be administratively managed by the IACUC administrator. In other words, the findings of a compliance liaison or a compliance coordinator may result in an action that impacts the entire program managed by the IACUC administrator. Consequently, the IACUC administrator is in the position to make the necessary changes to the program and to directly interact with the IACUC when necessary. Examples of common organization staffing charts that were provided and discussed at recent BP meetings are provided in Appendices 4 and 5.

Indirect Reporting Lines

In addition to the direct reporting lines, it is often important for an organization to develop indirect reporting lines. The BP meeting attendees believed it was important to identify reporting lines when the IACUC administrator, for example, is making changes to or developing portions of an animal care and use program that requires input from another area director and also requires agreement and confirmation. Knowing relationships and responsibilities makes communication clearer and more effective.

Consider the following scenario to illustrate the need for clear direct and indirect lines of authority: After conducting an analysis of the organization's occupational health and safety program, the IACUC administrator determines that the personal protective equipment (PPE) in the animal cage sanitation area is insufficient or maybe is not being used as prescribed. As a result, the compliance coordinator requires additional PPE in the form of, for example, safety glasses and hearing protection. The IACUC administrator bases the decision on an observation made during a facility semiannual inspection. He believes that, since it is the responsibility of the organization to minimize the personal risks associated with animal use and he is acting on behalf of the organization, he has the authority to make the change for enhanced PPE. In this scenario, this decision is communicated to the director of Environmental Health and Safety (EHS) and the director of the ARP, and tempers flare. Both directors demand to know the basis on which the decisions were made. In fact, the EHS director indicates that the area was recently accessed and that the sound levels were well below the Occupational Safety and Health Administration (OSHA) threshold for requiring hearing protection. She also indicates that safety glasses are provided, but since there are no direct risks that could lead to eye injuries they are optional.

The message is: Decisions often require multiple buy-in; it behooves the manager to identify the indirect reports (i.e., the directors of EHS, ARP, Public Relations, and Occupational Medicine) to avoid confusion and maintain a teamwork environment, as well as to ensure that a good working relationship between departments within the organization.

PERFORMANCE DATA

Directors often need to establish methods to access the effectiveness of efficiency of the programs they are managing. For example, an organization may choose to document the number of noncompliant events detected per year. These data could help to determine if the level of compliance is increasing or decreasing from year to year, or it could be a measurement of the increasing or decreasing efficiency of the ability to detect noncompliance. For example, an 80 percent decrease in noncompliant events over a 5-year period might suggest that the PAM program is effectively ensuring and improving the organization's level of compliance with the federal mandates. A second data set a director may choose to monitor relates to protocol reviews. For example, the organization may document the number of protocols received from the researchers that are approved upon review versus those that need to be returned to the PI for modifications before they can be approved. If the number of protocols requiring modifications prior to approval decreases over a 5-year period, those data could be indicative that the organization's PI training program on preparing complete IACUC submissions is effective and is improving program efficiency. The director may collect data on the length of time required for the IACUC to review and approve a submission. Once again, if the trend decreases (i.e., a typical approval goes from 30 to 10 days in a 3-year period), that would indicate the processes practiced within the program are improving the overall efficiency of the office.

IMPROVING EFFICIENCY AND WORKING SMART

Frequently, institutional budgets prohibit hiring multiple staff members to run a research compliance office. An efficiency that may be engaged is to identify those groups within the organization that have responsibilities that overlap the research compliance. For example, if the organization has a safety unit and an IACUC administrative office, establishing a partnership between the two offices may result in the sharing of workloads and eliminate the need for the IACUC to conduct the safety checks necessary for protocol approval—a simple "cleared" from the safety office may be sufficient. IACUC members partnering and performing semiannual and safety inspections of laboratory satellite facilities at the same time is another example of maximizing the value for the organization and minimizing distress to the researcher.

Another simple tool is to share the workload. In situations when the number of research scientists using animals is significant in comparison to the number of compliance staff, organizations enlist more scientists to serve on the committee to review protocols, thereby decreasing the number of submissions pre-reviewed by compliance staff. At some organizations, the staff members from EHS are appointed to the IACUC and participate in semiannual facility inspections representing both the IACUC and Institutional Biosafety Committee (IBC). Administration asks the EHS member not only to inspect the animal facilities during the walk-through, but also to conduct at least an annual environmental risk assessment of the facility. This practice frequently satisfies one of their current responsibilities and also satisfies an occupational health and safety requirement.

In conclusion, there is no single model or even a preferred model to address the management issues of animal care and use programs. However, there are several successful models and many successful tools that have been employed at organizations of all sizes. Organizations are most successful when they do not try to create an exact duplicate of another organization's program on their own campus, but rather when they see what has worked and choose the tools that would serve effectively on their campus. There is no one-size-fits-all management schema for an organization's program of animal care and animal use.

CHAPTER 3

The IACUC Administrator's Role and Responsibilities

SCENARIO

Small Eastern University (SEU), until recently, had only one faculty member who used live vertebrate animals in her Public Health Service (PHS)-funded research. Dr. Smith collaborates with colleagues at nearby Great Eastern University (GEU). Her animals are housed in GEU's animal facility where the animal work is conducted. Accordingly, GEU staff care for Dr. Smith's research animals, her animal work is covered under GEU's assurance with the Office of Laboratory Animal Welfare (OLAW), and her experiments are reviewed and approved by the GEU IACUC.

Recently, SEU hired a prominent investigator who conducts research using an animal model. Recruitment efforts are ongoing and the SEU president anticipates they will soon have additional researchers who use animal models. The president thought it would help recruitment if SEU had its own animal program and so he directed the vice chancellor for research to get a program started. The vice chancellor designated a room in the basement of the Science Building for mouse housing space. Then he considered his administrative staff roster and identified Leo as the person on his staff that had the most experience with federal oversight of research—in this case biohazard regulation oversight. The vice chancellor speculated that this made Leo the most qualified individual to establish an animal care and use program. So Leo was promoted to the dual position of IACUC/Biosafety Committee administrator. He did not receive a salary increase but was given a free parking space in a university garage.

Leo initiated the process of developing an animal care and use program by attending IACUC 101 and 201 training to become familiar with the components of an animal care and use program and the policies and regulations that govern animal research activities. He also attended a best practices (BPs) meeting where he developed professional relationships with experienced colleagues, some of whom work in organizations near SEU. At the BP meeting, he collected protocol application shells and samples of best practices for conducting semiannual inspections, operating an occupational health and safety program (OHSP), and conducting protocol reviews. At the BP meeting, Leo made sure to sit with the Association for Assessment and Accreditation of Laboratory Animal Care (AAALAC), OLAW, and US Department of Agriculture

(USDA) representatives. These colleagues provided advice and resources for establishing an animal care and use program including links to their organizations' websites, which had answers to questions frequently asked by IACUC professionals and templates for accreditation and assurance documents. Leo also attended Public Responsibility in Medicine and Research's (PRIM&R's) annual IACUC conference, where he was able to network with a number of experienced IACUC administrators and gather additional information on setting up an animal care and use program. He could have attended a Scientist Center for Animal Welfare (SCAW) workshop or an American Association for Laboratory Animal Science (AALAS) meeting, but he did not have time or the budget to attend so many conferences.

Leo soon realized that the responsibilities associated with managing an animal care and use program were too extensive to perform in addition to his Institutional Biosafety Committee (IBC) coordinator duties. Therefore, he asked the vice chancellor to hire an additional staff member to meet the responsibilities of managing and operating both research compliance programs. The vice chancellor said that since there were not a lot of faculty members or animals at this time, he would not be able to commit additional personnel resources to the program. He suggested that Leo establish the program first and promised that he would assess the administrative needs of the program and reconsider the additional staffing request after the program was established and faculty had been hired. Against his better judgment, Leo set out to establish the SEU animal care and use program.

His first priority was to establish an IACUC. Leo believed that if he could identify faculty and staff willing to serve on the IACUC, they would be a free resource to assist him in developing the animal care and use program. So he recruited scientists from the SEU Biology and Psychology Departments, an administrator from the university's accounting department, and his neighbor, who was a retired pastor. Since the organization did not employ a veterinarian, he contracted the services of a local veterinarian to serve on the IACUC as the attending veterinarian (AV).

Then Leo and his newly formed IACUC went to work developing a set of guidelines, standard operating procedures (SOPs), and policies that were subsequently implemented to establish the program infrastructure. The contract veterinarian developed a plan for the veterinary care program. Leo sought staff members from the Environmental Health and Safety (EHS) and Occupational Medicine (OM) offices to assist in establishing an OHSP. After the programmatic components were established, Leo continued working diligently. He continued to check one thing at a time off his list, but it seemed as if two new items were added each time one was completed. As Leo had suspected earlier, the intricacies of operating an animal care program were not a part-time job or an ancillary position. Leo was now living proof that an IACUC administrator wears many hats, often needs to be in two places at one time, and is expected to manage multiple facets of a complex program.

Federal regulatory and guidance documents frequently refer to the organization's responsibilities. Since the IACUC is generally volunteer and an "additional duty," and other program participants are employed full-time for their primary role (e.g., Attending Veterinarian overseeing animal care), many organizations have developed an IACUC administrator position. The organization frequently expects the IACUC administrator to manage inclusively the overall intricacies of the animal care and use program. The administrator is expected to maintain a compliance program, by providing regulatory expertise to IACUC members, ensuring that an operating and compliant OHSP, providing required training opportunities, and serving as the organization's overall program assessor and general process coordinator.

THE IACUC ADMINISTRATOR

An organization's animal care and use program is usually managed through an IACUC administrative office directed by an individual who manages, organizes, and coordinates the daily activities of the program, such as providing regulatory expertise, coordinating inspections, scheduling IACUC meetings, and facilitating protocol reviews.

Organizations use multiple titles to describe the individual who coordinates the activities of the animal care and use program. Common titles used by organizations across the country identified during past BP meetings include program administrator, IACUC coordinator, compliance coordinator, and compliance liaison. However, the most commonly used title is IACUC administrator. Discussions proved that no matter what the organizations called the individuals, the duties of the animal care and use program administrators were consistently similar. Historically programs have spoken of a three-legged stool as the pictorial representation of a program's senior leadership (e.g., IO, IACUC, AV), but the increasing complexity and regulatory seriousness of animal care and use noncompliance is morphing the historical three-legged stool into a new piece of furniture which is the four-legged platform (e.g., IO, IACUC, AV, IACUC Admin).

The IACUC Administrator's Experience and Background

BP meeting attendees accumulated and later identified the common qualifications of IACUC administrators during meetings conducted from year to year. Although there were a lot of variables on the list, commonalities included the following: Most individuals had earned a bachelor's degree in a science-related field and had a general understanding of the regulations governing research animal use. Ideal IACUC administrators also had at least 1 year of experience in conducting animal research. In addition, a senior IACUC administrator who supervises other staff members and is responsible for making final programmatic decisions often had training in program and employee management.

Some organizations have selected a veterinarian or a veterinary technician as their IACUC administrator. These organizations are typically looking for someone capable of reading and understanding the regulatory components of the animal procedures as well as the particulars of the experimental designs. For example, some organizations expect their IACUC administrator to be familiar with the details of surgical procedures and the intricacies of analgesia and anesthesia use, and knowledgeable about methods of blood collection.

In October of 2001, PRIM&R began offering a professional certification, known as certified professional IACUC administrator (CPIA), that validates an individual's level of expertise. Candidates for CPIA status must demonstrate proficiency with federal regulations and policies governing animal care and use activities by passing a comprehensive examination. To be eligible to sit for the examination, candidates must have a bachelor's degree and, within the last 4 years, have 2 years of relevant animal care and use experience or 4 years of relevant experience in the past 10 years.

In addition to the CPIA certification, some organizations also value the certified manager of animal resources (CMARs) and the animal technician certifications (i.e., certified laboratory animal technician/technologist) offered through AALAS, which are also earned by the candidate successfully passing a comprehensive exam. Organizations and hiring managers are routinely seeking CPIAs since they have been demonstrated to be knowledgeable and experienced IACUC professionals.

The Regulatory Role of the IACUC Administrator

IACUC administrators frequently are responsible for ensuring that an organization complies with all federal standards governing the care and use of animals in research, teaching, and testing. They must have a working knowledge of the regulations and program guidance. The IACUC administrator must have expertise with the Animal Welfare Act regulations (AWARs), the PHS policy, the *Guide for the Care and Use of Laboratory Animals* (henceforth referred to as the *Guide*), and other relevant documents (such as the American Veterinary Medical Association [AVMA] *Guidelines on Euthanasia of Animals* and the *Guide for the Care and Use of Agricultural Animals in Research and Teaching*) to ensure that these rules are appropriately applied at his or her organization.

The IACUC administrator is frequently one of the people at an organization responsible for ensuring that the terms and conditions of National Institutes of Health (NIH) grants that relate to animal care and use, as specified in the NIH guide for grants and contracts, are met. USDA veterinary medical officers (VMOs) conduct inspections of USDA-registered research programs. IACUC administrators frequently escort the VMO through their facilities and facilitate the USDA inspection by providing all the requested documents and answering the VMO's questions. The IACUC administrator frequently completes and submits all required reports to governing agencies and to AAALAC International. He or she commonly contacts federal agencies to discuss programmatic issues and to acquire advice on how to address specific situations within his or her organization.

The IACUC Administrator and the IACUC

IACUC administrators are the "operational arm" of the IACUC, working under the authority of IACUC decisions. They routinely partner and work closely with the IACUC chair to ensure that critical programmatic components are satisfied. The IACUC administrator is frequently a full-time employee responsible only for managing IACUC and animal care and use program responsibilities. The IACUC chair usually holds a full-time position as a faculty member at the organization, fulfilling IACUC duties on a volunteer or part-time basis. The IACUC administrator and the IACUC chair discuss IACUC functions (e.g., a new or modified policy), specific protocols, or proposal amendments (e.g., whether it is appropriate to review a proposal by designated member review [DMR]). For example, the IACUC administrator may distribute submissions to IACUC members to ensure that each has the opportunity to request that a submission be discussed during a full committee meeting. In the event a full committee review is not requested, the IACUC administrator may suggest DMR assignments to the IACUC chair. Providing that the proposed DMR assignments are accepted by the IACUC chair, the IACUC administrator notifies the IACUC members of their DMR assignments. In addition, the IACUC administrator may discuss issues of noncompliance with the IACUC chair and conduct semiannual program reviews.

A key role of the IACUC administrator is to interact directly with the IACUC members and advise the committee on regulatory affairs. They may monitor OLAW announcements and regularly update the committee on policy announcements and regulatory guidance documents. The IACUC administrator often drafts needed documents and policies that are further developed by the IACUC and ultimately become committee practices or institutional policies.

IACUC administrators frequently serve as the principal investigator's (PI's) liaison to the IACUC, and they routinely interact with the PI when protocol modifications are required. For example, they frequently gather and collate modifications the IACUC expects to be made before a specific protocol can be approved. They then assist the PI with modifying the proposal and resubmitting it to the IACUC for review.

The IACUC Administrator and the AV

Effective and successful IACUC administrators and AVs work together to combine their expertise and efforts. The IACUC administrator often serves in an advisory role to the AV when regulatory issues arise. For example, when a noncompliant event occurs, the IACUC administrator may advise the AV on oversight agencies' reporting requirements.

The IACUC administrator can help to preserve the AV's reputation as an advocate for the PI. The AV frequently serves on the IACUC, but must also be notified by the PI when animal health issues arise. If a PI reports an animal health issue to the AV that results in a noncompliant event, as a committee member they must report the concern to the IACUC for investigation. If the report results in negative repercussions for the PI, they may equate self-reporting and communication with the AV with negative IACUC interactions. For example, if a PI reports sick mice to the AV, who determines that the PI was noncompliant with IACUC-approved humane end points, an IACUC investigation will occur. The investigation may result in the PI being required to perform corrective action training. As a result, the PI may begin to believe it is in his or her best interest to not inform the AV when animal welfare issues arise. This cascade of events, would not support the ultimate goal of the program, which is humane animal care and use (well-being).

The IACUC administrator can effectively partner with the AV and assist with maintaining good researcher–AV relationships. One BP meeting member organization has developed a procedure to distance the AV from regulatory discussions and minimize the severity of the noncompliance to the extent possible while maximizing a positive outcome and assure future animal well-being. They describe the process that follows as effective but not easy. The process is initiated by the AV, who notifies the IACUC administrator when the PI makes an animal adverse incident report. Upon receipt of notification, the AV and the IACUC administrator meet with the PI at the animal facility. Once the AV has resolved the health-related issues, he or she departs, leaving the IACUC administrator and PI to discuss whether potential noncompliant issues exist. If the IACUC administrator identifies a noncompliant issue that must be reported to the IACUC, he or she can explain the process to the PI. The IACUC administrator may discuss the regulations that require reporting and provide guidance on how the incident is likely to be resolved. The two may discuss the investigation process, federal reporting requirements, and any other relevant points. The IACUC administrator may discuss whether the negative ramifications of the incident could result in additional noncompliant situations, and whether the PI could take any self-initiated action to resolve the current problem and prevent recurrence.

The IACUC Administrator and the IO

IACUC administrators often serve as the committee's liaison to the institutional official (IO). They routinely explain IACUC issues to the IO and, when necessary, arrange for the IO to discuss relevant issues with the IACUC.

The IACUC administrator may facilitate meetings between the IACUC chair and the IO to present documents such as the semiannual report and explain the details contained in the report. For example, the IACUC may identify an insufficient heating and ventilation system (HVAC) at an animal housing facility. The committee may decide that the problem must be rectified within a defined time frame if the facility is going to continue to house animals. In a situation such as this, the IACUC administrator and the IACUC chair may explain to the IO that a properly functioning HVAC system is required in all animal housing facilities. The discussion may include alternatives such as relocating the animals or taking the facility out of service. The IACUC administrator may discuss the consequences of each potential decision. For example, delaying corrective measures may delay research projects until animal housing facilities are available and require identifying alternative housing space. As part of the discussion, the IACUC administrator may present a budget and a plan that outlines the time frame and potential costs associated with repairing the system. At some organizations the IACUC administrator reports directly to the IO, serving as the IO's operational entity and keeping the IO informed on program issues or concerns.

The IACUC Administrator and Other Administrative Units

To ensure that the organization satisfies the animal care and use program regulatory requirements, the IACUC administrator must forge a close professional working relationship with staff members from other departments within the organization. For example, each organization must establish a functioning OHSP to protect its animal handlers from work-related hazards. The IACUC administrator will need to establish a working relationship with representatives from EHS and OM to successfully develop this program. The IACUC administrator and his or her colleagues should collectively develop the components (e.g., risk identification and assessment and the use of appropriate personal protective equipment [PPE]) of the OHSP. Once the team establishes the OHSP, the IACUC oversees it. The IACUC may delegate to the IACUC administrator the responsibility of ensuring the program remains compliant should regulatory requirements change. In addition, the IACUC administrator may work with Human Resources (HRs), the Office of the Physical Plant (OPP), or other units on required programmatic components.

The IACUC Administrator and Regulatory Agencies

The IACUC administrator may serve as the organization's point of contact (POC) and establish working relationships with OLAW and USDA staff members. At some organizations, the IACUC chair or the IO may serve as the POC for regulatory agencies. The federal mandates establish a system of self-governance for institutional animal care and use programs. This expectation requires organizations to establish an IACUC to oversee the programs and file reports with federal agencies (OLAW and USDA) when necessary. As a routine activity, IACUC administrators contact professional staff at OLAW or the USDA when guidance clarification is needed. For example, during an IACUC meeting, committee members may discuss when it is appropriate to use reagent grade medications in animal research activities. Prior to making a final decision, the committee may ask the IACUC administrator to consult OLAW staff about this issue and report the findings to the IACUC at a future meeting. In addition, the IACUC administrator may consult with OLAW staff to determine whether an incident under investigation by the IACUC is a noncompliant event that truly needs to be formally reported to OLAW.

The IACUC Administrator and AAALAC International

At AAALAC-accredited organizations, the IACUC administrator is frequently the AAALAC point of contact, receiving requests for information and coordinating interactions with critical organization representatives, such as the IO, the IACUC chair, and the AV. For example, the IACUC administrator would receive and respond to an "annual report due" notice from AAALAC and would be a primary participant in preparing for and coordinating the 3-year AAALAC International accreditation visit, including preparing and submitting the program description, working with the site team to prepare the visit itinerary, and scheduling appropriate meetings between the site visitors and institutional representatives. The IACUC administrator is frequently the one who explains the reason for accreditation and the site visit process to institutional personnel such as animal care staff, scientists, and safety professionals.

The IACUC Administrator and Principal Investigators

The IACUC administrator also works very closely with PIs and other members of the research team such as students and post-docs. The IACUC administrator is frequently the first point of contact for new PIs. At most organizations, new PIs meet with the IACUC administrator prior to submitting their first proposal to the IACUC. The IACUC administrator typically discusses the organization's submission and approval processes and time lines and may assist the PI with writing the first protocol since the administrator often understands the IACUC's expectations. The administrator commonly develops an understanding of the PI's proposal so that he or she may advocate on behalf of the PI during the IACUC meeting. As an advocate for the PI, the IACUC administrator may provide guidance that will resolve problematic issues and facilitate approval of the project.

Should the IACUC Administrator Be a Voting Member?

BP meeting attendees discussed the advantages and disadvantages of having the IACUC administrator serve as a voting member on the committee. Fifty percent of the represented organizations have their IACUC administrator serve as a voting member, which helps to facilitate certain administrative components of the program such as administrative reviews or subcommittee inspections. For example, when IACUC administrators are committee voting members they can contribute to the meeting quorum, satisfy the regulatory requirement of having two committee members present during the inspection of facilities housing USDA-covered species, serve as a DMR, and participate as a committee member in the semiannual program review. In the official voting member capacity, the IACUC administrator is also able to officially discuss programmatic concerns and express minority opinions if necessary.

On the other hand, there are some disadvantages when the IACUC administrator serves as a voting member on the IACUC. For example, in addition to the IACUC administrator's duties of providing regulatory guidance and support to the IACUC, he or she must also fulfill the responsibilities of a committee member, possibly complicating the professional relationship with PIs.

WHAT ARE THE COMMON DUTIES OF THE IACUC ADMINISTRATOR?

Coordinate and Conduct Protocol Prescreening and Reviews

IACUC administrators frequently are the first individuals to receive IACUC submissions. At that time, they review the proposal for completeness. This includes prescreening the document to confirm that the PI has responded to all of the questions. Frequently they also verify that personnel listed on the protocol have completed their required training and are appropriately enrolled in the OHSP. The IACUC administrator often checks the accuracy of the number of animals being requested for the study and verifies the inclusion of appropriate scientific justification of the species and numbers of animals. In some cases, for example, the IACUC also authorizes the IACUC administrator to check the completeness and dates of literature searches for alternatives to painful and distressful procedures.

The objective of the administrative pre-review is for the proposed animal use activities to be complete and accurate, thus facilitating a higher level review by the IACUC. During the pre-review process, the IACUC administrator may execute specific protocols according to predetermined IACUC policies, such as sending all surgery protocols to the AV for veterinary pre-review or scheduling all proposals involving painful or distressful procedures to be reviewed by the full committee.

The IACUC administrator may also conduct administrative reviews as directed by the IACUC. For example, the Committee may delegate reviews such as changes in personnel (with the exception of change in PI) and funding sources, minor animal number adjustments (<10 percent in rodent breeding colonies and large aquaria of fish), and changes in approved housing location to an IACUC administrator. The IACUC administrator may be directed to add new personnel after verifying that each has completed all required training and is enrolled in the OHSP. The IACUC administrator may also make, for example, small adjustments (<10 percent) in the number of mice used for a particular research project. Once the protocol is administratively complete, the IACUC administrator schedules it for IACUC review. After administrative pre-review has been completed, the protocol is reviewed by either the full committee review (FCR) or DMR. Some organizations have policies that specify the review method to be used. For example, at some organizations the policy may be for all painful or distressful procedures to be reviewed at the full committee level, with all other submissions being eligible for the DMR process providing IACUC members are given prior opportunity to call for FCR. In the event that the DMR process is used, the DMRs must be qualified and appointed by the IACUC chair. A frequent practice is for the IACUC administrator to preselect DMRs for approval by the IACUC chair.

If the protocol involves biohazardous materials, the IACUC administrator may send the protocol to either the biosafety committee or an OHSP employee for review. In situations such as this one, it may be the IACUC administrator's responsibility to ensure that all required reviews and approvals are conducted before IACUC approval is granted. The IACUC administrator may check the protocol for any unique potential hazards to ensure that appropriate training has been provided. For example, if a PI plans to work with nonhuman primates, the IACUC administrator may confirm that the PI and all relevant personnel receive herpes B-virus safety training. If staff members are working on a farm delivering sheep, the IACUC administrator may confirm that they have Q-fever training. If the research involves the use of the hepatitis virus, a vaccination may be recommended.

The IACUC administrator frequently is the POC for all PIs submitting a protocol for review and also works with the PI to ensure that annual reviews of protocols occur on time. He or she helps the PI ensure that protocols remain current and accurate. The administrator monitors approved protocols to ensure that *de novo* reviews of PHS-funded projects are conducted every 3 years and that projects involving USDA-covered species are reviewed at least annually.

Organize and Schedule Meetings, Inspections, and Program Reviews

IACUC administrators are frequently responsible for scheduling meetings and coordinating inspections and reviews. In addition to coordinating protocol reviews, the IACUC administrator is usually responsible for scheduling IACUC meetings, which are held monthly at most organizations. When scheduling meetings, the IACUC administrator typically contacts committee members to determine who will attend the meeting. A quorum of the IACUC is required to conduct official IACUC business. If a quorum (simple majority) of voting members is not able to attend, it is necessary to arrange for alternates to attend to ensure that a quorum will be present or to reschedule the meeting.

The IACUC administrator usually coordinates the biannual inspections of the animal facilities, procedure rooms, and other relevant spaces that require IACUC oversight. The IACUC administrator coordinates inspections with the facility managers. Once the manager confirms attendance, the IACUC administrator identifies at least two IACUC members to participate in the inspection when USDA-covered species are maintained in the facility. The PHS policy provides some flexibility to organizations by permitting the IACUC to identify and use consultants to conduct the inspections. In either situation if the IACUC administrator is a voting member, he or she may serve as one of the two inspectors. Other parties may also need to be scheduled to participate in the site visits; for example, an EHS, veterinary, physical plant, OM, or compliance staff personnel member may also attend. The same process is followed when scheduling inspections of procedure areas and locations (e.g., satellite facilities) that are not part of the main vivarium. An IACUC administrator typically maintains a list of procedure areas requiring inspections and uses it to ensure that all relevant areas are visited at least biannually by the IACUC.

The IACUC administrator also schedules, for example, IACUC member retreats, continuing education activities, and new member training as well as any other IACUC-related activities.

Prepare and Distribute Reports

IACUC administrators must also prepare and distribute reports (i.e., for the AAALAC, IO, OLAW, and USDA). Every organization with an animal care and use program must prepare a report to the IO every 6 months. This report, often called the IACUC or IO report, includes a summary of the IACUC's animal care and use program review (e.g., the adequacy of specific programmatic areas, a list of *Guide* departures and a copy of the facility inspection report). It is necessary to discuss serious or continuing program deficiencies in the report and to include the plan and schedule for resolving these issues. The IO report may include requests for funding when the resolution of issues requires financial support. In addition, it includes facility inspection information (e.g., a list of facilities inspected, deficiencies noted, and the needed resolutions including plans and schedules for correction). The IO report also includes minority opinion statements, which are written opinions expressed by a single member or a minority representation of the IACUC on programmatic issues (e.g., relating to veterinary care or personnel qualifications and training). The IACUC administrator maintains records of minority opinions and also includes them in other reports such as the OLAW and AAALAC International annual reports.

If an organization is maintaining an animal welfare assurance through OLAW, the IACUC administrator files an annual report to OLAW. To ensure that the IACUC administrator has the current and accurate information needed to complete this report the administrator usually maintains a current roster of IACUC members. The roster may include each member's name, contact information, credentials, appointment term, and role (e.g., AV, scientist using animals, nonaffiliated member) on the committee. Since OLAW permits organizations to list committee member ID numbers on the annual report to OLAW except for the IACUC chair and AV, some organizations may also choose to list committee member ID numbers on the roster. In this situation, rather than listing the specific committee member names on the report, the ID numbers are then provided. However, the names of the AV and IACUC chair must be provided in the annual report to OLAW. In addition, the IACUC administrator maintains records of when the facility inspections were conducted and when the program review was completed. This critical information is also provided on the report to the IO and OLAW.

If the organization is a registered research facility with the USDA, the IACUC administrator also files an annual report with the USDA. To ensure that the IACUC administrator has correct and accurate information to complete this report, he or she generally maintains records of animal use according to the USDA-covered species and the "pain/distress categories." For example, if an organization used fifteen rabbits in research activities and ten were used for noninvasive procedures with five being used for surgical procedures conducted under anesthesia, ten animals would be considered used in category C (no or minimal pain/distress) and five in category D (pain/distress relieved by anesthetics or analgesics). A record of total category E procedures (painful/distressful procedures performed without relief) conducted on site and a justification for each category E activity explaining why pain/distress could not be alleviated must be included as part of the report. In addition, a list of animal housing areas where USDA-covered species are maintained must be part of the report.

If an organization is AAALAC accredited, the IACUC administrator prepares the annual report to AAALAC International. He or she typically maintains a list of housing areas with the total square footage of each area. If the facility includes farms, a list of acreage used to support animal housing is also maintained. In addition, a list of the species and number of animals used at each facility is maintained. The IACUC administrator also maintains a list of noncompliant events and significant changes to the animal care and use program. He or she maintains contact information for key program representatives (e.g., AAALAC contact person, IACUC chair, AV, animal program director/IACUC administrator). The described information is transposed into a complete AAALAC International annual report.

A critical narrative in each agency report is significant changes in the animal care and use program. For example, changes in key personnel, additions of new vivaria, and the OHSP must be included in annual reports to OLAW and AAALAC International.

IACUC Record Maintenance

The IACUC administrator maintains records (e.g., agendas, minutes, and approval notices) of activities relating to the animal care and use program. It is the IACUC administrator that usually creates the meeting agendas. To complete this process effectively, the IACUC administrator often maintains not only a list of protocols for review, but also a list of potential IACUC discussion points. These discussion points may come from PI conversations with the AV, from OLAW webinars and IACUC administrators' BP meetings, or as a result of something heard during discussion with another IACUC administrator. For example, during a facility inspection, the AV may have had a discussion with a PI on the use of non-pharmaceutical-grade drugs for animal research. As a result of the conversation the IACUC administrator may identify the use of non-pharmaceutical-grade drugs in animal research as a discussion point for the next IACUC meeting. Alternatively, the IACUC administrator may have recently attended a BP meeting and discovered that GEU has a policy that would be useful at SEU.

The IACUC administrator maintains records of meetings (minutes). He or she may prepare an audio recording of the meeting and reference it while preparing the written document or may actively use a computer or pencil and paper to document activities during the meeting. The meeting minutes will include records of meeting attendance and votes and note the committee's major discussion points for each agenda item. In some cases the IACUC administrator may need to encode the minutes to maintain confidentiality (e.g., the names of PIs or the location of animal housing facilities). In such a situation, the IACUC administrator will need to maintain a code list that can be used to decipher the code if requested by a federal agency. For example, the code list may identify committee member "xyz" and the location of the animal work being conducted in facility "ABC." Once all of this information is secured, the IACUC administrator will prepare the meeting minutes, which will be discussed during the next IACUC meeting.

The IACUC administrator establishes a method to maintain IACUC records for the required time periods. For example, he or she will maintain records of IACUC-approved animal research projects for at least 3 years after the project is completed. In addition, relevant reports, meeting minutes, and project approval records are maintained for 3 years. Some organizations maintain paper files, but many retain digital files (e.g., scanned or formatted documents) using either a specific database or an organization's networking system.

OLAW (PHS) Assurance, USDA Registration, and AAALAC Accreditation

The IACUC administrator typically prepares and revises the organization's OLAW assurance, USDA registration, and the AAALAC International-accredited animal care and use program description.

To complete an OLAW assurance document, the IACUC administrator provides the required information according to a sample document posted on OLAW's website. In order to successfully prepare an assurance document for OLAW review, the IACUC administrator must have a thorough understanding of the organization's animal care and use program. He or she must verify that the organization's program complies with, for example, the *US Government Principle for the Utilization and Care of Vertebrate Animals Used in Testing, Research, and Training*, the PHS policy, and the *Guide*. Among other things, IACUC administrators must be able to describe the administrative structure of the organization's animal care and use program. They should document the reporting lines of the key personnel administering the program and their relationship with the IACUC. They should also identify the reporting structure of the AV and the IACUC—for example, that they report directly to the IO. They must list the credentials of those associated with the program and describe, for example, the AV's responsibilities. They must demonstrate that the IACUC is appropriately staffed and that the members were appointed by the CEO or CEO-appointed IO. The administrator must verify that the organization's IACUC fulfills regulatory obligations. It must also be determined whether the organization is a category 1 (AAALAC accredited) or category 2 (non-AAALAC-accredited) organization. The IACUC administrator typically gathers and collates this information into the organization's assurance document and submits it to OLAW for consideration.

When scientists within an organization are using USDA-covered species in research activities, the organization must maintain a current registration with the USDA. The registration type must be defined by the applicant and is valid for a 3-year period. For example, an organization involving animals in research activities is a class "R" activity. The registration application also needs to note the type of funding received by the organization, whether it is awards, contracts, grants, or loans. The registration documentation must also include a list of the organization's primary administrative leaders. For example, a university may list its president, a pharmaceutical company its CEO, and a private corporation its owner. In addition, the registration form must include a list of the organization's animal facilities. The list typically includes the building name, location, and address.

Once the relevant information is gathered, the registration application is frequently completed by the IACUC administrator. Since the document requires the signature of the IO, the IACUC administrator often explains the document to him or her while obtaining the required signature. To guarantee that appropriate mailings are sent directly to the IACUC administrator (so they are not lost in a desk or drawer), he or she must ensure that the mailing address is accurate and current. In addition, the USDA prints the certification number directly on the registration form and, when appropriate, renewal notice, which is often needed by scientists to verify that their organization is USDA registered.

In situations where an organization is interested in applying for or maintaining AAALAC accreditation, the IACUC administrator routinely takes the lead role in preparing for the accreditation visit. During the visit, representatives from AAALAC are invited to access the organization's program of animal care using federal guidelines and documents (e.g., the *Guide*, AWAR, and PHS policy) as the comparable standards. All of the organization's information is entered into the appropriate location of the program description template shell. In order for this document to be completed, relevant information must be gathered from facility managers, physical plant staff, and EHS and OM staff. In addition, the program description must include the details of the organization's OHSP. For example, this section must discuss risk identification and assessment. The document must also summarize how personal risk assessments are conducted and discuss how PPE is used to protect staff from any animal-related hazards. The program description must also include comprehensive information about the organization's animal facilities—for example, detailed information about each unit's heating and ventilation system, facility dimensions, and water supply sources.

Once the required information is gathered, the IACUC administrator completes and e-mails the program description to AAALAC International. An AAALAC council member is then assigned to work with the IACUC administrator on the logistics of the accreditation visit.

Conduct Training and Orientations for IACUC Members and PIs

IACUC administrators often have a key role in identifying, orienting, and training new IACUC members. In addition, they frequently find themselves conducting training and orientation sessions for new faculty members planning to use animals for research. These sessions are conducted in a variety of ways.

At some organizations, PIs are required to meet one on one with the IACUC administrator before they are permitted to submit a proposal to the IACUC. At this time, the two frequently discuss the particulars of the IACUC process. For example, the IACUC administrator may summarize and discuss the details of the DMR and the FCR processes. He or she may provide some specific information on each review process, including potential review time lines, the difference between the two methods, and why the DMR is often timelier than the FCR process. In addition, the two may review and discuss the protocol template, which could include offering specific suggestions and providing guidance on particular questions.

The IACUC administrator may discuss the expectations of the approval—in other words, that approved protocols are equivalent to a contract between the PI and the organization and that the PI is expected to follow the terms of the contract. For example, the PI must only conduct those procedures approved in his or her protocol, and any modifications to the study must be reviewed and approved by the IACUC before any new processes are conducted. The IACUC administrator may remind the PI of the terms and conditions of the granted funds and the obligation to communicate with the grants manager. IACUC administrators may use this opportunity to remind the PI that if a protocol expires, grant funds cannot be utilized to support the project.

In addition, the IACUC administrator may also take the opportunity to discuss issues relating to animal welfare. For example, he or she may describe the organization's veterinary care program and may discuss the fact that the veterinarians ensure that the health and well-being of the organization's animals. The

IACUC administrator may advise the PI to immediately contact the AV any time that he or she finds animals ill or suspects that they are experiencing undue pain or distress.

For new IACUC members, the IACUC administrator may discuss all of the same issues described for new faculty members plus the expectations of a committee member. For example, new committee members may receive an intense training on the regulations governing the use of research animals; they may discuss the fact that IACUC members will need to review and provide comments on a set number of protocols per month and may talk about the requirements of IACUC members participating in semiannual facility inspections. They may also discuss the role IACUC members play on subcommittees and at regular IACUC meetings. For many organizations, the IACUC member obligations are documented in an approved IACUC member expectation policy, or committee members' agreement (**Appendix 6**), which is frequently provided to each new member during his or her orientation.

Oversee Training Programs and Maintain Training Records

The IACUC administrator at some organizations is also responsible for maintaining records that document PI, animal care technician, and IACUC member training. The records typically include written documentation (a requirement of the *Guide*) that the individual has completed specified regulatory and compliance training. Training records should include the type of training that was completed, the date it was completed, and whether it was a webinar, online training, or a didactic session (and should include the name of the instructor). When refresher training is required, the record should indicate when the training certification expires and when the refresher training must be completed.

When protocol approvals are contingent upon personnel completing required training, the IACUC administrator is frequently the "gatekeeper." In other words, once the IACUC approves the protocol, the IACUC administrator works with the PI to ensure that all personnel working with animals as part of the protocol are trained before the research can be initiated by the PI. The IACUC administrator may also send out training notices to animal users to ensure that they receive training prior to working with animals.

Advise the IACUC and IO on Relevant Regulations and Policies

As the veterinarian is the expert in animal welfare and health concerns, the IACUC administrator is frequently the organization's resident expert on regulatory issues. The IACUC administrator typically spends a significant amount of time keeping current on regulatory issues. For example, IACUC administrators frequently attend activities such as the BP meetings, PRIM&R, and other IACUC training workshops. The IACUC administrator regularly monitors, reads, and interprets website announcements made by organizations such as OLAW and AAALAC to ensure that he or she is able to provide the most current regulatory advice to the AV and the IACUC. Frequently IACUC administrators develop a professional relationship with staff members at OLAW and AAALAC, and they routinely contact regulators and colleagues when questions arise.

During protocol review, the IACUC administrator regularly ensures that protocols comply with the federal standards while the IACUC, especially scientific committee members, reviews the projects for scientific merit or humane care or use issues. IACUC administrators frequently conduct educational events to discuss regulatory expectations at IACUC meetings.

Assist PIs with Preparing Protocols and Satisfying Regulatory Requirements

Many organizations have the IACUC administrator meet with PIs, especially new PIs, to discuss protocol development. BP meeting attendees noted that these sorts of interactions generally result in much "cleaner," more complete, and better prepared proposals; the purpose is to provide the IACUC all of the information necessary for it to make an informed decision. The advantage of having the IACUC administrator pre-review submissions is that the administrator usually knows the specific information that must be incorporated in the protocol, has a good sense of protocol specificity, and recognizes a response that will generally be accepted by the IACUC.

Some organizations require their PIs to complete an application and send it first to the IACUC administrator for prescreening. The IACUC administrator screens the document and makes suggestions that may improve the quality or completeness of the protocol. In addition, the IACUC administrator may contact the PI directly to discuss the points of contention in the project.

Another method frequently used is for the IACUC administrator to sit down and talk with the PI during the proposal development phase. During the discussion, the IACUC administrator may take notes, give advice on what to include in the protocol, and offer to prescreen the proposal prior to IACUC review. On occasion, the PI provides a draft document to the IACUC administrator for supplemental prescreening, in which case it is typically returned to the PI with comments in preparation of finalizing the proposal for IACUC review.

A critical component of the protocol preparation may be for the IACUC to support the decisions of the IACUC administrator. For example, if the IACUC administrator tells the PI that question five needs to be revised and the PI does not make the revisions, it can be valuable for the IACUC to send the submission back so that question five can be revised. This process helps to maintain the creditability of the IACUC administrator and ensure that IACUCs are not getting incomplete applications for review. Although the organization can benefit from the IACUC and the IACUC administrator collaboration, the IACUC is not obligated to support the IACUC administrator's decisions.

Facilitate Communication

The IACUC administrator typically facilitates communication and discussion. An effective measure that has been successfully used is to first bring the interested parties (e.g., the IO, IACUC administrator and members, AV, PIs, and any others) together. Some IACUC administrators have indicated it is important for the group to meet in a neutral environment (i.e., away from the workplace). The IACUC administrator usually assumes the lead role in the discussion and opens by elaborating on the fact that as the IACUC administrator, he or she is there to support the needs of the committee, the AV, and the IO. The administrator continues the presentation by recognizing the often overwhelming task of managing an animal care and use program and concludes the background by briefly describing the responsibilities of each individual and how he or she complements a complete animal care and use program. The discussion frequently includes dialogue that defines where duties overlap. For example, if a perceived animal welfare incident is discovered by the AV, the veterinary staff would provide immediate care to the animal and notify the IACUC administrator to investigate the potential for noncompliance. This process allows the AV to remain focused on animal health activities and shifts the primary burden of regulatory compliance to the IACUC administrator, thereby preserving the PI and AV relationship. The IACUC administrator continues to facilitate the discussion, taking notes and highlighting critical points until a resolution is defined.

The IACUC administrator may also serve as a facilitator for an initial animal care and use management retreat and/or refreshers as the program matures. Overall, the IACUC administrator frequently establishes ongoing meetings to facilitate constant communication between the interested parties (the AV, IACUC administrator, and IO).

Informational Resource to the AV, IACUC Chair, IO, and PIs

The IACUC administrator also interacts with all those having a role on the IACUC. Consequently, the IACUC administrator frequently acts as the liaison between committee members or as a conduit of communication between the AV, IO, PI, and the IACUC.

In addition, the IACUC administrator frequently serves as the point of contact when submissions must be modified prior to the IACUC approving them. For example, if a DMR reviews a protocol that involves blood collection in mice and it is not clear that the PI is appropriately trained, the DMR may ask that the PI provide his experience in collecting blood before the protocol can be approved. In this scenario, the IACUC administrator may contact the PI using e-mail with the PI ultimately addressing the concern directly in the e-mail. The IACUC administrator forwards the additional information to the DMR, who subsequently approves the submission. The advantage of having the IACUC administrator in this liaison role is that it provides interaction and responsiveness for additional information necessary for proper decision making. It also ensures that the identity of the reviewers is confidential through the entire review and approval process as well as facilitates the requirement of maintaining accurate and complete records.

Serve as Long-Term IACUC Participant and Provide "Institutional Memory"

Some BP meeting attendees discussed the fact that the IACUC administrator is often the "institutional memory." This means the IACUC administrator has a role in monitoring IACUC decisions over time and noting consistency. At many organizations, IACUC members may change every 3 years or so, but the IACUC administrator frequently serves for 10 or more years and therefore provides a connection to the IACUC's past. The IACUC administrator in some cases maintains an IACUC administrator log and notes major decisions made by the committee over the years. Other IACUC administrators develop policies or SOPs that can be used to document decisions and/or procedures followed by the IACUC. For example, if the IACUC decides that greater than 15 minutes of physical restraint is a category E procedure for one protocol, it needs to ensure that the same practice is followed for future proposals; a policy or procedure plan is one way to remain consistent over time and across studies.

Manage IACUC Investigations

The IACUC administrator typically has a lead/key role in investigating incidents of noncompliance and anonymous reports. For example, if an alleged incident of noncompliance is identified, the IACUC administrator typically contacts the relevant party and any others that can provide pertinent information. He or she compiles the information into a report and presents it to the committee for discussion and consideration. If the PI and/or any other individuals need to attend a meeting to discuss the event, the IACUC administrator and chair collaborate to coordinate the logistics.

The IACUC administrator follows the noncompliance issue through the process and satisfies the required steps associated with noncompliance. For example, if the IACUC confirms noncompliance, the IACUC administrator typically reports the problems to the relevant parties (e.g., the AAALAC, OLAW, and USDA). The IACUC administrator prepares the written report, obtains the IO's signature, and submits the document to the relevant parties. In addition, when the IACUC establishes sanctions that must be satisfied by the PI, the IACUC administrator monitors the required corrective measures to assure satisfactory completion. For example, the committee may require refresher training before additional research can be conducted. The IACUC administrator would establish methods to ensure that folks are retrained and no new animal procedures are conducted until that time.

In addition to investigating noncompliance, the IACUC administrator is frequently responsible for initiating investigations into anonymous reports of noncompliance or animal welfare concerns. Once a report is received in the administrative office, the IACUC administrator may first employ his or her knowledge of the regulations to determine if an infringement of the regulations has occurred. Once a regulatory concern is identified, the IACUC administrator summarizes the concern for the IACUC for investigation. If no obvious regulatory incident has occurred, the IACUC administrator may contact the AV or the IACUC chair to determine if the incident has led or could have led to an animal welfare incident. When animal welfare incidents are identified, the IACUC administrator may contact the PI and inform him or her of the concerns and ask that the AV be immediately contacted. In the event follow-up is required (e.g., result of investigation letter), the IACUC administrator would typically draft the letter and forward it to the appropriate individuals.

Program/Facility Deficiencies and Corrections

When program reviews are conducted and deficiencies are identified during the process, the IACUC establishes correction measures and time lines, which are outlined in the IO report. The IACUC administrator may contact the relevant parties to discuss the concerns. He or she will inform the identified party of the problem, discuss why it is a problem, and identify when the corrective actions must be complete. If an issue must be addressed before a facility can be used, the IACUC administrator communicates the necessary information. Once the discussion occurs, the IACUC administrator typically documents the conversation by sending an e-mail to the responsible individual, summarizing the conversation and reiterating the correction expectations. The IACUC administrator usually follows up as necessary until the issue has been resolved.

Conduct Protocol and Grant Congruency Reviews

The IACUC administrator may conduct protocol and grant congruency reviews and frequently establishes procedures to ensure that the protocol is congruent with the grant before funds are released to the organization. For example, if a grant is not congruent with the protocol, the IACUC administrator under the direction of the organization may not issue an approval letter to the PI and consequently delay the distribution of the granted funds to the PI.

Develop and Maintain IACUC SOPs and Policies

The IACUC administrator may drafts SOPs and policies that include the relevant regulatory guidance verbiage and memorialize committee practices. The IACUC administrator facilitates the review and approval of a document draft by the IACUC, introduces it to the animal users' group, and implements measures to ensure that the practice is observed.

Once the documents have been approved by the IACUC, the IACUC administrator typically is the custodian of the documents and ensures that they are re-reviewed and reapproved regularly by the IACUC.

Develop and Maintain Websites

Many IACUC administrators also maintain the content on the IACUC website. They typically update the relevant materials, post announcements, and make appropriate changes to the website as required.

Author and Distribute Newsletters

IACUC administrators may prepare and distribute relevant announcements to PIs and IACUC members using newsletters or other means of communication. For example, if a USDA VMO is concerned about the overall effectiveness of an organization's search for alternatives, the IACUC administrator may write a memo on how to conduct a thorough literature search. The article may discuss the problems identified by the VMO and the corrective actions the organization must take. Articles such as these may appear in a newsletter, which may also include a calendar of events.

Monitor Animal Rights Activism and Initiate Appropriate Responses

The IACUC administrator may monitor animal rights activities across the country. This process often involves a collaborative relationship with his or her colleagues as well as the organizations' law enforcement officers and, on occasion, the local FBI office. He or she typically notifies the IACUC, PIs, and other relevant parties at the organizations when threats of animal rights activities are imminent. The discussion may focus on such topics as a guest speaker from an animal rights organization to a scheduled protest in front of the organization's main vivarium.

Serve as a Point of Contact for USDA Officers and AAALAC Visitors

The IACUC administrator frequently serves as the initial point of contact for USDA VMOs and AAALAC site visitors. When the USDA visits an organization to conduct an inspection, it may be the IACUC administrator that is the first contact. The IACUC administrator typically accompanies the VMO during the inspection. He or she may make arrangements for the AV or a facility manager to accompany them during the inspection or during a visit to a researcher's laboratory. The IACUC administrator prepares copies of meeting minutes, the last two IO reports, and specific protocols for the VMO to review during the visit. The IACUC administrator typically sits with the VMO during the records review session and addresses his or her questions.

Once the inspection is completed the IACUC administrator reviews the USDA inspection report with the VMO. He or she makes sure the VMO's concerns are understood, assures the clarity of any deficiencies, and, when needed, ensures that appropriate clarifications appear in the final report. The IACUC administrator reports inspections to the IO, including any commendations that were made during the inspection as well as deficiencies that must be addressed.

Develop and Maintain Emergency Disaster Programs

Under the direction of the IACUC, the IACUC administrator may ensure that a disaster plan is in place that covers all of the facilities (e.g., all central and satellite facilities) where animals are held and that it has IACUC-approved methods for mass depopulation, should such measures be necessary. The IACUC administrator typically ensures that the document is updated as needed and that the relevant parities are involved in conversations relative to updating the components of the document.

CHAPTER 4

The Animal Care and Use Program

SCENARIO

While earning his PhD at GEU, Dr. Alan Marx developed a sweetener from plant extracts. The sweetener is as palatable as table sugar and has zero calories. He established Sweet Solutions Inc. near his home in Maryland as a conduit for conducting additional laboratory tests on his product. A candy manufacturer interested in producing weight- and diabetes-control products provided financial support for product testing and development.

Since Sweet Solutions did not have an animal care and use program, Dr. Marx began the process of establishing a program for his organization. As only part of the overall program, he would need to establish (1) animal husbandry and veterinary care, (2) training, (3) occupational health and safety, and (4) record maintenance and oversight (i.e., IACUC) programs. Dr. Marx soon realized the cost to establish a program was significant. For example, he would need to hire a veterinarian to manage the veterinary care program, hire an IACUC administrator to manage records maintenance, and contract the services of a health care provider to manage the occupational health and safety program.

After Dr. Marx conducted a careful review of the Public Health Service (PHS) policy and the Animal Welfare Act regulations (AWAR), he reconsidered the need to establish an animal care and use program. He learned that the PHS policy only applied to organizations receiving funding from PHS agencies, and that research involving *Mus musculus* (i.e., laboratory mouse) was exempt under the AWAR. Considering that Sweet Solutions would not receive fiscal support from PHS agencies and that only laboratory mice would be used for animal testing, Dr. Marx decided that a formal program was not required, and he planned to start his animal experiments.

Dr. Marx would conduct taste preference studies using laboratory mice as the experimental model. He would prepare a variety of experimental diets containing different concentrations of artificial sweetener. A selection of two different experimental diets would be offered to each mouse to identify its taste preference. The trial data would be collated and analyzed to identify the concentration of sweetener to be used in the candies.

Dr. Marx hired Dr. Howard Oprah, a research scientist with expertise in taste testing research using a mouse model. Dr. Oprah was eager to start his research and inquired about the process for submitting animal experiments to the IACUC for review and approval. Dr. Marx indicated that IACUC review and approval was not required in this situation and that he might order mice and begin the experiments. As a previous member of an IACUC, Dr. Oprah indicated there is an ethical obligation to have experiments involving animals reviewed, approved, and overseen by an IACUC.

Dr. Marx agreed that there is an ethical obligation to ensure that the welfare of the experimental animals, but he did not believe an IACUC was necessary for that process to occur. He indicated that based on the species and funding source, the primary regulations and policy (i.e., the AWAR and the PHS policy) are not applicable. He further explained that none of the projects to be conducted threaten the welfare of the animals. The animals will receive healthy food containing artificial sweetener, and scientists will identify which samples the mice prefer. Dr. Marx indicated the standards of care for the mice would be based on the regulations, but an IACUC would not be established to review, approve, and oversee animal care and use activities. Dr. Oprah reluctantly agreed and began the taste preference studies.

As described in the scenario, there are circumstances when an organization may establish an animal care and use program that is not overseen by an IACUC. However, organizations deciding to conduct animal care and use activities without IACUC oversight should consider the ramifications. Regulatory standards governing the care and use of animals in research, teaching, or testing were established or adopted by federal agencies to ensure that the health and well-being of animals. They guarantee society that the animals receive exceptional care, are not permitted to experience unnecessary pain, and are not unnecessarily used. Organizations fully implement, for example, the AWAR and the PHS policy not to simply have a compliant animal care and use program, but most importantly to demonstrate to society their commitment to humane animal care and use. Organizations choosing not to publicly validate their commitment to humane animal care and use by implementing the federal standards are at a higher level of public scrutiny. For example, they are identified as organizations that are not eligible to receive public funding (e.g., from the National Institutes of Health [NIH], Food and Drug Administration [FDA], and Centers for Disease Control and Prevention [CDC]) since they have not implemented the federal requirements that ensure animal well-being. In addition, scientists frequently publish their findings in professional journals. The publication not only strengthens the reputation of the scientist, but in the case of the scenario also serves to validate findings associated with Dr. Marx's artificial sweetener. An increasing number of professional journals are declining to publish work that was not reviewed or approved by a duly constituted IACUC.

In addition to implementing the federal requirements, some organizations seek Association for Assessment and Accreditation of Laboratory Animal Care (AAALAC) accreditation to increase the level of organizational commitment and public confidence. AAALAC does not govern the organizational process, but does provide an external assessment that the program and process meet the minimum standards of animal care, experimental use, personnel safety, and institutional oversight of animal activities. AAALAC assessment is a peer review process facilitated by AAALAC-identified colleagues who perform a confidential assessment of the program of animal care and use, using the *Guide for the Care and Use of Laboratory Animals* (henceforth referred to as the *Guide*), a federally adopted standard, as a primary reference standard. The AAALAC assessment is an example of an organization's peers validating that animals are being used humanely and according to the standards governing animal care and use at the organization.

Yes, Dr. Marx was correct; in his case there were no federal regulations to have a functioning animal care and use program. He was not required to comply with the AWAR or the PHS policy. However, there remained several important business reasons why such external oversight may be beneficial to the company. Public confidence that the organization is committed to the health and welfare of research animals instills community comfort with and support for the company. Although it is possible for an organization to conduct animal research without regulatory or IACUC oversight, the organization should carefully consider the consequences.

Regulatory Requirements and Guidance Documents

- **PHS policy (Section IV [A] [3] [a], p. 11)** requires an organization's chief executive officer (CEO) to appoint an IACUC to oversee the organization's program of animal care and use.
- **Animal Welfare Act regulations (Section 2.31[a], p. 14)** require organizations to appoint an IACUC that will prescribe methodologies that will administer the activities of the animal care and use program.

INTRODUCTION

The animal care and use program (ACUP) is a collection of dynamic processes that collectively safeguard the health and well-being of personnel and research animals. It includes, for example, the organization's procedures for ensuring the health and welfare of personnel and animals, the expertise of personnel that accomplish the associated tasks, and effective measures to assess the quality of the program.

The program is established by the organization at the level of senior management. The CEO, who is the most senior individual at the organization, initiates the process, for example, by requiring key staff members with expertise in animal care, safety, compliance, and research to participate in the development of the program. The CEO's ability to directly interact on a regular and supportive basis may not be possible or practical. In this case, the CEO may appoint an institutional official (IO) to serve as the functional head of the ACUP. Whether the CEO is also IO, or the CEO appoints the IO, the IO is legally authorized to commit on behalf of the organization that the requirements of the PHS policy and AWAR will be met. The IO must also be authorized to commit the required funding and resources to ensure that the ongoing maintenance of a functional and compliant ACUP. Since the regulations governing animal use activities require organizations to oversee their own programs, the IO initiates the process of self-governance by appointing the IACUC. The IACUC acts on behalf of the organization to facilitate the ACUP and to ensure that it remains compliant with the regulations and policies. A properly constituted, trained, and resourced committee is critical to the successful performance of the ACUP.

Maintaining a compliant and functioning ACUP also requires commitment from other departments within the organization. For example, the occupational health and safety (OHS) office must actively participate in ACUP activities. The OHS will establish processes that ensure that animal activities are performed in a manner that mitigates risk for members of the research team. For example, OHS staff will develop standard operating procedures for wearing protective clothing such as gloves and lab coats when manipulating research animals.

The organization must also establish a process that ensures that animal handlers have access to medical services. Consequently, health care professionals (i.e., physicians and nursing staff) are an essential component of an ACUP. The medical surveillance portion of the ACUP may be coordinated through a hospital or medical specialty unit. The health care professionals ensure the health and safety of animal handlers by quantifying the level of risk each may experience and mitigating that risk to acceptable levels. The physician will conduct a health evaluation of each animal handler to ensure that a preexisting illness will not preclude him or her from working with animals. For example, an individual that is severely allergic to rodents would not be able to care for laboratory mice.

The organization must also establish the equivalent of the Department of Animal Resources (DAR). The DAR staff routinely includes veterinarians (either full or part time) and animal care and veterinary technicians. DAR personnel are an integral component of an ACUP. Veterinarians provide quality health care to research animals and assist principal investigators (PIs) with aspects of their research—for example, when potentially painful procedures are proposed. Animal care staff members maintain a clean and suitable living environment for research animals. For example, they feed and water animals daily and routinely clean their cages. The DAR typically maintains inventories of medications, feed, animal bedding, and caging to ensure that research animals can be appropriately cared for.

In addition, the Office of Physical Plant, Public Relations, Police Services, Human Resources, and senior administration frequently dedicate staff members to support the ACUP. For example, senior management may hire a staff member (i.e., the IACUC administrator) to oversee and manage research compliance. The Office of the Physical Plant may dedicate a team of staff members to perform routine maintenance on the vivarium and its heating and ventilation system. Police Services and Public Information are frequently available and aware of all activities associated with the vivarium and ACUP, which enables them to act should disasters (e.g., fire or activity from animal activists) arise. In addition, Human Resources can help senior management to identify individuals that are qualified and suitable to be employed as animal handlers.

Most ACUPs consist of a core group that provides program leadership (e.g., IO, IACUC chair, attending veterinarian, IACUC administrator) and serves, for example, as liaisons to other area leaders in the Human Resources, OHS, Occupational Medical, Police Services, and Physical Plant offices.

KEY MEMBERS OF THE ORGANIZATION'S ANIMAL CARE AND USE PROGRAM

1. **Chief executive officer (CEO).** The CEO is the most senior individual at the organization. He or she assumes responsibility for ensuring all animal care and use activities are conducted according to regulatory standards. The CEO appoints IACUC members and ensures the terms and conditions for receiving and utilizing funds from external sources (e.g., the NIH or National Science Foundation [NSF]) to conduct animal activities are satisfied. The CEO may also serve as the IO, but in many situations the CEO delegates the IO responsibilities to an individual who has direct oversight of reseach activities within the organization—for example, the vice president for research at an academic organization.

2. **Institutional official (IO).** Since the IO is defined in the regulatory documents as the individual that is authorized to legally commit to regulatory compliance on behalf of the research organization and therefore the individual bearing ultimate responsibility for the program, he or she has been affectionately deemed by the research community as the "go to jail" guy. The IO, either the CEO or serving on behalf of the CEO, is responsible for ensuring the effective performance of the ACUP.

    Best practices (BP) meeting attendees discussed and compared terminology and the functional engagement of the CEO compared to the IO. The regulatory documents define the IO as having ultimate responsibility for the ACUP, but both the AWAR and the PHS policy indicate the CEO must appoint the IACUC members. To ensure that the IO also has the authority to appoint in writing IACUC members (Appendix 7), most organizations exercise the option for the CEO to delegate in writing (Appendix 8) to the IO the authority to appoint IACUC members.

3. **IACUC administrator.** The regulations do not mention the position "IACUC administrator" or define his or her responsibilities. However, those duties typically performed by the IACUC administrator are essential if an organization wants to maintain an organized and compliant ACUP. The IACUC administrator typically manages the daily operations of the IACUC administrative office. He or she develops and maintains the required annual US Department of Agriculture (USDA), the Office of Laboratory Animal Welfare (OLAW), and AAALAC reports. The IACUC administrator accepts written proposals of animal use activities and provides them to the IACUC for review and approval. He or she ensures that the IACUC has the information necessary to make informed decisions regarding the use of animals. The IACUC administrator provides technical assistance to researchers on protocol development, relevant training options, and regulatory requirements. The IACUC administrator may also serve as the IO's representative to the ACUP.

4. **IACUC chair.** The IACUC chair's job description varies among organizations. Organizations typically appoint, for example, a senior scientist with exceptional leadership skills as IACUC chair. Many BP meeting attendees indicated organizations should also focus upon the following qualifications when trying to identify a good IACUC chair. The ideal IACUC chair should have

a. Significant animal-based research experience, be detail oriented, and possess excellent written and verbal communication skills
b. A graduate degree (PhD, MD, DVM, etc.)
c. Previous experience serving on an IACUC (viewed as exceedingly helpful to manage efficient meetings)

At many organizations, a senior scientist is asked to also perform the duties of IACUC chair. For example, a university may ask a tenured faculty member managing an animal research program and teaching graduate-level courses to chair the IACUC. In this situation, the IO is asking a member of the faculty to fulfill the responsibilities of the IACUC chair in his or her spare time. This scenario results in a part-time IACUC chair that often relies heavily upon the IACUC administrator to manage and lead the program. Organizations with larger, complex ACUPs appoint an individual with similar qualifications. However, the appointee is often relieved of a large portion of his or her responsibilities, which allows active involvement in the daily operations of the ACUP. For example, a professor appointed as IACUC chair may be relieved of his or her teaching responsibilities and asked to dedicate 75 to 80 percent of his or her time to managing and leading IACUC functions. In this scenario, the IACUC administrator and the IACUC chair work collectively to ensure that the ACUP is compliant and efficiently operating to serve the needs of the organization and the researchers. The time IACUC chairs dedicate to IACUC-related activities often depends on the size of the organization's program, but in all cases the job description is similar.

The IACUC chair ensures that animal use activities are conducted in compliance with the AWAR, PHS policy, the *Guide*, the organization's animal use policies, and IACUC-approved animal care or use protocols. This responsibility is satisfied, for example, by leading the IACUC through a combination of protocol amendment reviews and approvals, semiannual program reviews, and investigations into incidents of potential protocol noncompliance and potential animal welfare concerns. The three common activities of the IACUC chair identified by BP meeting attendees are

a. Protocol review process
   i. In collaboration with the veterinarian, pre-review new protocols before they are considered by the IACUC.
   ii. Assign committee members as designated member reviewers (DMRs) based on their qualifications.
   iii. Serve as a DMR and, when necessary, interact with PIs to ensure that concerns are identified and resolved before approval.
b. Program oversight
   i. Be current and lead discussions on business items, compliance concerns, and proposed animal use activities as IACUC chair during full committee meetings.
   ii. Provide leadership to the IACUC during the semiannual program review process.
   iii. Provide leadership to the IACUC during noncompliance and animal welfare investigations.
   iv. Participate in the inspection of new and existing animal use facilities.

       v.   Oversee and participate in the preparation of official communications that are a result of IACUC actions (e.g., minutes, protocol results, enforcement activities, etc.).

      vi.   Oversee and participate in the preparation and submission of all regulatory reports (e.g., USDA and PHS annual reports, reports on protocol deviations and animal welfare concerns, etc.).

   c. Compliance oversight and enforcement

       i.   Direct, in consultation with the organization's compliance official if one has been established, investigations of any reported or suspected protocol deviations and animal health or welfare concerns.

      ii.   Lead emergency IACUC meetings to address animal welfare concerns that may be time sensitive.

     iii.   Lead the IACUC in difficult issues such as addressing animal welfare concerns, sometimes on an emergency and/or time-sensitive basis.

     iv.   Lead the IACUC when such issues as sanctions (e.g., protocol suspension) to correct significant protocol deviations or animal welfare concerns are needed. The IACUC chair must be cognizant of the consequences of such an action on the PI research productivity and funding status, for example.

5. **IACUC vice chair.** The governing standards do not require an IACUC vice chair. However, BP meeting attendees indicated that identifying one may be beneficial. Some organizations use an official appointment letter (Appendix 9) signed by the IO to identify the vice chair. The letter indicates when the vice chair acts as the IACUC chair and identifies his or her delegated responsibilities. Placing individuals in the role of vice chair can serve to train the IACUC chair's replacement.

6. **Attending veterinarian (AV).** The PHS policy requires organizations to appoint a doctor of veterinary medicine to the IACUC. The policy indicates they must have "training or experience in laboratory animal science and medicine, and have direct or delegated program authority and responsibility for activities involving animals at the institution" (PHS policy IV. A.1. c). The AWAR also have a requirement to appoint a properly credentialed veterinarian (Section 1.1 p. 7) to the IACUC and identify him or her as the AV. The AV has no more authority than any IACUC member, but does have special responsibilities (e.g., potentially painful/distressful protocol consultation).

    The AV must be employed by the organization either as a full- or part-time employee or through a formal agreement—for example, as a consulting veterinarian. An academic organization with a large, complex ACUP may choose to hire a full-time AV, and a contract research organization (CRO) with a small ACUP may employ a veterinary consultant through a formal contract. If the AV is part time or a consultant and the organization uses USDA-covered species, then the formal arrangements for veterinary care must include a written program of veterinary care and a list of regularly scheduled visits to the research facility.

    The organization's AV must assure there is a program of adequate veterinary care that is in compliance with the organization's expectations, the regulatory stipulations, and the funding agency's or the accreditation agency's requirements. The AV must have appropriate authority to ensure the provision of adequate veterinary care and to oversee the adequacy of other aspects of animal care and use.

    The AV or other qualified staff members must be available at all times, including weekends and holidays, to provide veterinary care to research animals. Organizations ensure that animals have veterinary care available at all times by implementing one of many commonly used practices. For example, those organizations with large, complex programs frequently employ two or more veterinarians with at least one being available by telephone at all times. The contract veterinarians at smaller organizations frequently train, for example, a full-time animal care or veterinary technician to recognize and treat common clinical problems associated with the species of interest. The AV, in conjunction with the IACUC, usually develops a standard operating procedure (SOP) that defines when a technician must telephone the AV for guidance and when he or she can treat the animal.

The AV is responsible for developing and managing the veterinary care program. Depending on the circumstances (e.g., a large, complex program vs. a small program), husbandry and veterinary technicians may report directly to the AV or to a vivarium manager. A vivarium manager typically supervises organization employees when the AV is contracted or serving in a part-time capacity.

THE IACUC

The IACUC has oversight responsibility for the ACUP. It ensures that research animals are being treated humanely and that the regulatory standards governing animal care and use are appropriately applied. The committee members are formally appointed by the CEO or the IO if the authority is formally delegated to the IO by the CEO.

Committee Membership

The IACUC is qualified based on the collective experience of committee members to oversee the ACUP and ensure the health and well-being of the research animals. Although IACUC members can be appointed based on the program needs and research profile, there are three primary regulatory documents that define the minimum number of members of the IACUC. The standard applied by an organization depends on the species of animals used and the external fiscal resources received. Organizations receiving PHS agency financial support for animal research activities must comply with the PHS policy. An organization using USDA-covered species must constitute its committee according to the AWAR, and an organization using traditional agriculture species for food and fiber research may apply the *Guide for the Care and Use of Agricultural Animals in Research and Teaching* (henceforth referred to as the *Ag Guide*). Many organizations are both PHS assured and USDA regulated. Where the requirements differ, these organizations must meet the more stringent regulation.

During past BP meetings, attendees discussed the challenge of ensuring that IACUC membership satisfies the regulatory standard(s). Attendees indicated they frequently must apply multiple standards when defining members to serve on the IACUC. For example, some organizations receive PHS funding to support research involving USDA-covered species, which requires organizations to apply both the PHS policy and the AWAR. In addition, other BP meeting attendees indicated they also use agriculture animals in food and fiber research, which typically results in the application of a third guidance document: the *Ag Guide*.

Attendees at past BP meetings discussed whether IACUC members had to possess special qualifications. Attendees agreed that the regulations identify members' role, but not specifically individual qualifications. For example, the PHS policy (Section IV, 3, [b]) states the committee shall consist of at least five members and include a doctor of veterinary medicine, a practicing scientist experienced in research involving animals, an individual whose primary concerns are in a nonscientific area, and an individual who is not affiliated with the organization. Consequently, the IO must appoint committee members according to the required roles, but he or she is free to identify individuals with the specific qualifications and expertise needed to oversee the organization's ACUP. Attendees reported that the qualifications of committee members depend on the organization's research portfolio. For example, an organization that conducts research using wild species, laboratory rodents, and nonhuman primates may appoint three different practicing scientists: one with expertise using wild species, one using rodents, and one using nonhuman primates. Organizations using only rodents may have one scientific member with expertise in rodent studies on their IACUC.

Many BP meeting attendees believe organizations with multiple administrative units should appoint IACUC members to represent each administrative unit that has scientists conducting animal research. For example, a large academic organization with active animal scientists in the College of Veterinary Science, the College of Health and Human Development, and the College of Science may appoint a scientist from each unit to the IACUC. Attendees indicated this practice is important to ensure that each unit has a representative that participates in policy development, program oversight, and protocol review on the committee. In addition, unit members provide a service to their specific administrative area by disseminating information to other scientists within their unit.

The Nonaffiliated Member

The primary regulations (e.g., the PHS policy, the AWAR, and the *Guide*) all require organizations to appoint a nonaffiliated member to the IACUC. The nonaffiliated member, also referred to as the community member, cannot have any affiliation with the organization. For example, an organization's nonaffiliated member cannot be an immediate family member of anyone employed by the organization. He or she cannot serve as the nonaffiliated member for any other organization committees such as the Institutional Biosafety Committee (IBC) or Institutional Review Board (IRB). The nonaffiliated member is the committee member that represents the interests of the local community. He or she helps to ensure that scientific members remain committed to the humane care and use of animals used in research.

BP meeting attendees agreed that locating an individual not affiliated with the organization who is willing to serve as a committee member is difficult, and training him or her can be just as challenging. Many meeting attendees had success identifying nonaffiliated IACUC members by contacting senior citizen organizations whose members have an interest in community service. Attendees indicated that many local public school systems support associations for retired school teachers whose members are eager to serve on the IACUC. Organizations have also found that local clergymen, attorneys, pharmacists, and private practice veterinarians can be excellent nonaffiliated members. Organizations have discovered that the background of the prospective nonaffiliated member is less important than his or her commitment and willingness to serve.

In almost all cases, BP meeting attendees reported that either the IACUC administrator or the AV thoroughly trains the nonaffiliated member. Many organizations invest in their nonaffiliated members by sending them to local, regional, or national meetings (e.g., IACUC 101 and IACUC administrators' BP meetings) used to train ACUP staff members. This type of training enhances their ability to meet the responsibilities of an IACUC member (e.g., conducting protocol and program reviews and facility inspections).

Alternate IACUC Members

Many organizations appoint regular and alternate (i.e., backup) members to the IACUC. Although organizations are not required to appoint alternate IACUC members, many choose to because it increases the efficiency of the IACUC by allowing the committee to conduct official business when a quorum of regular members is not available. Alternate members should be formally appointed (Appendix 10) by the CEO or the IO with delegated authority. The alternate member can serve as a voting member any time a voting member serving in the same role is unavailable. For example, an alternate nonaffiliated member can serve for the voting nonaffiliated member when that member is traveling or ill. An alternate practicing scientist may serve for any voting practicing scientist when he or she is unavailable. However, an alternate veterinary member cannot serve as a voting member for the unavailable nonaffiliated or scientific member.

Appointing alternate members helps the organization to ensure that a quorum can be established when, for example, an emergency IACUC meeting needs to be convened or during the regularly scheduled monthly meetings. Alternate IACUC members can also participate in facility inspections and program reviews when voting members are unavailable. OLAW is firm on the alternate not being used to "share the workload" of the voting member, but that he or she only be used when the voting member is unavailable. It is critically important to always inquire about the primary member's availability prior to asking the alternate to serve in an active role.

The alternate may attend all IACUC meetings and other functions open to voting members. He or she may participate in discussion, but may not cast a vote to determine the outcome of an issue except in cases where the voting member is not available.

Properly Constituted IACUC and Quorum

A properly constituted IACUC can only conduct official business during a convened meeting of a quorum of the IACUC. The PHS policy defines a quorum as majority of the PRIMARY members of the IACUC. A properly constituted IACUC is one that includes all of the members required by regulations.

Although the AWAR, PHS policy, and the *Ag Guide* define the required IACUC members differently, there are some consistencies. For example, all guidance documents (i.e., the AWAR, PHS policy, *Ag Guide*, and *Guide*) require organizations to appoint a veterinarian with training in the species being used (i.e., training or experience in lab animal science) and a nonaffiliated member to its IACUC. The discrepancies among the guidance documents can create challenges when establishing the IACUC. To help clarify the requirements, consider the following scenarios.

Scenario 1

The scientists at Farmer's University conduct research that is intended to improve the quality and production of human food and fiber. Traditionally known as "food and fiber research," research activities intended to improve nutrition, breeding, and the production efficiency of traditional agriculture species (e.g., cattle, swine, and poultry) are overseen by the IACUC. Farmer's University does not receive federal funds to support research. The scientists conduct nutrition studies that maximize the efficiency of production and the quality of food products produced from production animals. In addition, researchers identify positive attributes in specific strains of beef cattle, for example, and breed animals to emphasize desired traits such as fast growing animals. Since Farmer's University is not receiving PHS funding and agriculture animals used for food and fiber research are not covered by the USDA, the organization may choose to only apply the *Ag Guide* when developing the constituency of its IACUC. Consequently, in addition to the nonaffiliated and veterinary committee member, Farmer's University must appoint committee members with expertise specific to agriculture animals. The *Ag Guide* permits organizations to identify a single individual to satisfy more than one membership role. For example, a nonaffiliated member may be a pastor whose primary concerns are outside science. In this example, the individual would fulfill the requirement of appointing a nonaffiliated and a nonscientific member to the committee. However, the organization needs to ensure that it satisfies the minimum requirement of appointing at least five members to the committee. In this situation and in order to have a properly constituted IACUC, Farmer's University would need to appoint individuals with the following expertise to its IACUC in addition to the nonaffiliated and agricultural animal veterinary member:

- A scientist with experience in agricultural research or teaching
- An individual whose primary concerns are outside science (e.g., clergyman, lawyer)
- A scientist with training or experience in the management of agriculture animals

After appointments have been made by the IO, the described Farmer's University's IACUC would be properly constituted. It includes the required roles and the minimum five committee members. In the event the organization decides to have a single individual satisfy the nonaffiliated and nonscientific role, it would need to appoint at a minimum one additional individual to the committee.

Scenario 2

GEU conducts research using only mice and rats. It also receives funding from PHS agencies (e.g., NIH, FDA, and CDC) to support animal research activities. Since GEU receives PHS financial support for vertebrate animal research and laboratory mice and rats are not overseen by the USDA, it would need to establish an IACUC with a membership that satisfies the PHS policy requirement. The PHS policy includes provisions for organizations to appoint a single individual to satisfy more than one membership role, but it also requires the appointment of at least five members to the IACUC. To establish a properly constituted IACUC under the PHS policy provisions, GEU will need to appoint the following members to the IACUC:

- A practicing scientist with expertise in laboratory animal research
- An individual whose primary concerns are outside science (e.g., clergyman, lawyer)
- A doctor of veterinary medicine, with training or experience in laboratory animal science and medicine, who has direct or delegated program authority and responsibility for activities involving animals at the institutions
- An individual who is not affiliated with the institution in any way other than as a member of the IACUC and is not a member of immediate family of a person affiliated with the institution
- At least one other individual to satisfy the minimum requirement of appointing five members to the IACUC

In this particular scenario, once the IO has appointed individuals to the required roles, he or she must identify and appoint at least one additional member to the GEU IACUC. For example, if GEU conducts infectious disease research, the IO may appoint an individual with expertise in biosafety, which may help to maximize the committee's ability to oversee the program. Once the appointments have been made by the IO, the IACUC will be properly constituted and include the minimum number of members as defined by the PHS policy. Frequently an organization's IACUC has more than five members. It is ideal to have an odd number of members, making it easier to convene a quorum (e.g., five members equal a quorum of three and six a quorum of four).

Scenario 3

Andy's Biologics is a small antibody production company that uses only rabbits and guinea pigs to produce, for example, custom antibodies, whole blood, and other relevant biologics for researchers. The company charges customers for the products it produces and receives no funding from external sources. The organization uses only USDA-regulated animals as defined in the AWAR (9CRF, Subchapter A, 1.1 definitions). Since Andy's uses only USDA-covered animals and receives no funding from PHS agencies, it must establish an IACUC that is consistent with the AWAR. Regarding members' roles, the AWAR also require organizations to appoint an experienced veterinarian and a nonaffiliated member to the IACUC. However, the AWAR only require organizations to appoint a minimum of three individuals to the IACUC.

In addition to the veterinarian and the nonaffiliated member, Andy's Biologics will need to appoint an individual to serve as the IACUC chair.

Scenario 4

Great Western University (GWU) has an extensive research program that involves the use of both agriculture and laboratory animals. A significant portion of research activities are funded by PHS agencies. The laboratory animal research activities involve both USDA-covered (e.g., rabbits, dogs, and cats) and non-USDA-covered animals (e.g., laboratory mice and rats). GWU also has operational farms with scientists conducting food and fiber research using cattle, sheep, and swine. Since the organization receives PHS funding to support vertebrate animal research, uses USDA covered species, and uses agriculture species for food and fiber studies, the committee's membership must satisfy the PHS policy, the AWAR, and the *Ag Guide*.

The three standards consistently require organizations to appoint a nonaffiliated member to the IACUC. They also require the appointment of a qualified veterinarian. Since GWU uses agriculture and laboratory species, the organization will need to either identify a veterinarian qualified to provide medical care to both types of research animals or appoint two veterinarians (i.e., one with lab animal and the other with agricultural animal expertise) to the committee. GWU must also appoint practicing scientists with expertise in agriculture and laboratory animals to the committee. Since the minimum number of members among the regulatory documents is five, a minimum of five individuals will need to be appointed to GWU's IACUC. GWU will also need to ensure that each role identified collectively among the three standards is satisfied. In addition, the standards provide the organization the opportunity to satisfy more than one member role on the committee with the same individual. Consequently, the following membership appointments must be made:

- PHS component of the program (PHS policy)
  - A practicing scientist with expertise in laboratory animal research
  - An individual whose primary concerns are outside science (e.g., clergyman, lawyer)
- Agriculture component of the program (*Ag Guide*)
  - A scientist with experience in agricultural research or teaching
  - An individual whose primary concerns are outside science (Note: since a nonscientific member was appointed under the PHS policy requirement, appointing a second nonscientist is not required)
  - A scientist with training or experience in the management of agriculture animals
- USDA component of the program (AWAR)
  - No additional appointments are necessary, but at least one of the appointed members must be identified as the IACUC chair according to the AWAR and the PHS policy

In this particular scenario, GWU has appointed six members. Once the organization identifies a committee chairman, it will have satisfied the membership requirements of all regulating documents. GWU's properly constituted IACUC has the five required members, including a veterinarian trained in both agriculture and laboratory animals, a nonscientific member, a nonaffiliated member, and three practicing scientists with the required expertise. If GWU chooses to appoint two veterinary members (agriculture and laboratory animals), the committee will increase to seven members.

IACUC Functions

The regulations require each organization to self-regulate and oversee its ACUP. While the IO and AV provide critical program resources and support, the regulations require organizations to establish an IACUC to oversee and assess its ACUP components and functions. The PHS policy and the AWAR, for example, define the specific functions of the committee as being the following:

1. Review, at least once every six months, the organization's ACUP. The IACUC uses the review to verify that the components of the program are compliant with the federal regulations and policies. During the program review, IACUC members may review for example the organization's polic[i]es on the care and use of vertebrate animals for research, standard operating procedures, application templates, and procedures;
2. Inspect, at least once every six months, the vivarium and other areas (e.g., PIs' laboratories or satellite housing facilities) used to house research animals, locations where animal activities are conducted and areas such as cage wash and feed storage that support the organization's animal facilities, including animal study areas;
3. Prepare and submit written reports of its program review and inspections to the IO;
4. Review and, if warranted, investigate concerns involving the care and use of animals at the organization. The concerns may be as a result of public complaints, allegations of noncompliance, or adverse events affecting the welfare of animals;
5. Make recommendations to the IO regarding any aspect of the organization's ACUP, for example, facilities, personnel training, or OHSP;
6. Review and approve, require modifications in (to secure approval), or withhold approval of those components of proposed activities related to the care and use of animals, as specified in paragraph (d) of this section;
7. Review and approve, require modifications in (to secure approval), or withhold approval of proposed significant changes regarding the care and use of animals in ongoing activities; and
8. Be authorized to suspend an activity involving animals in accordance with the specifications set forth in paragraph (d) (6) of this section. This authority exceeds that of the CEO or IO. In the simplest terms, "if the IACUC suspends or disapproves an animal care and use activity no one can override that decision." However it is important to note that the CEO or the IO may overrule the IACUC and disapprove any animal use activity approved by the committee.

Conflicts of Interest

During the course of conducting official business, an IACUC member may encounter a situation that disqualifies him or her from taking official action. The

PHS policy (Section IV [C] [2], p. 14) and the AWAR (Section 2.31 [2], p. 24) indicate that no member "may participate in the IACUC review or approval of an activity in which that member has a conflicting interest (e.g., is personally involved in the activity)...." Consequently, IACUC members must report potential or perceived conflicts of interest (COI) concerning all business conducted by the IACUC.

Attendees at best practices meetings indicated that they consider IACUC member participation in any of the following categories as being a COI:

1. The review and approval of an activity in which the member is the PI or co-PI
2. The review and approval of an activity of which the member is a financial sponsor
3. The review and approval of an activity from which the member is receiving funds
4. The review and approval of an activity of which the PI is the member's supervisor
5. The review and approval of an activity in which the member is a family member of the PI
6. The review and approval of an activity of which the IACUC member is listed as personnel on an accompanying grant proposal
7. An activity in which the member has a financial interest (as defined by the GEU's conflict of interest policy) in the sponsor funding the project (Note: Peripheral knowledge or indirect involvement in certain activities may not rise to the level of a COI but might result in member abstention from a vote or discussion.)

If a member believes that he or she would be conflicted if involved in a particular discussion or deliberation, that member should make the conflict known prior to the committee beginning the discussion. This can be accomplished by the concerned member notifying the IACUC chair or the IACUC administrator before the meeting or discussion begins. Committee members having a specific conflict are obligated to leave the meeting room prior to the committee's deliberation on the conflicted issue. If the COI is recognized during discussion of an issue, the member should declare a potential conflict exists, and depart the meeting room for the remainder of the discussion and subsequent vote. The conflicted member can provide information for the committee's consideration, but should not be part of the decision process of the committee.

To provide evidence of the IACUC's adherence to clear and nonconflicted review activities, some organizations read a conflict statement at the beginning of each IACUC meeting. For example:

If you are aware of a conflict of interest involving any investigator, protocol or business item that is under consideration today, you must leave the room before the IACUC's deliberations and vote occur. Should you realize during a discussion that you have a conflict of interest, you should declare it and depart the meeting room for the remainder of this discussion and vote. The actual nature of the conflict does not need to be disclosed. Members unsure of whether a particular situation is a potential conflict may present the issue to the IACUC for consideration during the meeting, or prior to the meeting they may present the issue to the IACUC chair or IACUC administrator.

Expectation of Member Confidentiality

The organization's IACUC members have a part in assessing the humane care and use of animals. To protect the integrity of the organization and its researchers, members must not disclose confidential or proprietary information (protocol- or investigator-specific information) to any non-IACUC member. They must not discuss, communicate, or disclose any details of IACUC business (e.g., protocol reviews, noncompliance discussion, subcommittee investigations or review, etc.) to third parties without the consent of the IACUC chair, IACUC administrator, or IO.

CHAPTER 5

Protocol Review and Approval

SCENARIO

The IACUC at GEU reviews and approves animal care and use activities before they are initiated by principal investigators (PIs). In addition, GEU's internal policy is for all IACUC submissions that include activities in which the animals die (i.e., death as an end point) as part of the research to be reviewed and approved by the full committee.

During a full committee meeting, the IACUC reviewed a protocol that included infecting mice with a methicillin-resistant *Staphylococcus aureus* (MRSA). The research objective was to determine how invasive the organism would be in the absence of antibiotic therapy.

During discussion, IACUC members considered information provided by the biological safety officer (BSO) and the university physician. Based on the information provided, committee members were satisfied that the MRSA would be appropriately contained, and that staff members handling the organism were adequately trained and protected from infection.

However, regarding the animal activities, two IACUC members recommended modifying the protocol before granting approval. The other committee members concurred, and they collectively generated a list of significant changes that must be made by the PI before IACUC approval would be granted.

Committee members identified two primary issues as being significant concerns. Although the PI had proposed death as an end point, members were not convinced that allowing the animals to die from the infection was necessary for the study to be successful. They agreed that the PI should consider developing both experimental end points (the point when most or all scientific outcomes had been achieved) and humane end points (the point when the distress to the animal is unacceptable and should be discontinued). The committee concluded that, once appropriate experimental end points were established, death as an end point would be unnecessary and that criteria (e.g., rapid weight loss, inability or difficulty to rise or ambulate, breathing difficulties, or dehydration) should be developed to indicate when an experimental animal must be removed from the study.

The proposal briefly described a process for infecting the animals with MRSA. The PI indicated the skin of the animal would be scratched, and MRSA would be directly applied to the wound using a cotton swab that was saturated with a MRSA suspension. The proposed activity was problematic for the IACUC since the description did not include the details of the procedure. For example, the description did not discuss wound location, specific details on how the wound would be caused (e.g., whether the skin would be shaved, whether a scalpel would be used, the depth of each wound, and what process would be used to assure consistency between animals), and the dose of MRSA each animal would receive.

The IACUC chair assigned the two IACUC members that originally suggested the need for refinements as the designated member reviewers (DMRs). The remaining IACUC members agreed the protocol could be reviewed using the DMR process with the expectation that the DMRs would ensure that the committee's concerns were adequately addressed.

The DMRs contacted the PI and discussed why death as an end point was necessary. After discussion, the DMRs agreed death as an end point was needed for scientific reasons. In addition, the inoculation procedure was described to the satisfaction of the DMRs. Realizing that securing National Institutes of Health (NIH) funding was contingent on the IACUC reviewing and approving the animal activities, both DMRs agreed that the PI had adequately revised the protocol, and it was approved by the DMRs.

Although the GEU IACUC established a policy requiring all protocols that use death as an end point be reviewed and approved during a full committee meeting, both DMRs justified approving the study since the review was initiated during a full committee meeting. In addition, they believed that, since the primary questions were their own and they had been assigned by the IACUC chair to serve as DMRs, referral back to the IACUC was not necessary and would only prolong the review period. Though the committee policy required studies involving death as an end point to be reviewed and approved by the IACUC during a full committee meeting, it was the DMRs that ultimately approved the project.

A few days later one of the DMRs questioned the protocol approval since it did not occur during a full committee meeting as the IACUC policy mandates. As a result, he discussed the issue with his reviewing colleague, and they decided it was necessary to advise the IACUC chair. After discussion, the IACUC chair indicated that since the reviewers did not refer the issue back to the IACUC as policy required, the protocol was not approved. He said that the protocol would need to be reviewed and approved during the next full committee meeting.

What is permissible and what is appropriate?

In this scenario, the IACUC chair's decision was not entirely correct. The protocol review and approval process did satisfy the federal regulations and policies. Federal guidelines allow designated reviewers to approve, require modifications to secure approval, or refer a submission back to the committee for review. Although deviating from the established GEU policy may result in the IACUC taking action; the federal regulations do not require DMRs to refer a submission to the full committee for review. According to federal regulation and policy, the DMRs are empowered to determine the protocol outcome with the exception of disapproving the protocol.

However, the IACUC chair's logic was not totally flawed in that any protocol may at any time be re-reviewed by the IACUC. Although the protocol was officially approved by the DMRs; it could also have been reconsidered by the full committee upon the request of any member including the IACUC chair.

Special procedural note: When a project is discussed at the full committee level, Public Health Service (PHS)/Office of Laboratory Animal Welfare (OLAW) guidelines only permit a protocol to be assigned to DMR if *all* of the voting members are in attendance and all agree that the DMR assignment can be made, or all members have previously signed an agreement that all members in attendance at an IACUC meeting (not necessarily all members of the IACUC) may agree by unanimous consent to send the document to DMR.

Regulatory Requirements and Resources

- **PHS policy (Section B (6) (7), p. 12)** requires organizations to review and approve the proposed use of animals in research before it is conducted.
- **Animal Welfare Act regulations (AWARs) (Section 2.31(8) (d) (i–x)** require organizations to review and approve the proposed use of animals in research before it is conducted.
- **OLAW frequently asked questions (FAQs); Section D (Protocol Review) Question 19** asks, "May an IACUC use designated member review (DMR) to review an animal study protocol subsequent to full committee review (FCR) when modifications are needed to secure approval?"

PRESCREENING PROTOCOLS (ADMINISTRATIVE AND VETERINARY REVIEWS)

Although a prescreening process is not a federal requirement, a large percentage of the IACUC administrators attending past best practices (BPs) meetings indicated their organization uses a protocol prescreening process. Submissions are prescreened by most organizations to ensure that they are complete before entering the formal IACUC review process. Many organizations reported that the prescreening process would ensure that committee members are provided all of the information they need to make an informed decision, which has led to shortened protocol approval times.

During BP meetings attendees identified and discussed two different types of prescreening practices and the benefits of each.

IACUC Administrator Pre-Review

The IACUC administrator often pre-reviews all submissions to the IACUC. During the prescreening process, he or she will ensure that the protocol is complete. The IACUC administrator will ensure that all of the submission questions were answered and all of the listed personnel have completed the required training and been authorized by medical services to work with animals. In addition, many

IACUC administrators ensure that the responses are consistent among questions. For example, the number of animals needed to conduct the experiments should equal the number of animals requested.

Several IACUC administrators reviewed the "Search for Alternatives" to potentially painful or distressful procedures to ensure that the appropriate keywords, databases, and search periods were used.

Frequently, IACUC administrators ensured that all procedures listed in the protocol are fully described. For example, if the PI indicated that he or she would be conducting stereotaxic surgery, but did not describe the procedure or failed to include patient preparatory steps, the IACUC administrator would request additional information. Select IACUC administrators ensured that reagent doses were provided in the protocol and that end points were appropriately defined and provided when necessary.

At many organizations, an administrative policy was developed requiring submissions to be prescreened by the IACUC administrator before IACUC review is initiated. At other organizations, the IACUC administrators were frequently considered an assistant of the researcher. They would help the PIs to prepare complete and clear protocols that would be easily understood and quickly approved by the committee. The IACUC administrator would also employ the necessary measures to ensure that the PI maintained a compliant research program.

Veterinary Pre-Review

In addition to an administrative pre-review, the organization frequently mandated a veterinary pre-review be conducted before a formal IACUC review was initiated. In almost all situations, the primary purpose of the veterinary prescreening was to help the PI to develop a submission that was complete and clear. This process would help to ensure that the committee could conduct a thorough review and make an informed decision on project approval.

As with the administrative pre-review, a separate veterinary pre-review is not required by policy or law unless the protocol includes procedures that have the potential to be painful or distressful to the animals. In that situation a veterinary review is required by AWAR. At some organizations, the veterinary pre-review is also the required veterinary review of projects involving potentially painful or distressful procedures.

When conducting the veterinary pre-review, veterinarians typically focus on procedures that may affect animal well-being (e.g., methods of euthanasia, surgical procedures, end points, and the appropriate use of anesthetics and analgesics).

At some organizations, the veterinarian will physically sign the protocol once he or she has completed the review. This process is occasionally used to document that the veterinary review and consent occurred particularly when the project included specific activities such as survival surgeries, the use of pain management medications, and euthanasia. At organizations using this process, the veterinarian's signature verifies that the procedures were medically satisfactory.

INITIATING THE IACUC REVIEW PROCESS

To initiate the protocol review process, organizations must provide to IACUC members a list of submissions for review and adequate time to call a protocol to FCR. Organizations develop different methods for distributing proposed animal care and use submissions to IACUC members.

Upon receiving the submissions, IACUC members must be given adequate time to review each submission and the opportunity to request it be discussed during a full committee meeting. Once each member is given the opportunity to request a submission be reviewed by the full committee, organizations initiate their internal review practices. For example, some organizations review all new submissions during a full committee meeting, while others predominantly use the DMR process. For the latter organizations, the IACUC chair can assign qualified DMRs to review a submission, providing no member has requested it be reviewed during a full committee meeting.

As expected, each organization defines "adequate time" to call for full committee differently. Although some organizations provide committee members 2 days and others 30 days; many BP meeting attendees identified 10 days as "adequate time" to decide if a full committee review is necessary. For example, let us assume an organization decides to give IACUC members 10 days to review each submission to decide if it should be discussed during a full committee meeting. The administrative office may adopt the following process:

- The IACUC administrator develops a submission distribution template that will be used to send weekly submissions to committee members. The template notifies committee members of the submissions being considered for review. The notice indicates that each committee has, for example, 10 days to notify the IACUC administrator and IACUC chair if it would like any of the listed submissions reviewed using the FCR process.
- The submissions are accepted throughout a business week. At the end of the week, a list of submissions received is prepared. The list may be titled, for example, "The January 29, 2015, List of IACUC Submissions." For each submission on the list, the PI's name, protocol number, and submission type (e.g., new, modification, or annual review) are provided with either a synopsis of the submission or access to relevant documents through a private server. The submission list is e-mailed to all IACUC members.
- Once the 10-day period has lapsed, those protocols identified as requiring an FCR would be discussed at the next regularly scheduled full committee meeting, and all others would be assigned to qualified designated reviewers by the IACUC chair.

BP meeting attendees provided additional guidance and indicated that each organization should establish a mechanism to validate that all members were given the opportunity to request FCR for each submission. For example, the use of e-mail can be used to effectively distribute submissions, but if an e-mail is not delivered to one committee member, the requirement to notify each member and allow time to call for full committee review is not satisfied. The use of an e-mail system that records when e-mails are received by the recipient was preferred. In addition, many BP meeting attendees indicated that a policy on how to perform these actions and what constituted the minimum period for calling for full committee review was developed.

Full Committee Review Process

Organizations often use the FCR process to review and approve animal activities. During FCR, a quorum (i.e., a simple majority) of the IACUC voting members discuss proposed animal activities in an open and impartial business session. At that time, committee members question and deliberate on the proposed activities. Once deliberations conclude, a voting member recommends that an official action (i.e., a motion) be taken. If the protocol is complete and the IACUC can make a final determination, members typically motion to either approve or withhold approval of the submission. Once a motion is made, the IACUC chair usually provides members a final opportunity for discussion. In the absence of discussion, a second voting member gestures positive confirmation (i.e., seconds the motion) of the proposed recommendation. The IACUC chair then asks committee members to vote either in favor of or against the motion with the understanding that any member may choose not to vote (i.e., abstain) on the motion. The motion is approved when a majority of the voting members vote in favor of the motion.

In certain circumstances, modifications to the protocol must be made before the IACUC can approve the submission. Organizations practice four primary processes to facilitate the full committee review process: (1) the use of a primary reviewer; (2) consultation with the PI during the FCR; (3) table the submission until the next FCR; and (4) the use of a DMR.

The Use of a Primary Reviewer

At some organizations, primary reviewers are used to support the FCR process. At organizations using the primary reviewer process, the IACUC administrator and/or chair assigns the most qualified committee member to serve as the primary reviewer. For example, a proposal investigating cancer treatments using a mouse model would be assigned to a scientific committee member experienced in cancer research using mice.

Once the primary reviewer is identified, he or she conducts a thorough review of the protocol. Prior to FCR, the reviewer and the PI work together to ensure that the protocol is ready to be approved by the IACUC. For example, in the event the reviewer believes the submission requires modifications before IACUC approval can be granted, he or she would contact the PI, and together they would modify the protocol. Conversely, if the reviewer believes the protocol is ready to be approved, he or she may forward the submission as written to the committee. Approximately 2 weeks (on average) before the FCR, the primary reviewer provides the protocol to the IACUC administrator, who distributes it to the committee as part of the meeting agenda.

During the meeting, the primary reviewer initiates the FCR by summarizing the protocol for the committee members. For example, the reviewer may highlight any painful procedures and discuss humane end points. The primary reviewer may also identify specific procedures within the protocol that require expert input from other committee members. The reviewer typically completes his or her review with a recommendation for approving the protocol or withholding approval. The IACUC then completes the FCR process.

Consultation with the PI during the FCR

Some organizations require PIs to present their protocols to the committee as part of the FCR process. Approximately 2 weeks (the average) before the FCR, the PI provides the protocol to the IACUC administrator, who distributes it to the committee as part of the meeting agenda. The PI initiates the FCR by summarizing the protocol at the IACUC meeting. The PI typically discusses the objectives of his or her research and identifies why it is necessary to use animals. The IACUC then conducts the FCR process after the PI leaves the meeting.

In the event the protocol requires modification to secure approval, the PI may be invited to attend the meeting if clarifications are necessary before the IACUC can grant approval. In this scenario the PI would be available to answer any questions posed by IACUC members. BP meeting attendees agree that this process also affords IACUC members the opportunity to introduce themselves and the IACUC process. It gives committee members the opportunity to explain their role and answer any questions the PI may have about IACUC oversight mechanisms.

Table the Submission Until the Next Full Committee Meeting

At some organizations, a protocol that must be modified before the IACUC can grant approval can be tabled. The *Merriam-Webster Dictionary* defines "to table" as "to decide not to discuss until a later time." Consequently, when an IACUC tables a protocol, the review of that submission occurs at a future meeting. The IACUC chair can decide to table a submission. An IACUC vote to table a submission is not required.

During protocol deliberation, the IACUC identifies modifications that must be made to the protocol before approval. The IACUC administrator notes the required modifications while the IACUC is deliberating. Once the deliberation process is completed, the IACUC administrator reviews the list of required modifications with the IACUC. The required modifications are then communicated (e.g., via e-mail or through a physical meeting or telephone conversation) to the PI. The PI is asked to make the necessary modifications to the protocol and submit it to the IACUC for a subsequent review. The submission is then reconsidered using the FCR process.

The Use of a Designated Member Reviewer

Frequently, a protocol undergoing FCR requires significant changes to be made before it can be approved. At some organizations, the IACUC chair assigns a DMR subsequent to the full committee review. The DMR then serves in an official capacity and approves the protocol, requires modifications be made to the protocol before approval, or asks that the protocol be sent back to the full committee for review. For DMR assignments to be made subsequent to FCR, all IACUC members must agree in writing (**Appendix 11**) that the members at an appropriately convened meeting can unanimously decide to employ the DMR process. If the signed agreement has not been established, for one or more DMRs to be assigned subsequent to full committee review, all full committee members must be in attendance and unanimously vote that the submission be reviewed using the DMR process. Most commonly, the chair served as the DMR, but in several cases, the IACUC administrator, when a voting member of the IACUC, was assigned.

Note: For a DMR to be assigned subsequent to an FCR when all members are not present, the stipulations in the OLAW FAQs (Section D, Protocol Review, Question 19) must be followed.

On occasion, an organization's committee recruits ad hoc specialists to assist with the review of complex issues. While non-committee members may not vote in the final disposition of the proposal, ad hoc specialists provide expert information during discussion that occurs prior to the IACUC's determination.

Although not federally required, some organizations have policies that require specified types of protocols to be reviewed by the full committee; these categories of proposed activities are not eligible for DMR. The criterion is often based upon the invasiveness or type of research being conducted. For example, some organizations require survival surgery and category E protocols to be discussed by the full committee. At other organizations, a broader classification is employed. For example, some organizations require all protocols involving US Department of Agriculture (USDA)-covered species to be reviewed by the full committee, and others require all new and triennial reviews (i.e., de novo) to be reviewed by FCR. Frequently, the IACUC also identifies special conditions that require full committee review (e.g., death as an end point or death losses during a protocol anticipated to exceed 10 percent).

Designated Member Review Process

In addition to the FCR process, the regulations and policies include provisions for conducting reviews using the DMR process. Organizations using the DMR process permit the IACUC chair to assign a member or members of the IACUC with the appropriate expertise to serve as the DMR and review and then approve the protocol. During the DMR process, the assigned reviewers act independently of the IACUC to review and approve the assigned protocol. They may or may not choose to consider the opinions of other members. The decision of approving the protocol belongs to the designated member(s) alone. The DMR has the authority to approve a project, require modifications to secure approval, or refer it back to the full committee for review. The DMR does not have the authority to disapprove a protocol.

In order to initiate the DMR process, certain conditions must be satisfied. Before a committee member can be assigned by the IACUC chair as the DMR, a list of proposed research projects to be reviewed must be provided to all IACUC members.

Written descriptions of the projects must be available to the IACUC and each member must be given adequate time to decide whether he or she would like to have the study receive FCR. In addition, the IACUC chair must decide which member has the appropriate expertise to act as the DMR. The IACUC chair can also assign multiple DMRs for the same submission, with each reviewing the same version of the protocol. If this scenario occurs, all DMRs must agree on the disposition of the protocol. For example, if a protocol requires modifications before final approval can be granted, the DMRs collectively must agree that the PI's modifications adequately address the concerns and that the project can be approved. If the DMRs disagree, the proposal must receive an FCR.

It is important to restate the philosophy of unanimous agreement. In the event that more than one member is assigned as the DMR to the same protocol, all must agree that the submission can be approved before final approval is granted. If there is more than one reviewer, it is critical that they do not "vote" on the disposition of the project. The DMR process is not a democratic majority, but rather an agreed unanimous conclusion. The decision of the reviewers must be unanimous; any decision other than unanimous results in the protocol receiving an FCR.

Assigning the Designated Member Reviewer

Federal policy requires DMR assignments to be made by the IACUC chair. During past IACUC administrators' BP meetings, attendees discussed various methods that have been used to satisfy the federal requirement. BP meeting attendees agreed that it may be more reasonable to have the IACUC administrator assign the designated member reviewers, but regulations require the appointment to be made by the IACUC chair. There were two methods commonly used by attendees to fulfill the regulatory requirements without the process being unduly burdensome.

In both cases, the IACUC administrator facilitates the process. One method is for the IACUC administrator to identify a qualified designated reviewer candidate to serve as the DMR for specific submissions once the "call for full review" period had lapsed. In this situation, the recommendations were e-mailed to the IACUC chair, who confirmed, by using the "reply" function of the e-mail, that the appropriate assignments had been made. The IACUC administrator then notified the appointed DMRs of the assignments. The other process, although similar, involves the IACUC chair e-mailing the DMR assignments to the IACUC administrator once the "call for full review" period has lapsed; the administrator then notifies the DMRs of their assignments.

Benefits of Using the Designated Member Review Process

BP meeting attendees indicated that effectively using the DMR process significantly increased the efficiency of their review process and decreased the workload for committee members. In most cases, organizations developed standard operating procedures (SOPs) which identified the circumstances and procedures for engaging the DMR process. BP meeting attendees reported that the DMR process generally occurred 10 days or less, while the FCR required between 30 and 45 days for completion.

Potential Pitfalls of the Designated Member Review Process

Colleagues from federal offices offered the following advice as it relates to assigning DMRs. The USDA representative indicated that assigning DMRs is not directly addressed in the regulations, but that the IACUC administrator should not make the formal assignment unless he or she has received written delegation from the IACUC chair. It was also recommended that the IACUC chair be copied on the message that advises committee members of their DMR assignments.

In addition, regulatory attendees reiterated that before a DMR can be assigned, all committee members must have an opportunity to review the protocols (i.e., a time period is not defined in the regulations, so each organization must develop reasonable time lines) and to request full committee review at their discretion. Prior opportunity to review means that each member must receive, at a minimum, a list of protocols with a brief but detailed synopsis (i.e., not just a list of PIs and protocol numbers) available at the request of a committee member. Some organizations simply provided the protocol rather than craft a new description of the details of the proposed activities and provide the complete documents (e.g., protocols)—for example, using a private server.

Occasionally, organizations have a rotating roster of DMR assignments developed and previously approved by the IACUC chair. In other cases, the IACUC vice chair is authorized in writing to assign the DMRs when the IACUC chair is unavailable. However, it is not the intent of the regulations or policies to have the duty routinely delegated to anyone other than the IACUC chair.

Conducting Protocol Reviews Using Video and Teleconferencing

Although the USDA and OLAW permit limited telecommunication for performing IACUC activities, almost all organizations limited this option to special circumstances. BP meeting attendees stressed the criticality of the regulatory phrases "without compromising the quality of deliberation and interaction" and "assuring teleconferencing provides the same or better opportunities for robust deliberation and interaction."

It is a best practice to conduct protocol discussions and reviews during face-to-face meetings. However, it is appropriate for a committee member working at a satellite facility, many miles from the core operations, to attend a meeting using teleconferencing equipment. Conversely, it is not acceptable that all members on the core campus meet by teleconference in preference to a face-to-face meeting. Face-to-face meetings provide the benefit of personal interaction and body language—actions that could have merit in the discussion of the proposal before the committee.

The provision to permit the use of video and teleconferencing for IACUC meetings should not be misused. These methodologies were established to allow committees to conduct protocol reviews in difficult or challenging circumstances and are not considered an acceptable replacement for routinely convened face-to-face meetings. Teleconferencing should not be used for convenience.

Full Committee versus Designated Member Review

In general, represented organizations equally used the DMR and FCR processes. However, an increasing number of organizations are maximizing, or expressed an interest in maximizing, the use of the DMR process. Attendees favorably considered the efficiencies offered by the DMR process, as compared to other review options. Attendees were in full agreement that if the DMR process were used, the organization must have clear and defined procedures for its engagement.

Receiving Proposed Animal Activities

BP meeting attendees described many different protocol templates, but they all had a common theme. The protocol template was designed to gather the information needed for IACUC to assess and approve animal activities proposed by researchers. Organizations designed protocol templates that gathered the information needed to safeguard animal welfare and determine that the proposed activities complied with federal regulations and policies.

BP meeting attendees identified many options for the submission of protocols, amendments, and annual reports. Although no process was proven to be superior, each fit the needs of and was unique to the organization.

Although the submission processes were unique, they did share some common characteristics that organizations identified as steps to improve the efficiency of the process. For example, many organizations no longer require PIs to physically sign their submissions. This change facilitated the use of e-mail allowing PIs to more efficiently submit protocols to the IACUC administrator. To ensure that PIs are engaged in the submissions process, most organizations require IACUC submissions to come directly from the PI's organizational e-mail address. The shift to accepting protocol submissions electronically also allowed the IACUC administrator to use e-mail to distribute protocols to IACUC members for review.

Using Approved SOPs as Part of the Protocol

Some organizations permit PIs to reference a particular IACUC-approved SOP in their protocol rather than requiring a complete description of a procedure. For example, the IACUC may develop and approve an acceptable process (i.e., SOP) for collecting tissue samples that will be used to genotype mice. At some organizations, rather than PIs providing the specific procedures in their protocol for collecting tissue samples for genotyping, they indicate that tissue samples for genotyping mice will be collected according to the IACUC's approved SOP on collecting tissue samples in mice for genotyping. In cases where SOPs were referenced, most organizations required the PI to include a copy of the SOP as part of the submission.

While this may be a method to make the process more efficient, most BP meeting attendees expressed reservations with this practice, as SOPs are dynamic documents that may be modified. Organizations using SOPs to supplement the protocol narrative must re-review SOPs on a frequency that assures congruence with the protocol and/or reviews/approves an amendment to the protocol when the SOP is changed.

Most BP meeting attendees agreed that the information should be put directly in the protocol; otherwise, the process becomes too automated and PIs begin to rely too much on SOPs and do not think about their protocol. If SOPs are employed, they should be reviewed periodically and if a "search for alternatives" is part of the SOP, it should be regularly updated. A system must be in place to inform PIs when an SOP is revised.

One BP meeting note was that IACUCs develop best practices or SOPs for a specific procedure, but that PIs still list how they will conduct the activity in their protocol. Caution should be used if a PI refers to a guideline in a protocol, because guidelines by nature are intended to provide guidance when developing procedures. It is important to note that since some protocols are only reviewed every 3 years, the IACUC must have a process to assure that the most current SOP is being used. Additionally, if the SOP is being employed on a protocol that involves USDA-covered species, then the SOP must be reviewed at least annually. In addition, when interpreting the AWAR, the PI must include in the protocol a summary of the search for alternatives, including for SOPs where potentially painful procedures will be performed. This summary must be done by the PI and included as part of the IACUC proposal.

If SOPs are employed by an IACUC at an Association for Assessment and Accreditation of Laboratory Animal Care (AAALAC)-accredited organization, the SOP must be included (in some manner) in the program review. Attendees noted that when referencing an SOP in the protocol, the IACUC must also consider how the SOP activity affects protocol outcomes.

Protocol Modifications

Most scientists discover at some time during the 3-year life cycle of their protocol that they have a new idea, wish to try a different procedure, or wonder how an alternate agent would impact the course of their research outcomes. While such adjustments are a characteristic of an engaged scientific investigation, the scientist is obligated to perform only those procedures, use those agents, or attempt those methods that have been approved by the IACUC. Since all animal care and use procedures must be approved by the IACUC prior to being performed, organizations have developed methods for modifying protocols. The instrument used to achieve the approval for modification is called an amendment.

Most BP meeting attendees indicated that their organizations accepted protocol amendments as supplemental documents. In this particular situation, several organizations indicated that a protocol could only be amended four or five times before a new protocol incorporating all of the amendment information was required. Once the organization's IACUC reviewed and approved the new submission, the other would be closed. Other organizations established policies that required PIs to include amendments directly into the text of their approved protocol, thereby creating a new and complete protocol. These organizations generally required the PI to identify the new text (i.e., the amendment) by either highlighting or changing the font to differentiate it from the rest of the protocol. In both cases, the amendments were reviewed and approved by the IACUC using either the full committee or designated member review process.

Significant versus Nonsignificant (or Minor) Modifications

Significant changes are defined by both USDA regulation and PHS policy (OLAW FAQ Section E, Program Review and Inspection of Facilities: Question 2). BP meeting attendees agreed that the more practical definition for a significant change was any action or outcome that may affect the health and well-being of the animals or humans engaged in the research activity.

BP meeting attendees agreed that only significant protocol changes (i.e., modifications requiring committee review and approval) must be reviewed and approved by the IACUC. However, the majority agreed that all protocol changes, including minor amendments, should be submitted through the IACUC's administrative office for the purpose of tracking and documenting.

For consistency purposes, each organization must establish a policy that defines significant protocol changes requiring IACUC review and approval—those changes that were not significant (minor) and could be processed by an alternate methodology. The policy should also identify types of protocol changes that may be reviewed and approved by the IACUC administrator, DMR, or IACUC chair.

Some organizations identified certain protocol modifications that can be reviewed and approved by the IACUC administrator (e.g., personnel additions, title changes, adding a new granting agency, a change in the method of animal acquisition and/or vendor). Many attendees noted that their organization did not require IACUC review of the minor modifications, but that all approvals (minor modifications) were reported to the IACUC at the next meeting. BP meeting attendees were clear that this type of approval process should not be referred to as an "administrative approval," believing that such terminology was misleading and could present problems with PIs understanding necessary approval requirements. Regardless of the process being used, the methodology should be included in the PHS assurance, for those organizations that are assured.

Personnel Changes

Over 50 percent of those attending BP meetings agreed that changes in protocol personnel, other than the PI, could be reviewed and approved by the IACUC administrator. Representatives from the USDA and OLAW confirmed that handling personnel changes administratively is appropriate, providing the individual in question is not the PI in the study. The regulatory concern associated with the addition of personnel to an approved animal study was that the IACUC is required to verify that protocol participants have the appropriate qualifications and skills to conduct the procedures they plan to perform. Several attendees noted that their IACUC had developed operating procedures for IACUC administrators to use when evaluating protocol personnel additions. Generally, the IACUC's SOP defined the minimum training each animal user must complete for certain types of activities. For example, when someone is added to a protocol that involves a surgical procedure in mice, the new protocol participant would be required to complete the online compliance training, the rodent surgery training module, the species-specific mouse module, and the Occupational Health and Safety Program (OHSP) training module. The IACUC administrator would review the folder to confirm successful completion of each module prior to granting the addition of the personnel to the protocol. If the individual wished to be approved based on current skills or desired not to complete the required modules, the IACUC administrator would forward the personnel amendment to the committee for formal IACUC action.

Annual and De Novo Reviews

While the USDA requires annual reassessment of approved activities, many organizations had developed an annual review form for all approved animal protocols (i.e., not just USDA-covered species) and required PIs to submit relevant information to the IACUC at least annually. The type of information typically requested included, whether any unreported adverse events occurred over the past year, a research progress report, the number of animals that were used, and a summary of any changes that were made since the last report.

In accordance with the PHS policy, a de novo review of each animal care and use protocol must be conducted by the IACUC at least every 3 years when the project is funded through PHS funds (e.g., NIH, Centers for Disease Control and Prevention [CDC], Food and Drug Administration [FDA]). During the de novo review, the IACUC must re-review the activities associated with an ongoing animal care and use protocol. The review must be consistent with the regulations and ultimately approved by the IACUC. BP meeting attendees reported that their organizations used either the DMR or full committee review process to conduct de novo reviews (3-year re-reviews). In summary, while most BP represented institutions perform both annual and triennial reviews on all IACUC-approved protocols, in fact, Annual Reviews are required by regulation only for USDA-covered activities and Triennial Reviews (De Novo) are required, by policy, only for PHS-covered activities.

Expedited Reviews

BP meeting attendees were using several different practices for managing reviews requiring a quick turnaround. The regulatory representatives in attendance expressed caution with the use of the term "expedited," as such often implied a shortcut or minimal approach to review and approval. BP meeting attendees agreed that the process was an accelerated review process rather than a shortcut review. All agreed that the use of the term "expedited" should be discouraged in preference for the term "accelerated."

Reviewing Proposals for Custom Antibody Production

Frequently a scientist conducting only in vitro research uses custom antibodies that must be generated in animals. When antibody production involves the use of animals, even if all of the other procedures are in vitro, IACUC oversight is required. In particular when in vitro studies are funded through a PHS grant or contract, the grantee needs to have a mechanism to ensure that custom antibodies are produced by an organization maintaining an animal welfare assurance with OLAW. An assurance is not required for the use of commercial or off-the-shelf antibody production.

These antibodies are often custom produced to satisfy the researcher's need. In the event that a research scientist is conducting PHS-funded research using antibodies that were custom made for his or her research needs, the scientist must ensure that the antibodies were made under the supervision of an IACUC. Some organizations satisfy this requirement by requiring PIs to complete a "custom antibody production" protocol application. The document typically asks the PI to provide the PHS assurance number, and USDA registration number of the organization producing the antibodies. Frequently, AAALAC-accredited organizations also ask the producer of the antibodies to provide the accreditation status.

The USDA considers custom antibody production to be research. Consequently, antibody production is conducted under the auspices of the USDA, and animals used for those purposes must be reported on the organization's annual report to the USDA. When antibodies are produced using commercial sources, the producer of the antibodies reports the animal use. Since commercial antibodies may be purchased off the shelf or from a published catalog (does not require IACUC approval), the IACUC needs to determine whether a PI is using custom or commercial antibodies. A simple test to determine if a protocol is necessary is to ask: Can the antibodies be purchased by anyone out of a publicly available catalogue? If yes (anyone could purchase the exact same antibody), then no protocol is required. If no (the antibody is being prepared according to the PIs specifications or using the PIs protein), then a protocol is required—these are defined as "custom" antibodies.

Reviewing a Protocol for Scientific Merit

The IACUC administrators discussed whether the IACUC is responsible for reviewing the scientific merit of a submitted protocol. Those in attendance agreed that if a protocol does not receive a peer review as part of the funding process, the IACUC should review the protocol for scientific merit (according to US government principle II and the eighth edition of the *Guide for the Care and Use of Laboratory Animals*). Some organizations request that the IACUC chair sign protocol approvals to verify that the proposal has scientific merit before the committee considers it for review. Conversely, if a project has been peer reviewed and funded, it would be expected that the funding agency considered the relevance of the scientific merit, even though an organization's IACUC may reconsider the merit.

OLAW expects IACUCs to uphold US government principal II: "Procedures involving animals should be designed and performed with due consideration of their relevance to human or animal health, the advancement of knowledge, or the good of society." Therefore, some merit review by IACUCs is needed.

CHAPTER 6

Protocol Noncompliance

SCENARIO

During a GEU IACUC meeting, the compliance coordinator reported three concerns to the committee. The compliance coordinator told committee members that she and the principal investigator (PI) had worked together to resolve all three concerns. She recommended that the IACUC take no further action.

The first concern was seven identical issues coupled into a single case. Seven different PIs allowed students to conduct animal use activities even though the IACUC had not approved them to work with animals. Since an IACUC-approved policy authorized the compliance coordinator to administratively approve personnel to work with animals, the coordinator asked each of the PIs to submit an amendment to their protocols to add the students. The coordinator ensured that the students completed the required and specific training (e.g., animal user and occupational health and safety program [OHSP] training) and ultimately approved the amendments. The coordinator had conducted a postapproval monitoring visit and concluded that no animal welfare or adverse care or use had occurred; it was only a matter of unapproved personnel working with the animals.

The coordinator continued her report by describing the second concern. She explained that Dr. Smith was approved to conduct survival surgery but the surgical records were incomplete. She indicated the records did not document that animals received postoperative analgesia. The coordinator reported to the IACUC that Dr. Smith indicated that he was "fairly sure" analgesia had been provided in all cases. The coordinator believed that Dr. Smith simply needed to be shown what type of information was required in the surgical record. She had provided Dr. Smith with a surgical record template and provided instruction to him and the laboratory staff on maintaining accurate and complete records. She felt sure that the concern was resolved.

The final concern involved a PI that was approved to collect blood from mice by nicking a tail vein and collecting the samples using a capillary tube. During the visit with the PI, the coordinator learned that the PI had been unable to collect enough blood using the tail nick method. Consequently, he decided to collect the blood by using a retro-orbital blood collection technique. The PI indicated that he was approved to do the same procedure as part of a previously approved protocol;

therefore, he did have the skills to correctly conduct the procedure. The PI indicated that using this method enabled him to collect the volume of blood required. The coordinator noted that retro-orbital bleeding was not IACUC approved. She assisted the PI with preparing a modification to add the new method of blood collection to his protocol, which was subsequently approved by the IACUC. She also reported to the committee that the PI stopped conducting retro-orbital blood collection until after the modification was approved by the IACUC.

The compliance coordinator concluded her report to the committee by reiterating that she had resolved all of the reported concerns and recommended that the IACUC require no further action. She noted that although the retro-orbital blood collection was done without IACUC approval, this particular PI had worked very well with the IACUC previously and always submitted to the IACUC for approval to conduct experimental procedures involving animals. It was her opinion that the primary concerns were identified and resolved.

The IACUC members acknowledged that the concerns were addressed by the coordinator and that the noncompliant issues were resolved. The members discussed the compliance coordinator's recommendation of requiring no further action. During the discussion, a committee member reminded his colleagues that considering and resolving cases of noncompliance was the responsibility of the committee and that they should, for example, consider the cases of noncompliance, report them to government agencies if necessary, and develop methods to ensure that the concerns did not recur.

After further discussion, the committee made the following decisions. Regarding the personnel concerns, the IACUC members agreed with the coordinator and required no further action. The members noted that the decision was based upon the PIs' rapid responses and completed training for the new personnel. The committee disagreed with the coordinator regarding the other two situations. Committee members noted that failure to maintain adequate records and performing a potentially painful procedure without IACUC approval were violations that exceeded the minimum threshold for reporting to the Office for Laboratory Animal Welfare (OLAW).

Regulatory Requirements and Resources

- **Public Health Service (PHS) policy (Section IV (B) (4), p. 12)** requires "the IACUC to review concerns involving the care and use of animals at the organization."
- **PHS policy (Section (IV) (C) (7), p. 15 and (IV) (F) (3), p. 18)** requires self-reporting through the IO to OLAW and PHS funding agencies when (1) serious or continuing noncompliance with the PHS policy occurs, (2) a serious deviation from *the Guide for the Care and Use of Laboratory Animals* (henceforth referred to as the *Guide*) occurs, or (3) when the IACUC suspends an activity.
- **Animal Welfare Act regulations (AWARs) (Section 2.31(c) (4))** "require the IACUC to review, and, if warranted, investigate concerns involving the concerns and use of animals at the research facility."
- **AWAR (Section 2.31(c) (3) and (d) (7))** require self-reporting through the institutional official (IO) to the US Department of Agriculture (USDA) and federal

funding agencies when (1) a previously approved activity is suspended, or (2) when a significant deficiency is not corrected within 15 business days of the established corrective action deadline.

INVESTIGATING AND REPORTING ANIMAL WELFARE CONCERNS

In the process of administering an animal care and use program (ACUP), the IACUC will encounter activities that may violate federal standards. These situations must be investigated and resolved by the IACUC.

Organizations must develop a process that ensures that alleged incidents of non-compliance are reported to the IACUC. For example, this type of incident may be encountered by compliance coordinators while visiting principal investigators; incidents may be reported by animal husbandry staff members or the general public, discovered by IACUC members during routine facility inspection, or concerns identified by the AV, IACUC Chair, or as an anonymous report.

All alleged violations of the animal welfare standards reported to the IACUC must be investigated. IACUC administrators attending best practices (BPs) meetings offered suggestions for conducting investigations.

Conducting Investigations

Method 1

Form an advisory group to gather relevant information and report the findings to the IACUC with the IACUC deciding the next step. When first advised of an allegation, the IACUC chair may ask an individual or individuals to gather additional information regarding the allegation. The information gathered will be communicated to the IACUC, which will then determine if sufficient credibility issues or questions exist to form an investigation subcommittee. In most instances, the subcommittee includes the IACUC administrator and other relevant research compliance staff members such as compliance coordinators and postapproval monitors. In other organizations, investigation subcommittees comprise a combination of administrative IACUC staff and IACUC members, while still others select a veterinarian and a committee member with a specific expertise (e.g., a scientist with experience in using mice for noncompliance involving rodents).

In most cases, the investigation is led by an individual in a research administrative role (e.g., the IACUC administrator, the IACUC Chair, a vice chair), senior IACUC member, or veterinarian not affected by the alleged event. Institutional representatives indicated that their noncompliance subcommittee members would familiarize themselves with the case and begin the investigation by interviewing relevant individuals. If an alleged case was reported by an animal care technician, for example, the subcommittee might first meet with the animal care technician unless the report was made anonymously. If the report was anonymous, the subcommittee members might review the written reports at their disposal. In the event that a non-compliant activity involved a PI, the subcommittee would interview the PI as well as relevant staff members associated with the concern.

The subcommittee then gathered and collated the relevant information and pre-pared a report for the IACUC. The report typically summarized and verified the legitimacy of alleged activity. In some cases, the subcommittee might try to resolve the issue before reporting to the IACUC. For example, the subcommittee might encourage a specific PI to voluntarily discontinue an animal use activity until the IACUC had completed the investigation (only the IACUC can suspend an activity).

The incident did not need to be reported until after the IACUC completed the investigation and determined whether noncompliance occurred. The subcommittee might help the PI prepare amendments to modify the PI's protocol or to help the PI address any other relevant concerns (e.g., develop complete records or add personnel to a protocol).

Since the IACUC is the body that deliberates and issues a decision relating to investigations of alleged incidents when the subcommittee has completed its inves-tigation, a recommendation is referred to the IACUC. A meeting of the IACUC is scheduled for final deliberations and to make a final decision. At some organizations, these styles of meetings (investigation reviews) are held during a regularly called IACUC meeting; at other organizations, they are held as a special IACUC meeting for the sole purpose of deliberating and deciding the outcomes as identified by the subcommittee.

Method 2

Alleged concerns of noncompliance are reported directly to the IACUC, and the investigation is conducted during a face-to-face meeting of the committee. BP meeting attendees noted that organizations following this practice depend heavily upon their IACUC administrator. The IACUC administrator will need to familiar-ize himself or herself with the allegation, schedule relevant staff members to appear before the committee, and arrange for collection of forms or documents pertinent to the discussion to be made available for the investigation.

Depending upon the organization culture, the IACUC administrator may start a discussion about an alleged case of noncompliance by summarizing the activity for committee members. Organizations typically allow the committee members a period of time to discuss the allegation, which is followed by interviews with the involved individuals. During the interview process, the IACUC typically asks ques-tions of the guests to clarify any activities relating to the event. During the meet-ing, the IACUC ultimately deliberates the issues to decide the outcome and whether sanctions to impede future occurrences should be implemented.

Suspensions

In some cases during the course of an investigation, the IACUC may find instances when circumstances threaten the welfare of animals. At times, the IACUC may also discover researchers conducting activities that have not been approved by the IACUC. Situations such as these may require the IACUC members to exercise their responsibility of suspending research activities. In addition, suspensions can also be initiated by the IO. It is important to recognize that the IO cannot reactivate a protocol that has been suspended by the IACUC.

In the event that the IACUC suspends an approved animal care and use activity, the suspension must be reported to the relevant funding and regulatory agencies. For example, if all or part of an approved IACUC activity is suspended and if the activity is funded by the Public Health Service (e.g., the National Institutes of Health [NIH]), the suspension must be reported to OLAW and the appropriate funding component. When reporting to funding agencies, it is appropriate to provide the grant number and the PI's name.

If the organization's PHS assurance indicates all animal care and use projects will be equally evaluated, whether the project is PHS funded or not, the suspension must be reported to OLAW.

If the project involves US Department of Agriculture (USDA)-covered species such as rabbits or guinea pigs, the suspension must also be reported to the USDA. If PHS-funded research activities are suspended, then PHS funds cannot be used to support the project during the suspension period. For example, the daily per diems for maintaining a rodent breeding colony cannot be funded with PHS monies during the suspension period.

BP meeting attendees suggested that the IACUC should suspend animal use activities only when all other options have been completely exhausted. One option suggested by the attendees was to suspend an individual's privilege to use animals rather than suspending the research activities. However, according to regulatory representatives, if the individual suspended from conducting animal research activities is the PI on the grant, then suspension of his or her animal use privileges must be reported to OLAW. In this case, the research activities may be continued if a qualified and trained individual is able and willing to assume the role of research activity performance during the PIs suspension (i.e., change the role of PI to another person by IACUC-approved amendment).

Reporting

Organizations have developed various methods for determining when and what type of incident should be reported to federal agencies. Some organizations delegate the responsibility of reporting noncompliance to an individual in the compliance office, which is frequently the IACUC administrator. At other organizations, the committee determines which agencies (e.g., the USDA, OLAW, Association for Assessment and Accreditation of Laboratory Animal Care [AAALAC]) need to be notified when noncompliance is confirmed. BP meeting attendees agreed that when there is doubt about reporting to OLAW, OLAW should be consulted. Organizations could also review the OLAW notice "Guidance on Prompt Reporting to OLAW under the PHS Policy on Humane Care and Use of Laboratory Animals" (http://grants.nih.gov/grants/guide/notice-files/NOT-OD-05-034.html).

In most cases, IACUC-suspended activities need to be reported to all agencies with the following clarifiers:

- If the organization is accredited, then significant deficiencies in the ACUP must also be reported to AAALAC. The notification can be made at the time of the infraction or in the annual report to AAALAC. It is a common practice for accredited organizations to copy AAALAC on deficiency reports sent to OLAW and/or the USDA.
- OLAW must be notified if an activity significantly deviates from the *Guide* if the activity is PHS funded or when the PHS assurance considers all animals equally.
- If a significant deficiency provided as part of the semiannual inspection report is not corrected within the IACUC-established time line, then the USDA must be notified within 15 business days, providing the deficiency involves USDA-covered species. If a noted deficiency is not significant but has been repeated as part of the last inspection report, the IACUC should change the status to "significant" based on the fact that an ongoing problem exists. A new plan and schedule for correction must be developed and adhered to or it will require reporting as an ongoing noncompliance.
- Organizations often file preliminary reports (phone or e-mail) by notifying the appropriate agencies once an alleged issue is identified. If a preliminary report is made and found to be unfounded, the preliminary report can be withdrawn by contacting the relevant agencies.

Once an organization's IACUC makes a final determination, a formal report can be filed with the appropriate agencies. The final report should include a summary of the incident(s) and the IACUC discussion, decisions, and any imposed corrective actions. Since reports to federal agencies are subject to the Freedom of Information Act, it is appropriate to use various codes in the report, providing a reference to identify the codes is maintained and understood by the agency receiving the report.

Corrective Actions

On occasion, the IACUC must develop corrective actions to address noncompliance. BP meeting attendees agreed that a tiered approach was most useful (except if the original incident was egregious). Organization representatives attending past BP meetings identified a common tiered approach.

Many times corrective actions for first offenses were simple. For example, the IACUC chair will often issue a notice to the individual who committed the infraction and copy that person's supervisor. Several BP meeting attendees noted that their organization required those found to be in noncompliance to undergo retraining and frequently that training was followed by a face-to-face meeting with the IACUC chair and/or the IACUC administrator. In select cases, the IACUC may provide a copy of a policy to the PI and ask that he or she verify in writing that they have received and read the policy.

In the case of second or subsequent offenses by the same individual, BP meeting attendees reported that IACUCs considered additional training, higher level supervisory advisement, meetings with senior members of the organization, restricted animal use privileges, termination of animal use privileges, protocol termination, and, in the most egregious cases, recommendations to the organization for employment termination. A second (or more) offense usually involves visitation to the PI's laboratory by the IACUC chair and/or other members and requirements for specific and occasionally extensive training.

CHAPTER 7

Protocol and Grant Congruency

SCENARIO

A GEU principal investigator (PI) was awarded a National Institutes of Health (NIH) grant to evaluate clinical treatments that would minimize scarring from injuries. The project was a 5-year study that assessed methods for treating skin lacerations in the first phase of the study (i.e., the first 3 years) and treatments for burn injuries in the second phase (i.e., the last 2 years). The PI submitted a research proposal to the IACUC that evaluated only the first phase of his study. He decided that he would submit the second phase (i.e., burn studies) when he prepared his triennial submission.

The IACUC administrator and PI met to prepare his triennial research proposal. The proposal continued to outline how the PI incised the skin of experimental animals. It also outlined the medical care that included novel treatments, which promoted healing and minimized scarring. By the end of the third year of research, the PI was able to demonstrate that his process for treating skin lacerations was therapeutic and the patients were left with minimal scarring.

The PI explained to the IACUC administrator that the next phase of his research was to evaluate his treatment regime on skin injuries resulting from second- and third-degree burns. The scientist indicated that he planned to anesthetize experimental animals, shave their hair, and use various methods to inflict second- and third-degree burns. He added that these experimental animals would receive the same medical treatments that proved to be successful in the laceration studies.

The IACUC administrator disclosed to the PI that she would help him prepare his triennial submission, but alerted him that the organization had a policy prohibiting burn-related studies that involved experimental animals. The PI replied that his research was funded by the NIH, and the research design was based on a 5-year study. He stated that the primary objective of the research was to develop new and innovative techniques for treating burn victim injuries, and that the burn studies were an essential component of the experimental design. The PI reiterated the success of his initial experiments and stressed that completing the burn study experiments was a condition of receiving the NIH award, for which the organization has already received the funds.

The IACUC administrator informed the PI that since the burn studies were a component of the NIH grant, those procedures should have been described in the initial submission to the IACUC. She explained that GEU adopted this practice to ensure that all of the animal activities described in a grant application are reviewed and approved by the IACUC prior to the organization receiving the funds. The PI replied that, although the details of the burn studies were not described in the IACUC submission, the entire vertebrate animal section of the grant had been provided to the committee for evaluation. He noted that GEU's policy indicates that IACUC reviews ongoing research every 3 years. He explained that based on the policy his initial application to the IACUC included the details of those procedures he would conduct during the first 3 research years with the understanding that the details of his remaining activities would be thoroughly described in subsequent IACUC submissions. He reiterated that a copy of his grant was part of his submission to the IACUC, which included his entire research plan. The PI indicated that through his submission the IACUC was made aware of the burn studies and that he expected the committee to contact him if there were any problems with his overall research design, including the burn study.

Since an organization's IACUC must review and approve animal care and use activities described in a grant submission to the Public Health Service (PHS) before accepting awarded funds, the organization must develop a process to document that this practice occurs. In this particular case, the PI submitted a protocol to the IACUC that listed the activities he would conduct during the 3-year term of his approved IACUC protocol. In addition to the protocol, he provided a copy of his grant to the IACUC, which discussed those procedures to be conducted in subsequent years. Since the entire research design was submitted to the IACUC for review, the PI believed the committee understood that burn studies would eventually be conducted upon review and approval by the IACUC. The organization did issue a concordance memorandum to the NIH based upon the original protocol. Consequently, the organization did not conform to the PHS policy since the burn studies were not approved during the initial IACUC review. In addition, if the burn studies are not approved by the IACUC and ultimately conducted, the PI will not be able to satisfy the terms and conditions of the NIH award. As a result of not being able to satisfy the terms and conditions of the award, the organization may be required to return the proceeds of the award to the NIH.

Regulatory Requirements and Guidance Documents

- **PHS policy** requires IACUCs to review and approve those components of PHS-supported grant applications and contract proposals that relate to the care and use of vertebrate animals to ensure that the proposed research complies with the PHS policy (Section IV.C(1)). In addition, the PHS policy requires organizations to provide written verification to the granting agency that those procedures relative to the care and use of animals described in a grant application have been reviewed and approved by an IACUC prior to receiving an award (Section IV.D(2)).
- **The NIH grants policy statement** includes the terms and conditions that an organization must satisfy to be eligible to receive the granted funds. Specific sections

of the grants policy statement discuss the organization's responsibility for ensuring that the IACUC has reviewed and approved the animal activities described in a grant application prior to receiving the funds.

- **Part II: terms and conditions of NIH grant awards (Subpart A: General—File 2 of 5)** indicates that the organization, as a condition of receiving funds, agrees that "the IACUC should ensure that the research described in the grant application is consistent with any corresponding protocols reviewed and approved by the IACUC."
- **OLAW (Office of Laboratory Animal Welfare) frequently asked questions (FAQs); Section D (Protocol Review), Question 10** asks, "Is the IACUC required to review the grant application?" The FAQ discusses the importance of ensuring that the information on the protocol is consistent with that provided on the grant application. Although neither the PHS nor the NIH grants policy requires a side-by-side comparison of the application and the protocol, some organizations use this process to ensure that their IACUC has reviewed and approved all of the procedures relating to animal care and use listed in grant proposals.

BACKGROUND

Research involving vertebrate animal activities must comply with federal regulations and policies; those experiments must be reviewed and approved by an IACUC before initiation.

IACUCs at most organizations use a protocol template to obtain the information they need to evaluate proposed animal care and use activities. The template includes a series of questions that will be completed by the PI and used by the IACUC to ensure that proposed animal use activities comply with federal regulations and policies. The protocol template is designed to ensure that IACUCs have all of the information required to make an informed decision.

The animal activities described in the protocol template and provided to the IACUC for evaluation should be consistent with those animal use activities describe in the PI's grant proposal. Upon accepting the granted funds, the organization has guaranteed that those activities and actions described in the grant proposal will be performed. Consequently, the IACUC protocol should thoroughly describe all of the live animal activities associated with the performance of the specific tasks or objectives described in the grant proposal.

Scientists frequently apply to organizations such as the NIH for funds to support their research activities. The application process usually involves completing and submitting a detailed proposal (i.e., grant application) to the funding agency. The grant application can be quite extensive, and it methodically describes all of the investigator's purposed research activities. The proposal may comprehensively discuss in vitro procedures, processes for performing biochemical assays, and methods for sequencing DNA. The proposal may also include discussions about past research and experimental results that justify or strengthen the research activities proposed in the grant application. In addition, the application may include budgetary information and the credentials of collaborating scientists, as well as animal research activities.

Since IACUC proposals contain different information and different levels of detail than grant applications, the most common practice is for animal care committees to review IACUC protocols rather than the grant applications.

Scientists transpose relevant information from their grant application to their IACUC protocol and therein exists the potential for some information provided in the grant application not to be transcribed accurately or completely into the IACUC proposal. If an IACUC does not review all of the animal care and use procedures in the grant application before the organization receives the financial resources, the organization is placed in jeopardy with the contractual language of the PHS policy and the grant's terms and conditions statement. As discussed, federal policies require grantees to verify that the IACUC reviews and approves all of the animal procedures fiscally supported by the PHS and the organization. Organizations typically choose not to have their IACUCs review and approve the PI's grant proposal. They often establish a process through administration that ensures that all of the animal care and use activities described in the grant proposal are included in the protocol submission.

A common method implemented by organizations is to assign to the IACUC administrator the task of confirming that all of the animal activities described in a grant proposal are IACUC approved. This activity is commonly known as a "protocol and grant congruency review." Depending on the organization's best practice, after a robust congruency review has been performed, the IACUC administrator may forward the protocol to the IACUC for review or issue a "concordance memorandum" to the PI. Subsequently, the concordance memorandum or perhaps an IACUC protocol approval letter will be forwarded to the organization's grant office and/or the funding agency; only then are the funds released to the grantee organization.

Since it is the organization's responsibility (PHS policy, Section IV.C (1)) to ensure that the IACUC conducts the relevant evaluations before accepting PHS-granted funds, many times the protocol and grant congruency review is conducted through the IACUC office (e.g., by the IACUC administrator). BP meeting attendees noted that 58 percent of the time, the protocol and grant congruency reviews are conducted by the IACUC administrator. Other organizations placed this responsibility with the sponsored program office staff, the attending veterinarian (AV), or the IACUC chair.

IS CONGRUENCE REQUIRED?

At a minimum, congruency reviews are required for all new PHS-funded animal research activities. Since federal regulations and polices do not necessitate a similar process for non-PHS-funded research, the decision to conduct congruency reviews on all funded research varies from organization to organization, and from granting agency to granting agency. BP meeting attendees have discussed whether organizations should conduct congruency reviews on all projects, or just those funded through federal agencies. In September of 2009, a survey of thirty-three IACUC administrators revealed that protocol and grant congruency reviews are conducted on all externally funded projects 73 percent of the time.

Contrary to that survey, a majority of IACUC administrators attending earlier BP meetings agreed that congruency reviews need only be conducted when a requirement must be satisfied, as in the case of PHS-funded research. The primary point of the discussion focused on the fact that a 5-year research design may change over the term of the grant. In fact, some procedures described in a grant application may only be performed dependent on the results of initial experiments. Consequently, as a matter of efficient research administration, BP meeting attendees agreed that IACUC administrators should only conduct congruency reviews when it is required by a federal or granting agency's policy. IACUC administrators also agreed that, when congruency reviews are conducted on PHS-funded activities, cursory explanations for procedures that will be conducted or are planned after the scheduled protocol 3-year de novo review date should be accepted.

BP meeting attendees agreed that the scientist's goal is to satisfy the overall objective of the grant by conducting what he or she has determined to be the necessary procedures, and the IACUCs role is to ensure that the animal procedures are conducted according to the regulations and policies governing the care and use of animals for research. Scientists have their animal use procedures approved by the IACUC before they are initiated; therefore, both the scientist and the IACUC fulfill their obligations for congruence through the IACUC protocol review process. A majority of BP meeting attendees agreed that a congruency review has no bearing on the health and welfare of animals. Consequently, the only obligation is to satisfy the NIH grants policy and there is no need to conduct reviews on all non-PHS-funded projects.

COLLABORATIONS AND PROTOCOL AND GRANT CONGRUENCY

On occasion, research grants are awarded to an organization, but the animal activities are subcontracted to a collaborating colleague from a different organization. The PHS requires the prime grantee to ensure that regulatory standards and terms and conditions of the grant are satisfied. In other words if grant funds are distributed, PHS expects that the primary grantee (i.e., the organization that accepted the funds from the PHS) maintain documented proof that the animal use procedures were approved by the IACUC. Accordingly, if the animal activities are to be conducted at the collaborator's organization, the grantee must establish a method to document that all of the animal care and use procedures are approved by the IACUC before the funds are accepted.

Consequently, when organizations are establishing a subcontract with collaborators to conduct the animal activities described in a grant, they require the collaborator to provide documentation that the collaborator's IACUC has reviewed and approved the proposed animal care and use activities. This process is followed before any of the granted funds are accepted by the grantee.

BP meeting attendees reported the common process of requiring collaborating organizations to provide a copy of their associated approved protocol, which undergoes a congruency review. An alternate method described by some organizations was for both the grantee and subcontracting organization to have an approved protocol. The grantee organization would then conduct a congruency review between the two protocols, requiring the collaborator's approved protocol to include equivalent information. The key concept for congruency is to ensure that all of the animal care and use activities described in the grant are included in a protocol that has been reviewed and approved by an IACUC. However, it is appropriate for the protocol to include more detail about procedures than is included in the grant.

Three common methods that have been identified by IACUC administrators for conducting protocol and grant application congruency reviews are summarized next.

COMMON PRACTICES

Method 1

- The organization develops a policy requiring PIs to submit a unique IACUC protocol for each awarded grant.
- The title of the IACUC protocol must match that of the grant, and a copy of the grant must be provided with the protocol submission.
- The PI must list all of the animal use procedures described in the grant application on the IACUC protocol, and once IACUC approval is granted, the approval must remain current throughout the funding period of the grant.
- As part of the administrative prescreening, the IACUC administrator does a side-by-side comparison of the grant and the IACUC protocol.
  - The IACUC administrator first records all of the animal care and use procedures listed in the grant application on a congruency review form.
  - He or she will then cross reference the congruency form with the protocol, checking off those procedures listed on the congruency form when identified in the protocol.
  - Any procedure remaining unchecked on the congruency form after completing the review must be added to the protocol through the protocol amendment process before the submission is provided to the IACUC for review.
  - The IACUC administrator contacts the PI and discusses the modifications that must occur before the protocol review can be initiated.
  - Once no procedures remain unchecked, the protocol and the grant are determined to be congruent.
  - Prior to releasing the granted funds to the PI, the IACUC administrator verifies that the congruency review was conducted and that the IACUC has approved the protocol.

From an administrative standpoint, BP meeting attendees agreed that this method was the most accurate. When a protocol and grant match from title to procedures, a direct side-by-side comparison can be effortlessly conducted. Since the IACUC administrator can readily ensure that all of the procedures in the grant are listed on the protocol, when the IACUC members review and approve the protocol, they are also approving what is listed in the grant. The IACUC administrator can easily ensure that the protocol remains active throughout the entire grant funding period. The disadvantages of using this process is that it can significantly increase the number of IACUC submissions a PI must prepare, as well as increase the number of protocols the committee must review and approve. In addition, when multiple grants are used to fund the same research activities, there is the potential that duplicate protocol submissions could be made to the IACUC. In summary, this process could significantly increase the workload of the IACUC administrator, the PI, and the IACUC, but will provide a greater assurance that the vertebrate animal procedures listed in the grant have been IACUC approved.

Method 2

Similar practices are followed when applying this method, but with one significant variation: Multiple funding sources may be covered under a single IACUC protocol.

- When funds from multiple sources are supporting a single project, a copy of each awarded grant submission must be provided with the protocol.
- The PI must describe all of the animal use procedures listed in each grant application on the same IACUC protocol, and once IACUC approval is granted, the approval must remain current throughout the funding period of all the associated grants.
- As part of the screening process, the IACUC administrator does a side-by-side comparison of the each grant application and the IACUC protocol.
- The IACUC administrator repeats the process of listing all of the procedures in each grant application on a unique congruency review form. This process is repeated for each grant application.
- The IACUC administrator cross references the IACUC proposal with the congruency form(s) checking off those procedures that are included in the IACUC proposal.
- If all of the procedures on each congruency form are not identified in the IACUC proposal, the IACUC administrator contacts the PI and discusses the required modifications.
- Once all of the procedures listed on the congruency forms (i.e., in the grants) have been checked off, the protocol and grants are considered congruent.
- Prior to releasing the granted funds to the PI, the IACUC administrator verifies that the congruency review was conducted and that the IACUC has approved the protocol.

Administratively, this method has been proven to be effective. A disadvantage is that increasing the number of funding sources listed under a single approval can increase the administrative burden—for example, the protocol expiration date and the funding period cannot be synchronized, and there is no easily identified link between grant submissions and protocols (i.e., the title of the grant does not match that of the protocol). However, from the standpoint of the PI and IACUC, this process is very efficient and can significantly decrease the number of submissions prepared by the PI and reviewed by the IACUC.

Method 3

BP meeting attendees identified a third process for conducting protocol and grant congruency reviews, but the majority of the BP meeting attendees and the regulatory agency representatives expressed extreme caution when applying this approach.

- Some organizations allow PIs to perform their own congruency reviews.
- Once the review is complete, the PI signs an assurance statement confirming that both documents are congruent.

The advantage to this process is efficiency. Since the PI is familiar with his research and the requirements, he or she is expected to ensure that the regulatory requirement is satisfied. The disadvantage of this method is the absence of accountability. In other words, there is no methodology in place to ensure that procedures listed in a grant were not provided in the IACUC protocol.

Regardless of the method chosen, the primary concern is to ensure congruence. Organizations also find it important to ensure that documentation exists that verifies that the congruence requirement was satisfied. In methods 1 and 2, the congruency form is used to document congruence. The form becomes a permanent part of the file and will serve as verification. In method 3, the assurance statement in the IACUC protocol documents the congruence.

As a concluding comment for this discussion, there are generally three ways in which protocols and grants may be associated; all three processes work, but each has its advantages and disadvantages.

The one-to-one direct relationship is the easiest to manage and extremely simple to verify, but can increase the number of submissions prepared by the PI and reviewed by the IACUC. In contrast, several grants on one protocol is the easiest method for PIs since only one protocol is required, but it is the most challenging for IACUC administrators because assuring congruency is difficult since proposed activities in grants change.

HOW DOES A GRANT MODIFICATION
AFFECT THE IACUC APPROVAL?

Some organizations develop a way to ensure that when the procedures in the grant are modified, the corresponding protocol is also modified. IACUC administrators attending past BP meetings indicated it is very helpful for the IACUC office to receive a copy of the PI's annual grant report to the funding agency. This report is then cross referenced with the protocol to ensure that appropriate protocol modifications are made. If the procedures listed in a grant are revised, the IACUC may not require a new submission, but may ask that the protocol be revised appropriately.

OVERSIGHT AT COLLABORATING ORGANIZATIONS

The prime grantee is responsible for ensuring that funds are used as described in the grant. If funds are sent to collaborating organizations, the grantee must ensure that the collaborator's IACUC oversees the animal activities, or the grantee's IACUC must provide oversight. If grant funds are distributed, PHS will hold the prime grantee responsible for ensuring IACUC oversight of animal activities. A written agreement or documentation (MOU) of defining IACUC oversight responsibilities has been successfully used as a best practice.

CHAPTER 8

Animal Care and Use Program Review

SCENARIO

The GEU IACUC conducted a program review during the October IACUC meeting. The IACUC administrator reminded committee members that each component of the program must be evaluated to confirm that it complies with federal regulations and policies. To ensure that the complete program was reviewed, the GEU IACUC used the Office of Laboratory Animal Welfare's (OLAW's) "Semiannual Program Review and Facility Inspection Checklist" (grants.nih.gov/grants/olaw/sampledoc /index.htm) for guidance and as a reference. The IACUC administrator informed committee members that the review would be conducted by systematically going through each point on the checklist.

The IACUC administrator initiated the review by reading the first section heading ("Animal Care and Use Program") on the checklist to IACUC members and asked if anyone had questions or concerns relating to the topics listed in Section 1. He followed this process through the end of the checklist. Any concerns expressed by committee members during the review were noted, discussed, resolved, and reported to the institutional official (IO), which completed the semiannual program review. GEU IACUC members were pleased that they had completed the program review in less than an hour.

The GEU IO wished to have an outside assessment performed as a tool to assist the GEU IACUC in their efforts to prepare for the initial accreditation assessment by the Association for Assessment and Accreditation of Laboratory Animal Care (AAALAC). The GWU IACUC chair, a member of the assessment team, asked GEU committee members to describe how they conduct a semiannual program review.

GEU IACUC members explained their program review process to the GWU assessment team. Concerned that GEU's process may be too cursory and as a result inadequate, GWU's IACUC chair focused on a specific program component, the emergency disaster plan, to better understand the process. To initiate discussion, he asked what topics were discussed during the evaluation of the emergency disaster plan. GEU committee members indicated that, during the meeting and subsequent program review process, the initial conclusion was that the overall program lacked an emergency disaster plan. However, they explained that the attending veterinarian (AV) indicated that, although IACUC members may not have been aware, a complete disaster plan was established and instituted by the organization. Satisfied that the emergency disaster plan was complete and in place, they continued to the next topic of discussion. The GWU assessment team asked GEU IACUC members if their program covered satellite animal housing facilities and included measures for emergency euthanasia processes and contingency plans for emergency backup power sources. As the assessment continued, the GWU AV asked GEU's committee members if new animals coming into the university were quarantined before they were placed in the established vivarium. Appearing puzzled, the member looked to the GEU's AV to respond to the question.

Ultimately, GEU IACUC members indicated that they were not aware of the specific program details. They agreed that their program review process did not involve assessing the details of program processes or components, but rather confirming that a system was established to address the points included in the OLAW program review checklist.

Although the records document that semiannual reviews are being conducted, whether the standards are being satisfied is questionable. Based on the sample assessments, committee members failing to review the organization's emergency disaster plan and a committee member being unfamiliar with a component (i.e., a quarantine program) of the program that he verified was established and functioning, GEU's process does not appear to adequately achieve the overall goal of a program review.

During best practices (BP) meetings, attendees cautioned that checklists should not be used to simply "check off" that program components are established. While checklist are useful, BP attendees cautioned about extreme adherence to a checklist. Such a dedication could prevent wider discussion and discovery of potentially important program processes which could benefit from focused consideration. They encouraged their peers to use checklists as a guidance tool to ensure that each programmatic component is adequately discussed and evaluated by the IACUC. Most attendees believed that a checklist can be a very valuable reference tool, but that it should not be the only tool used by the organization to review the program. Organizations should also consider information from other entities on campus (e.g., safety, medical surveillance, grants, legal, etc.) to fully and completely assess the program's completeness and performance.

Regulatory Requirements and Resources

- **Public Health Service (PHS) policy (Section B [1], p. 12)** requires organizations to review their animal care and use program at least every 6 months to ensure that it complies with the *Guide for the Care and Use of Laboratory Animals* (henceforth known as the *Guide*). The first three chapters of the *Guide* outline the components of what an organization's program should include.
- **Animal Welfare Act regulations (AWAR) (Section 2.31[c] [1])** require organizations to review their animal care and use program at least every 6 months, using Title 9, Chapter I, Subchapter A—Animal Welfare Act, as a basis for the evaluation.

- The **OLAW semiannual program review checklist (http://grants.nih.gov/grants /olaw/sampledoc/cheklist.htm)** has been provided as a resource. Organizations are not required to use it.

THE PROGRAM REVIEW

The PHS policy and the AWAR require the IACUC to review the organization's program of animal care and use every 6 months. To conduct thorough reviews, IACUC members devise a system to validate that the program is complete and that the processes used to achieve the overall goals comply with the regulatory standards.

An effective review occurs when committee members confirm that a requirement is satisfied. They systematically scrutinize the process used to achieve the standards. For example, imagine that the GEU IACUC is reviewing the program during a convened meeting. As part of the review, the IACUC administrator validates for the committee that the organization's program includes provisions for training IACUC members. With the training component documented as satisfied, the IACUC chair asks the IACUC administrator to describe the training program for IACUC members. The IACUC administrator explains that the *Guide* (p. 17) requires the organization to establish methods for training IACUC members to fulfill their responsibilities and understand their roles as committee members. The IACUC administrator describes a comprehensive training program to the IACUC that involves (1) a formal orientation for new members to discuss GEU's program; (2) a comprehensive training on regulations, policies, and guidelines; and (3) opportunities to attend ongoing training such as webinars, meetings, and workshops. The IACUC chair concludes the conversation by asking committee members how well the training prepared them for their role as an IACUC member. The IACUC administrator finalizes the review by asking committee members if they believed the IACUC member training program continued to satisfy the regulatory expectations. The IACUC administrator also asks committee members for suggestions to improve the program. With no concerns or comments expressed, the review of the IACUC member training program is documented as being compliant with federal directives.

Preparing for the Program Review

In order for IACUC members to conduct a thorough program review, they must have the materials and knowledge necessary to engage in the process. In other words, committee members conducting program reviews must be informed and involved. In preparation of the program review, the IACUC administrator frequently distributes program materials to committee members; for example, the last IO report, guidelines, application templates, standard operating procedures (SOPs), and policies are provided as program review resources. In certain situations, satellite facility managers prepare semiannual reports that are distributed to committee members. These reports summarize, for example, a satellite facility's program components such as husbandry practices, the veterinary care process, and the emergency disaster plan. As part of the program review, committee members may also assess departmental SOPs that involve the care and use of animals. For example, agriculture animal husbandry and aquatic facility SOPs are often assessed during the program review. For organizations that are AAALAC accredited and PHS assured, copies of the most current program description and PHS assurance can be provided. In addition, IACUC administrators at US Department of Agriculture (USDA)-registered research facilities frequently provide a copy of the latest USDA annual and inspection report as resources. Although using the OLAW-provided checklist is optional, organizations frequently use the checklist to ensure that all components of the program are evaluated. The most commonly used checklist can be accessed in the resource section (grants.nih.gov/grants/olaw/sam pledoc/index.htm) of the OLAW website. Queried BP meeting attendees indicated that they provide the necessary materials to committee members at least 1 month prior to the scheduled program review. Many BP attendees noted that while they start with the OLAW checklist, they then modify the checklist to focus their review to their program (e.g., no primate? remove the question regarding monkeys).

Proposed Best Practices for Conducting Program Reviews

Past BP meeting attendees offered many effective ways for conducting program reviews. They agreed that there are various parameters that impact how a program review is conducted. For example, some organizations have small, simple programs (e.g., mice in only one vivarium and less than twenty protocols). Organizations with small programs may conduct the program review as an agenda item during a regularly convened monthly meeting.

Conversely, organizations with large, complex programs (e.g., multiple species, multiple buildings, and several hundred protocols) have developed other successful practices. Some organizations use subcommittees, with each focusing on specific components of the program (e.g., the occupational health and safety program or veterinary care) with the subcommittee reporting to the IACUC at a full committee meeting. Others have a meeting called solely for the purpose of conducting the program review. A few organizations also conduct the program review with updating their PHS animal welfare assurance or AAALAC-accredited animal care and use program document.

BP meeting attendees agreed that there are three general processes for conducting an adequate program review, which is primarily based on the size of the program: very large and very small programs. Most programs would fall somewhere along the spectrum and could choose aspects of each that would most effectively assess their program.

Program Review Best Practice 1

Organizations conducting their program reviews during regularly scheduled monthly IACUC meetings have used the following common practice:

1. **Hold a program review training session.** During the IACUC meeting (directly preceding the scheduled program review meeting), the IACUC administrator provides a training/refresher training session for committee members. During the training, the IACUC administrator may review:
   a. The federal regulations and policies requiring the IACUC to conduct a program review at least every 6 months
   b. The overall goal of ensuring that the organization's program complies with the applicable federal standards (e.g., the *Guide,* the *Guide for the Care and Use of Agricultural Animals in Research and Teaching* [henceforth known as *Ag Guide*], PHS policy, and AWAR)
   c. The program review process using, for example a checklist (e.g., the OLAW checklist) or an appropriate reference document, to identify the programmatic components that must be reviewed
   d. Other factors (e.g., whether there are an adequate number of information technology and administrative staff members for program support and enough committee members to routinely review protocols and inspect facilities) having the potential to affect the quality and operations of the program
   e. The process for identifying and resolving programmatic deficiencies, including the requirement for the IACUC to identify deficiencies as significant or nonsignificant and establish resolution plans with the action being formalized by a majority vote of a quorum of the IACUC
   f. The process for identifying recommendations for program improvement, which may include, for example, an IACUC recommendation to increase the number of IT staff to ensure that adequate human resources are available to support the program (also formalized by a majority vote of the quorum of the IACUC)
   g. The reasoning and process for an IACUC member to express a minority opinion, including a discussion that a single committee member may not agree with the IACUC's consensus and choose to write and substantiate his or her personal opinion into the reports (e.g., IO and OLAW reports)
2. **Provide the program review resources to IACUC members.** Upon completion of the training, the IACUC members are e-mailed the instructions and resources for initiating the program review. The e-mail to committee members may include:
   a. A request for each committee member to review the components of the program and to e-mail any identified comments or concerns to the IACUC administrator in, for example, 2 weeks
   b. A copy of the previous IO report, which includes the details of the last program review conducted
   c. A copy of the OLAW program review checklist for reference purposes
   d. Copies of or website links to the organization's guidelines, standard operating procedures, policies and training modules, for example
3. **Prepare for the program review meeting.** The IACUC administrator then prepares and e-mails (approximately 10 days before the meeting) the monthly IACUC meeting agenda and materials relevant to the program review:
   a. A copy of the OLAW program review checklist is consistently provided to ensure consistent, complete, and thorough review.
   b. IACUC members' programmatic comments are collated according to the relative section of the program. For example, concerns directed toward the IACUC member-appointing process would be categorized as "IACUC membership and

functions" issues and concerns associated with methods of euthanasia or analgesia use as "veterinary medical care" issues.

4. **The meeting and the program review.** Information reported at past BP meetings
suggested across organizations that either the IACUC chair or the IACUC administrator (50 percent of the time) facilitate the program review using the following
process:

a. The OLAW checklist is prominently displayed (e.g., projected on a screen or
using the distributed paper copies) and used to guide the IACUC through the
program review process.

b. The facilitator begins the process by bringing committee members' attention
to Section 1, "Animal Care and Use Program," on the checklist.

c. Focusing on this section only, the facilitator identifies point 1 under Section
1 (*responsibility for animal well-being is assumed by all members of the program*). He or she highlights the concerns cited during the last program review
and whether or not the correction plan has been satisfied, as well as those comments e-mailed by committee members in preparation of the review.

d. Committee members are given the opportunity to comment on the specifically
identified topic:

  i. When no concerns from the past review or e-mailed comments are noted,
  and committee members have no other topics for discussion, the facilitator continues to the next topic on the checklist following the same review
  practice.

  ii. When a deficiency correction plan from a past review is not completed,
  the IACUC discusses the concern, decides if it constitutes ongoing noncompliance, and may revise the correction plan to ensure that the problem
  is corrected; then it continues to the next topic on the checklist following
  the same review practice.

  iii. When a new concern is expressed (e.g., either through the e-mailed comments or during the ongoing discussion), the IACUC discusses and identifies the problem, notes why it is a problem, determines whether the
  concern is significant (i.e., whether the problem has the potential to have a
  negative impact on the health and well-being of an animal), and develops
  an appropriate plan for correction; then it continues to the next topic on the
  checklist following the same review practice.

e. The IACUC administrator finalizes the review by reading each IACUC-
identified concern and correction plan (if any), asking the committee to
motion and second that the review be accepted, and registering that an
acceptable review was completed (reviews that are not voted acceptable must
continue until the IACUC determines that a satisfactory review was completed) based on a majority vote of a convened meeting of a quorum of the
IACUC.

f. The IACUC administrator reminds committee members of their right to
express a minority opinion in the event that they disagree with a committee
decision and asks if anyone would like to communicate a minority opinion to
the IO or OLAW through the required reports (IO report and OLAW annual
report).

5. **Finalizing the process.** The IACUC administrator informs the IACUC members
that the details of the review will be included in the IO report, which will be provided to them for review and approval at the next convened IACUC meeting.

Program Review Best Practice 2

Many organizations with large, complex programs frequently choose to conduct their program review using subcommittees of the IACUC or during a full committee meeting scheduled specifically to conduct the review. Those organizations using IACUC subcommittees have used the following common practice:

1. **Hold a program review training session.** Sixty days preceding the scheduled IACUC program review meeting, the IACUC administrator conducts a training/refresher training session for committee members. During the training, the IACUC administrator may review:
   a. The federal regulations and policies requiring the IACUC to conduct a program review at least every 6 months
   b. The overall goal of ensuring that the organization's program complies with the applicable standards (e.g., the *Guide, Ag Guide*, PHS policy, and AWAR)
   c. The program review process using, for example, a checklist (e.g., the OLAW checklist) or an appropriate reference document to identify the programmatic components to be reviewed
   d. Other factors (e.g., whether there are an adequate number of information technology and administrative staff members for program support and enough committee members to routinely review protocols and inspect facilities) having the potential to affect the operations and quality of the program
   e. The process for identifying and resolving programmatic deficiencies, including the requirement for the IACUC to identify deficiencies as significant or minor and establish a plan and schedule for correction with the action being formalized by a majority vote of a quorum of the IACUC
   f. The process for identifying recommendations for program improvement, which may include, for example, an IACUC recommendation to increase the number of IT staff to ensure that adequate staffing is available to support the program (also formalized by a majority vote of the quorum of the IACUC)
   g. The reasoning and process for an IACUC member to express a minority opinion, including a discussion that a single committee member may not agree with the IACUC's consensus and choose to write and substantiate his or her personal opinion into the reports (e.g., IO and OLAW reports)
2. **Assigning IACUC subcommittees to conduct program reviews.** Upon completion of the training session, the IACUC chair identifies IACUC members and, in certain situations, ad hoc consultants to serve on program review subcommittees. The IACUC chair may utilize the following practices when establishing subcommittees:
   a. A qualified chair to lead the subcommittee is identified, with his or her appointment being based on expertise—for example, the IACUC vice chair may be selected to lead the review of the "IACUC Membership and Functions" section, and the IACUC administrator to lead the subcommittee on "Programmatic Records and Reporting Requirements."
   b. Ad hoc consultants may be identified (depending on the section to be reviewed) to provide technical expertise—for example, the organization's biosafety officer and an occupational medicine physician may be asked to serve on the subcommittee reviewing the occupational health and safety program.
   c. Time lines for completion (typically 30 days) are established.

3. **Subcommittee support.** The IACUC administrator then prepares and e-mails to subcommittee members the materials needed to conduct the program review, which may include:
   a. A copy of the OLAW program review checklist for reference purposes
   b. A copy of the previous IO report, which includes the details of the last conducted program review
   c. Copies of relevant website links to the organization's guidelines, standard operating procedures, policies, and training modules, for example

4. **Subcommittee activities.** Members of the subcommittee review only the portion of the program that they have been assigned, observing the following practices:
   a. The subcommittee chair schedules and coordinates meetings.
   b. During meetings, members consider each program requirement specific to their assigned section.
   c. Subcommittee members evaluate each point appearing on the OLAW checklist by determining if and how the organization complies with a listed standard (e.g., whether the requirement for conducting semiannual program reviews is satisfied and, if so, what process is followed).
   d. The subcommittee decides if a standard practice is compliant and efficient.
   e. The review (i.e., findings and recommendations) is summarized in a report that is provided to the IACUC administrator (at least 10 days before meeting) and later presented to the IACUC during the meeting.

5. **Prepare for the program review meeting.** The IACUC administrator then prepares and e-mails (approximately 10 days before the meeting) the IACUC meeting agenda and materials relevant to the program review to IACUC members, including:
   a. A copy of the OLAW program review checklist, which is provided to ensure a consistent, complete, and thorough review
   b. Copies of each subcommittee report

6. **The meeting and the program review.** Information reported at past BP meetings suggested across organizations that either the IACUC chair or the IACUC administrator (50 percent of the time) facilitate the program review using the following process:
   a. The OLAW checklist is prominently displayed (e.g., projected on a screen or using the distributed paper copies) and used to guide the IACUC through the program review process.
   b. The facilitator begins the process by bringing committee members' attention to Section 1, "Animal Care and Use Program," on the checklist.
   c. Each subcommittee chair initiates an IACUC discussion on that section of the program that he or she was asked to review by identifying perceived deficiencies and recommendations for improvement.
   d. Committee members are given the opportunity to comment on the subcommittee's recommendations and/or identify other concerns specific to the topic under discussion:
      i. When no subcommittee concerns are noted and other IACUC members have no additional topics of concern, the facilitator continues to the next topic on the checklist following the same review practice.
      ii. When a deficiency correction plan from a past review is not completed, the IACUC discusses the concern, decides if it constitutes ongoing

noncompliance, and may revise the correction plan to ensure that the problem is corrected; it then continues to the next topic on the checklist following the same review practice.

iii. When a new concern is identified (e.g., by the subcommittee), the IACUC discusses and identifies the problem, notes why it is a problem, determines whether the concern is significant (i.e., whether the problem has the potential to have a negative impact on the health and well-being of an animal), and develops an appropriate plan for correction; it then continues to the next topic on the checklist following the same review practice.

7. **Finalizing the review and reporting.** The IACUC administrator finalizes the review by

a. Reading each IACUC identified concern and correction plan (if any), asking the committee to motion and second that the review be accepted, and registering that an acceptable review was completed (reviews that are not voted acceptable must continue until the IACUC determines that a satisfactory review was completed) based on a majority vote of a convened meeting of a quorum of the IACUC

b. Reminding committee members of their right to express a minority opinion in the event that they disagree with a committee decision and asking if anyone would like to communicate a minority opinion to the IO or OLAW through the required reports (IO report and OLAW annual report)

c. Informing the IACUC members that the details of the review will be included in the IO report, which will be provided to them for review and approval at the next convened IACUC meeting

Program Review Best Practice 3

As discussed for best practice 2, organizations with large, complex programs often use a subcommittee process to conduct a program review. However, some organizations elect to use subcommittees and conduct an ongoing program review in the interest of time management. Those conducting continuous program reviews frequently utilize the following practice:

1. **Hold a program review training session.** The IACUC administrator conducts a training session similar to that described in best practices 1 and 2 every 6 months.
2. **Subcommittee activities.** Subcommittees are assigned, supported, and conduct their reviews as described in best practice 2.
3. **The program review meeting.** A schedule for reviewing each section of the program every 6 months at a monthly IACUC meeting is developed (i.e., one-sixth of the program is reviewed at each monthly meeting with a repeat review of the same section occurring 6 months later).
4. **Prepare for the program review meeting.** The IACUC administrator then prepares and e-mails (approximately 10 days before the meeting) the IACUC meeting agenda (routine business items and a section dedicated specifically to reviewing a portion of the organization's program) to IACUC members, which includes:
   a. The section of the OLAW program review checklist corresponding to that portion of the program to be reviewed (e.g., if programmatic records and reporting

requirements are to be reviewed, only those sections of the OLAW checklist are provided to the committee)

b. A copy of the subcommittee report corresponding to those portions to be reviewed during the meeting

5. **The meeting and program review.** The program review is conducted as described in best practice 2, with the exception that only those sections (a portion of the program) identified as part of the meeting agenda will be reviewed.

6. **Finalizing the review and reporting.** The IACUC administrator finalizes the review by

a. Reading each IACUC identified concern and correction plan (if any) for the sections reviewed, asking the committee to motion and second that the review of that section be accepted, and registering that an acceptable review of that specific section was completed (reviews that are not voted acceptable must continue until the IACUC determines that a satisfactory review was completed) based on a majority vote of a convened meeting of a quorum of the IACUC

b. Reminding committee members of their right to express a minority opinion in the event that they disagree with a committee decision and asking if anyone would like to communicate a minority opinion to the IO or OLAW through the required reports (IO report and OLAW annual report)

c. Informing the IACUC members that the details of the review will be collated into the IO report after every section of the program review has been completed (i.e., at the end of the 6-month time period), which will be provided to them for review and approval at that time

ILLUSTRATING THE ANIMAL CARE AND USE PROGRAM REVIEW PROCESS

During IACUC administrators' BP meetings, attendees selected sections of the animal care and use program and conducted mock reviews to demonstrate commonly used practices for conducting program reviews. Sample discussions are outlined next.

1. **IACUC membership and functions (Section 5 of the OLAW program review checklist).** The following illustrates how an IACUC may engage in the review of the "IACUC is comprised of at least 5 members, appointed by CEO" section:

a. The IACUC administrator opened the discussion by saying, "Let's begin our review by considering our IACUC membership and the committee member appointment process—point 1."

b. The IACUC administrator began by reviewing an issue identified during the previous program review and stated the following: "After reviewing the minutes of our last program review, I noticed we identified that our committee member appointments were being made by the IO, who is not the CEO of the company. At that time, the IACUC asked that the CEO delegate in writing the IO's responsibilities, including appointing individuals to the IACUC, which has occurred. Do any of you have additional concerns or require further clarifications on this particular matter?"

c. The IACUC administrator continued by asking IACUC members if "anyone had any other concerns they wish to address regarding this particular section of the program."

d. A committee member said, "I am concerned that the organization has not appointed an alternative committee member to serve for the nonaffiliated member in the event he is incapacitated for an extended period of time." To facilitate the discussion, the IACUC administrator projected the IACUC roster on a screen for the committee to review and discuss. The IACUC administrator reminded committee members that having alternates is not a federal requirement, but also recognized that if the community member was not able to serve as a voting member for an extended period of time the committee would not be duly constituted and therefore unable to conduct official business.

Considering all of the relevant information, the IACUC agreed that the situation did not create a program deficiency. However, the committee agreed the program could be improved if another community member was identified and appointed to the IACUC as an alternate member. Although the programmatic expectations were satisfied, the committee formulated a plan for identifying an alternate community member. The details of the plan and all relevant points of the discussion were documented in the meeting minutes and presented to the IO in the semiannual report.

e. The IACUC administrator continued the process by again asking IACUC members if anyone had any other concerns they wish to address relating to this particular section of the program.

f. Hearing no additional concerns, the IACUC administrator continued to the next section of the program.

2. **Personnel qualifications and training (Section 9 of the OLAW program review checklist).** The following illustrates how an IACUC may engage in the review of the this particular section:

a. The IACUC administrator initiates the discussion by saying, "As committee members we need to decide whether our training program satisfies the regulatory expectations by asking ourselves the following questions: (1) Does it include all of the required content? (2) Does it ensure all animal care and use personnel are trained? (3) Does it offer continuing education opportunities? (4) Does it ensure animal users are trained to conduct the procedures proposed to the IACUC? (5) Does it include a process for documenting the training?"

b. The IACUC administrator then begins a discussion on each of the preceding five points by asking the committee members how each of the five points is satisfied:

   i. When asked, "How do we include all of the required content in our training programs?" the committee members discuss the web-based regulations training, the hands-on training offered by veterinary staff, and the safety training offered by EHS (Environmental Health and Safety). Having confirmed the training program is complete; the IACUC administrator continues.

   ii. "How do we document the completion of all training requirements?" The IACUC discusses the training database maintained and managed by the training coordinator. They consider the fact that all training records are submitted to the coordinator and entered in the database before a research protocol is approved. Having confirmed the program includes effective measures for documenting training, the IACUC administrator continues to facilitate the review of each subsequent section.

DOCUMENTING THE PROGRAM REVIEW

The IACUC administrator is frequently responsible for documenting the details of the program review. The details of the review are routinely documented in the meeting minutes and semiannual report to the IO. The documentation commonly includes a list of committee members participating in the program review, those areas of the program that were evaluated, and any noted program deficiencies.

When program deficiencies are identified, the IACUC must identify them as either significant or minor deficiencies and summarize each in the report to the IO. In addition, the review must include a written plan and time line for resolving each major deficiency. Once the report is completed it is reviewed and signed by a quorum of the IACUC and then submitted to the IO.

CHAPTER 9

Semiannual Facility Inspection

SCENARIO

The GEU IACUC established a process to oversee the organization's decentralized animal housing facilities. The facilities include four laboratory animal vivaria that are located on the main campus and three agriculture animal research stations fifty miles north of campus. In addition, GEU scientists maintained three animal housing facilities throughout the continental United States. An ichthyologist established facilities for housing research salmon in Alaska. A wildlife biologist maintained a herd of mule deer in Colorado, and an immunologist housed laboratory rats for Public Health Service (PHS)-supported research in San Diego, California.

The IACUC administrator easily scheduled at least two IACUC members to physically inspect the on-campus vivarium and agriculture stations semiannually. The IACUC administrator discovered that inspecting the remote housing locations required extensive travel and a considerable time commitment from IACUC members. In addition, the expenses associated with the inspections were prohibitive. Consequently, the IACUC administrator found scheduling the physical inspection of the remote housing facilities impractical. As a result, the IACUC established alternative methods for conducting the inspections.

During a full committee meeting, the GEU IACUC approved a standard operating procedure (SOP) for inspecting remote animal housing facilities using audio and video technology. The SOP specifically described how the inspection video must be prepared. The videographer was asked to begin recording upon entering the main door of the animal room. He or she was asked to first pan the entire room including all four walls, the ceiling and the floor, which helps committee members to assess the materials' suitability and cleanliness of the walls, floor, and ceiling. To assist the committee with evaluating the cleanliness of vents, drains, and sinks, the videographer was asked to provide close-ups of each and then to provide close-up views of all relevant room records—for example, temperature and humidity, daily animal check, and health records. Finally, he or she was asked to slowly pan over every animal housed in the room and provide close-up views of at least 25 percent of the caged animals. The videographer was also asked to narrate the inspection and provide to the committee relevant details of the entire process. To conduct the inspection, the committee members reviewed the recording during a full committee meeting, discussing it with the facility manager through conference call.

This scenario includes a species (i.e., mule deer) governed by the Animal Welfare Act regulations (AWAR), one (i.e., laboratory rats) by the PHS policy, and another (i.e., salmon) regulated under other applicable standards such as the *Guide for the Care and Use of Laboratory Animals* (henceforth referred to as the *Guide*). Therefore, the IACUC must develop an SOP that satisfies the AWAR, the PHS policy, and all other applicable guidance documents. Both the AWAR (p. 22, Section 231(c) 3) and the PHS policy (p. 12, footnote 8) permit the IACUC to use its discretion for determining the best means for conducting semiannual facility inspections, but the AWAR requires at least two IACUC members to participate. Consequently, the GEU IACUC determined the best way to inspect remote housing facilities was to use audio and video technology; however, to comply with the AWAR, a crucial step in the process was to conduct the virtual inspection by reviewing the recording during a properly convened full committee meeting.

US Department of Agriculture (USDA) representatives participating in past best practices (BP) meetings have indicated that the AWAR do not prohibit the use of digital technology (e.g., video or photography) to conduct inspections. However, they did suggest that the most effective process would be to transmit the video directly to the IACUC during a full committee meeting to elicit an interactive discussion during the inspection process. Office of Laboratory Animal Welfare (OLAW) representatives also agreed that remote facility inspections can be conducted using digital technology. In both cases, federal colleagues indicated that digital technology should not be used out of convenience, but rather in situations when visiting housing sites is prohibitive or impractical.

Regulatory Requirements and Reference Resources

- **PHS policy (Section IV. B. 2, p. 12)** requires an organization's IACUC to "inspect at least once every 6 months all of the organization's animal facilities (including satellite facilities) using the *Guide* as a basis for evaluation."
- **OLAW frequently asked questions, Question 1, Program Review and Facility Inspection (http://grants.nih.gov/grants/olaw/faqs.htm#prorev_1)** provides clarification on the IACUC's responsibility for inspecting satellite facilities (defined by PHS policy as a containment outside a core or centrally managed area in which animals are housed for more than 24 hours), animal study areas (defined by the AWAR as areas where USDA-covered animals are housed for more than 12 hours) and areas where surgical manipulations occur.
- **AWAR (Section 2.31(c) (2), p. 15)** require an organization's IACUC to "inspect, at least once every six months, all of the research facility's animal facilities, including animal study areas, using Title 9, Chapter I, Subchapter A—Animal Welfare, as the basis for evaluation."

THE FACILITY INSPECTION

The PHS policy (p. 12) and the AWAR (p. 15) require the organization's IACUC to inspect those facilities associated with the animal care and use program at least every 6 months. As part of program oversight, the IACUC must ensure that the associated facilities (e.g., animal housing and procedure areas) conform to the policies and regulations governing the care and use of vertebrate animals in research, teaching, and testing. To ensure that a complete and thorough inspection occurs, the IACUC administrator must devise a system that ensures that the IACUC inspects all areas requiring inspection.

A key to establishing an effective process requires the IACUC administrator to first identify those areas that must be inspected by the committee. A common practice is for IACUC administrators to continuously maintain a list of areas that must be visited and inspected. Although some organizations accomplish this task using an elaborate commercially available electronic data management system; a best practice for IACUC administrators is to maintain a simple database (e.g., using Excel or Access) that can be referenced when scheduling facilities for inspection. The list typically includes all areas used for animal housing, surgical manipulations, and support areas (e.g., cage wash, pharmacy, and feed storage).

What Facilities Must Be Inspected Semiannually by the IACUC?

Housing Facilities

Perhaps the most obvious facilities requiring IACUC inspections are animal housing facilities. In general, an animal housing facility is defined as any structure or device used to confine animals or to shelter them from environmental extremes such as winter storms or summer sun. The list of housing facilities should include both primary and secondary housing units. For example, a dairy farm housing facilities list will not only include the barns, but also stalls and field shelters. Agriculture animal facilities frequently include parcels of fenced-in land (pastures) to maintain animals. Since the fencing is used to confine animals, it and the land would be considered a housing facility and require IACUC inspection.

The IACUC must also inspect the organization's satellite facilities (PHS policy) and study areas (AWAR). The AWAR define a study area as an area outside the core facility used to house animals for periods greater than 12 hours. The PHS policy defines a satellite facility as an area outside the core facility that is used to house animals for periods greater than 24 hours. Many BP meeting attendees indicated that their IACUC had developed and adopted a policy that required prior IACUC approval before any animals could be held outside an established vivarium for periods greater than 12 hours, regardless of species. The goal of this policy is to apply an equivalent standard to all species of vertebrate animals, whether they are species covered by PHS policy or the USDA.

Surgery Areas

In addition to housing areas, the IACUC must also inspect all of the organization's surgical areas. Although the AWAR indicate that survival surgery must be conducted aseptically and in a dedicated suite, they do not specifically indicate that the IACUC should inspect that facility on a semiannual basis. The PHS policy, on the other hand, indicates that areas where surgical manipulations (major, minor, survival, or nonsurvival) occur must be inspected semiannually by the IACUC. As in the case of satellite facility housing, the IACUCs at most organizations have established and implemented a conservative policy. IACUCs inspect all surgical facilities at least semiannually and before an initial surgery is conducted, which ensures that the site and facility can provide the appropriate conditions needed to conduct surgery. This practice ensures compliance with both the AWAR and the PHS Policy.

Specialized Animal Laboratories

Organizations maintaining specialized animal laboratories must have those facilities inspected by the IACUC. For example, veterinarians, facility managers, and principal investigators (PIs) identify specific areas in animal facilities to conduct specialized procedures. The attending veterinarian (AV) may identify a specific area for radiology and necropsy. The facility manager may identify a specific location for food preparation and storage. PIs routinely establish areas within their labs to conduct euthanasia and procedures requiring the use of specialized equipment. Areas such as those described should be inspected by the IACUC at least semiannually.

Support Areas

The IACUC must also inspect animal facility support areas. For example, committee members must inspect animal facility storage areas such as rooms or areas used to store materials—for example, food and bedding, pharmaceuticals, biologics, and waste—as well as supplies such as mouse cages and water bottles. They should inspect the animal care technicians' support areas and locker rooms.

Equipment Areas

Committee members must also inspect equipment support areas such as the clean and dirty cage wash areas. Organizations should also make provisions for inspecting animal transport vehicles such as livestock trailers, automobiles, and vans. Vehicles moving animals from place to place are "mobile animal holding facilities" for the period of time that animals are inside. While the period of holding in this mobile facility may be short, it still behooves the IACUC to assure that the animals' best interests are addressed during transport, especially if the transport is a non-institutional vehicle (e.g., PIs car). In addition to issues regarding animal care, private vehicles inappropriately used may result in potential health impacts to persons riding in the vehicle after the animals have been removed.

HVAC and Emergency Monitoring Systems

IACUC members should also evaluate the organization's environmental control and emergency monitoring systems. For example, the committee should

ensure that the air exchange requirements published in the *Guide* (p. 46) are satisfied in each animal room. They may also ensure that light/dark cycle timers
and computerized monitoring systems are functioning appropriately. Unique to
the Department of Veterans Affairs (VA) animal care and use programs (ACUPs),
the heating, ventilation, and air conditioning (HVAC) system must be recertified
by the IACUC during inspections, assuring that it meets the minimum standards
as outlined in its primary guidance document, the *VA Handbook* 1200.7. Many
other organizations require an HVAC reassessment at a least annually to assure
continuing compliant environmental conditions. All Association for Assessment
and Accreditation of Laboratory Animal Care (AAALAC)-accredited organizations must have the HVAC system assessed within 12 months of the site visit to
maintain their programs' accreditation.

Preparing for the Program Review

Training/Refresher Training

In order for IACUC members to conduct a complete and thorough facilities inspection, they should be regularly trained to conduct such inspections. To conduct the
training, the IACUC administrator often prepares a semiannual facilities inspection
checklist that committee members can use as a reference when conducting the inspections. Although each organization's checklist may vary, a commonly used but optional
checklist can be accessed (http://grants.nih.gov/grants/olaw/sampledoc/checklist_html
.htm#2a) in the resource section of the OLAW website. Queried BP meeting attendees
indicated that they provide the checklist to committee members and briefly review the
inspection criteria at least 1 month prior to the scheduled facility inspections.

Establishing the Inspection Schedule

An organization's inspection schedule often depends on the size of its program.
For example, some organizations' programs may include several hundred facility
locations that must be inspected every 6 months. Frequently, these organizations
decide to conduct inspections throughout the 6-month semiannual cycle. In a situation such as this one, an organization may decide to inspect approximately 20 percent
of its facilities in each of a 5-month period. The findings from each month's inspections are discussed at a monthly meeting. During the 6-month cycle, the IACUC
administrator consolidates the past 5 months' findings into the semiannual facilities
inspection report that is presented to the IACUC for approval and to the institutional
official (IO). The process is repeated at the beginning of the next semiannual period.
This ongoing inspection process allows organizations to more efficiently manage
human resources and the workload associated with facility inspections.

Alternatively, organizations with smaller programs frequently conduct their
facility inspections over a 3- to 10-day time period twice a year. For example, facility inspections may be conducted the first week of March and the first week of
September each year. This process maintains specifically defined inspection time

periods, which ensures compliance to federal standards and gives committee members the opportunity to plan for the inspections.

Coordinating the Inspections

During past BP meetings, attendees raised the concern that facility inspections are often difficult to coordinate since they cannot occur unless IACUC members participate. Many IACUC administrators indicated that it is difficult to get committee members to participate in the inspections due to personal conflicts or obligations associated with teaching and research. Several attendees indicated that scheduling the inspections at the same time every year helps committee members plan time to participate. They indicated that it was also helpful to advise new committee members that they must participate in facility inspections at least twice a year. Most IACUC administrators noted that they require committee members to dedicate a minimum of 2 days every 6 months (most actual inspections last 2–3 hours) to facility inspections. The IACUC chair and the IACUC administrator must also emphasize the importance of the facility inspections and emphasize that an ACUP cannot operate if the inspections are not conducted.

Who Should Serve on the Inspection Teams?

To coordinate facility inspections, the IACUC administrator begins the process by establishing inspection teams. A consistent practice throughout the IACUC administrative community is to have the inspection teams composed of at least two committee members and the IACUC administrator. Although not required under the regulatory standards, more and more organizations are beginning to identify other individuals as key personnel to participate in the inspection process.

As required, the IACUC must establish processes to conduct facility inspections. Since the AWAR provide a regulatory requirement of having at least two IACUC members assigned to each inspection team when USDA-covered species are involved, it is common for IACUCs to establish a process that includes active participation by committee members. Consequently, many BP meeting attendees indicated that they have adopted a policy at their organization requiring each inspection team to include at a minimum two IACUC members. BP meeting attendees agreed that such an approach would enhance the review and inspection process, whether involving USDA-covered species or not.

IACUC administrators at past BP meetings agreed that the PHS policy provides IACUCs the latitude to determine the best means for evaluating the organization's animal facilities. The PHS policy permits the use of qualified consultants to inspect animal facilities. As a result and under the direction of the IACUC, some organizations ask postapproval monitors, IACUC administrators, hired consultants, and/or veterinarians to conduct the inspections when USDA-covered species are not involved.

Many IACUC administrators attending past BP meetings generally believed that the primary goal of the inspection team is not only to evaluate the facilities using the regulations as the standards, but also to ensure that the health and welfare of the animals are being satisfied. As a result many BP meeting attendees thought it necessary to also include a veterinarian on each inspection team to answer any questions relating to the health and welfare of research animals.

Although many IACUC administrators indicated that it was important for a veterinarian to participate in each facility inspection, they discussed whether having the AV act as an official member of the inspection team created a conflict of interest. The primary concern was that the AV would be inspecting areas that he or she managed or, at a minimum, indirectly oversaw. For example, the inspection process should include an evaluation of the veterinary care program, which is one of the AV's primary responsibilities. In the event questions arise as to the adequacy of the veterinary care program, concern would be raised if the AV were officially involved in determining whether a program that he or she managed was not compliant with federal expectations. Consequently, some organizations do not permit their veterinarians to participate in the inspection of areas they manage. One suggestion was that the laboratory animal veterinarian could be used to inspect agriculture facilities or PI-managed care facilities and the agriculture animal veterinarian or a local veterinarian serving as a consultant could inspect the laboratory animal facilities.

Most IACUC administrators agreed that if the AV always attended inspections, especially those managed by him or her, a serious question of process integrity could develop. Therefore, the best suggestion was that the AV (or designee) should participate in the IACUC inspection to answer questions regarding animal health, but not as a member of the inspection subcommittee.

In addition to veterinarians and IACUC members participating in facility inspections, the IACUC administrator (at almost 100 percent of the organizations sending staff to BP meetings) or one of his or her administrative subordinates participated in all of the facility inspections. A primary role of the IACUC administrator was to document the inspection process, record findings, and communicate essential information to facility managers, PIs, and the IO. The IACUC administrator (or designee) also ensured that every facility was visited by the IACUC inspection team.

Although BP meeting attendees recognized the advantages of minimizing the number of people involved in the inspections (e.g., decreased risk of spreading disease), they indicated that it is helpful when staff members from supporting units participate in the facility inspections. For example, some organizations include a representative from the Office of the Physical Plant (OPP) on the inspection team. IACUC administrators agreed that many deficiencies are associated with facility maintenance and that having an OPP representative on the team can expedite the deficiency correction process. IACUC administrators that utilize OPP members in the inspection process noted that the OPP representative was often able to assist the IACUC in identifying and developing efficient and timely deficiency correction plans. For example, the OPP member could take photographs of a problem, specifically identify the location of the issue, and ask pertinent questions relating to the deficiency during the inspection and then help the committee establish an appropriate correction plan on site. IACUC administrators indicated that, at times, the problem was immediately resolved or often before the end of the day.

In addition to an OPP representative, some organizations also have an Environmental Health and Safety (EHS) staff member participate in the process. BP meeting attendees emphasized the importance of identifying someone from EHS with an industrial hygiene background. The goal of including this individual in the inspection was to completely address concerns relating to the occupational health and safety program (OHSP). During the inspection process the EHS office representative is able to conduct environmental risk assessments, thereby satisfying the OHSP requirement of conducting periodic facility risk assessments.

In addition, the facility managers were identified as key individuals on the inspection teams. Some organizations ask that the facility manager of the unit being assessed participate in the inspection. Having the unit facility manager involved in the visits significantly increases the efficiency of the visit. For example, a perceived deficiency such as lack of paperwork may be easily resolved by the facility manager since he or she maintains the records in his or her office.

The Size of the Inspection Teams

For various reasons the size of the inspection team should be limited to the number of individuals and the expertise needed to conduct a thorough review. Large groups touring the facilities can compromise biosecurity and be a distraction, and in some cases a smaller vivarium may not be able to accommodate larger groups; consequently, it is prudent to keep the size of the inspection team to a minimum.

Since the number and expertise of staff members differ between organizations, each must develop its own efficient process. For example, some organizations choose to appoint someone (e.g., biological safety officer) from their EHS office to serve on the IACUC. In a situation such as this one, the biological safety officer can satisfy the OHSP requirement of conducting periodic environmental risk assessments and serve as an IACUC member conducting facility inspections. Other organizations have appointed administrative professionals (e.g., IACUC administrators and Grants and Contracts personnel) to the IACUC, which allows them to conduct administrative responsibilities (e.g., reporting) while conducting inspections as an IACUC member.

Announcing the Inspection

Once the core team and schedules have been established, the next step is to announce the inspections. Since the AWAR require that all IACUC members be given the opportunity to participate in semiannual inspections, many IACUC administrators announce the inspection to committee members using some form of electronic technology, such as e-mail. The e-mail effectively reminds the core team members of the inspection dates, invites all other IACUC members to participate, and documents that the regulatory requirement (i.e., inviting all members to participate) has been satisfied.

One question organizations managed differently is whether inspections should be announced or unannounced. Many organizations have tried both methods and ultimately decided the only way to maximize efficiency (i.e., assure lab personnel are present) was to announce the inspections. When attempting the unannounced visits, inspection teams would frequently find the PI and his or her staff to be out of the lab. Consequently, the inspection needed to be rescheduled.

Many organizations scheduled the inspections within a range of times (e.g., Thursday morning between nine and noon) to accommodate the inspection team's schedule and yet maintain a modicum of unannounced nature. In addition, most organizations required that the PI or one of his or her staff members be available at a specific time to participate in the inspection.

Preparing for the Inspections

Since IACUC members are essential for the inspection process, with others having ancillary roles (e.g., OPP staff identifying needed maintenance issues), BP meeting attendees agreed that it is critical to regularly train IACUC members on the inspection process. Prior to facility inspections, many IACUC administrators conduct brief retraining sessions on facility inspection processes. In addition, they provide the previous relevant facility inspection reports to the inspection team. These reports remind IACUC members of previously identified issues, allowing them to verify that the issues are not ongoing repeat problems. In addition, some IACUC administrators discuss deficiency trends during the IACUC meeting. For example, the IACUC may discuss trends in physical plant maintenance issues and ways to mitigate them.

Inspectors visiting a PI's procedure area should be given a copy of the protocol(s). One suggestion to facilitate the process was to e-mail copies of the protocols to the inspection team 30 days before the inspection and ask that each subcommittee member provide a couple of questions to ask the PI.

To facilitate the inspection of PI-managed animal care and use areas (e.g., satellite housing facilities, surgery areas, and specialized animal laboratories), the IACUC administrator notifies the responsible PI of the inspection at least 30 days prior to the visit. The initial notification is typically done using e-mail and simply states that the IACUC will be conducting semiannual facility inspections on "date" and asks that either the PI or a designated lab member, who is using animals, be present to facilitate the inspection. In addition PIs are frequently provided a checklist, which lists the items the IACUC reviews during facility inspections and any concerns that were documented during previous inspections. The PI is given approximately 10 days to confirm that he or she will be available for the inspection. If confirmation is not received by the end of 10 days, the IACUC administrator frequently telephones the PI until confirmation is acquired.

CONDUCTING SEMIANNUAL INSPECTIONS

Inspecting Animal Housing Areas

IACUC members generally find that different housing situations offer unique challenges. For example, inspection team members visiting agriculture animal housing facilities have different criteria to apply than those visiting laboratory animal facilities. An important first step is the selection of the appropriate applicable standards. For example, when laboratory or biomedical animal housing facilities are visited, the *Guide* and, when appropriate, the AWAR would be applied; when farms housing agriculture animals being used for food and fiber research that is not funded by PHS agencies are visited, the *Guide for the Care and Use of Agricultural Animals in Research and Teaching* (henceforth referred to as the *Ag Guide*) is used. However, in all cases, the primary concern for IACUC members is the welfare and safety of the animals. BP meeting attendees noted that one of the most effective ways of learning about a particular area in the course of the inspection is to talk with one of the animal handlers from that area.

Laboratory Animal Facilities

Although BP meeting attendees stressed the importance of ensuring that housing facilities comply with the relevant regulations, the group agreed that concerns relating to animals' welfare are the overarching priority for the inspection team. For example, the inspectors must ensure that the animals appear to be healthy, that their cages are clean and dry, that they are not overcrowded, and that appropriate enrichment devices are being used. Inspection team members must be trained in the needs and well-being of the species of animals that are maintained in those areas they are inspecting.

In many situations the inspection team may not be able to look at every cage in a vivarium, but members should visit each room and look at a sampling of the cages from each room to reach a satisfaction level that the animals are healthy and housed according to the relevant regulations.

Once the inspection team is comfortable that the animals are healthy and being cared for appropriately, the team can turn its attention to necessary documentation such as daily visitation and facility sanitation records. The team should also consider the condition of the physical plant. For example, room surfaces should be impervious to moisture and easily sanitized. They should be free of chipping paint and rust that could compromise the ability to effectively disinfect the room.

The inspection team may choose to talk with one of the animal care technicians to learn how veterinarians are notified when animals appear to need clinical care. For example, what methods are employed for notifying the veterinarian? How does the technician know when a veterinary follow-up has occurred? What process is used to track the health status of animals under veterinary care? In addition, IACUC members can also take this opportunity to assess the OHSP. For example, ask technicians if they are enrolled in the OHSP and if they know of any biohazards in their work area.

The information gathered by the inspection team during this process should then be assessed to ensure that it adheres to the relevant regulations.

Inspecting a Laboratory Animal Room

An IACUC member inspecting a mouse housing facility frequently utilizes the following best practices:

The animals. Upon entering the room the inspector observes the animals and inspects their respective cages.

- The mice should exhibit the physical characteristics of healthy animals. They should be active and moving about the cage, engaged in nest building or grooming activities, eating and/or drinking.
- Mice are considered social animals, and they should be housed in socially compatible groups of two or more unless an exception has been made through an IACUC-approved protocol.
- The bedding in each cage should be clean and dry. The amount of fecal matter should be within the organization's acceptable limits; remember that the cages should be changed every 2 weeks (the *Guide*, p. 70).
- The food and water supply should be clean and free of fecal matter.

Enrichment. The inspector verifies that enrichment expectations are satisfied.

- When practical, devices that enrich the environment of the animals (e.g., Nestlets™, PVC pipes, or other devices that can be used as hiding places) should be placed in each cage.
- If animals must be housed singly or enrichment devices cannot be used for scientific reasons, provisions for the exception should be included in an approved IACUC protocol.
- The animals should not be overcrowded, and particular attention should be paid to ensure that pups are weaned from nursing dams at the appropriate interval.

The environment. Upon entering the animal room, the committee member should pay attention to the room's environmental conditions.

- A properly functioning HVAC system will maintain appropriate room-to-hallway air pressure differentials, stable room temperature and humidity, and ensure that the room receives an adequate supply of fresh air.
  - The room temperature and humidity levels should be within the acceptable parameters. The acceptable temperature range for mice is 68°F–79°F (20°C–26°C), and the range for humidity is 30 to 70 percent.
  - The inspection team should determine the appropriate room air pressure differentials prior to conducting the inspections. This information is usually available in the organization's program description, especially when the program has been accredited by AAALAC International. In situations when the room pressure should be positive to the hallway (e.g., immune-suppressed animals), the air pressure in the animal room should be higher than that of the hallway. For negative pressure areas (e.g., quarantine, infected animals, most rodent spaces), the air pressure in the animal room should be less than that of the hallway. A simple test to verify the direction of airflow is to hold a tissue or paper towel near the crack of a slightly opened door. A positively pressured room will blow the tissue toward the hallway while a negatively pressured room will pull the tissue into the room.
  - To ensure that each animal room receives an adequate supply of fresh air, there should a minimum of 10 air room exchanges per hour. Upon entering the room,

if the inspector detects an intense ammonia smell or experiences a slight eye irritation, the ammonia level may be too high, suggesting that the fresh air exchange to the room is too low.

- The inspector also ensures that the light intensity and cycle are appropriate for mice.
  - The lighting cycle for mice is generally maintained at the optimum of 12 hours of light and 12 hours of dark unless experimental parameters dictate otherwise. This time cycle promotes normal murine behaviors including breeding activities and the routine intake of food and water. The inspectors often check to ensure that the lights and timers are functioning properly.
  - Light intensity has been shown to cause phototoxic retinopathy in albino rodents such as traditional laboratory mice. Consequently, in this situation the inspector should ensure that the light intensity adheres to the *Guide* (p. 49) recommendation of 130–325 lux (at cage level) to minimize the risk of mice experiencing the problem.

The physical plant. The inspector then observes the physical condition of the room.

- The room floor, ceiling, and walls should be clean. In addition, the floor should be free of clutter.
- The doors, walls, floor, and ceiling should be free of chipping paint and, if metal, free of rust.
- All equipment (i.e., tables, chairs, garbage cans, countertops, etc.) utilized and maintained within an animal room should be easily disinfected and made of impervious material (e.g., no cloth chairs or cardboard should be stored in the room).
- The integrity of the electrical system should be validated (i.e., no frayed wires and waterproof electrical receptacles with a ground fault interrupt system).

Records review. The inspector should review all of the relevant records to ensure that they are current and include the required information.

- **Sanitation records.** The inspector should review the sanitation records that verify that the facilities are routinely disinfected.
- **Cage cards.** The inspector should be able to identify the protocol number and a contact person when reviewing the cage cards.
- **Daily visitation records.** The inspector should be able to identify through documentation that the animals are being checked each day, including weekends and holidays. For example, the visitation log (or electronic entry report) should confirm that staff members were in the facility looking at the animals every day including weekends and holidays.
- **Clinical care (veterinary) records.** Veterinary records should be maintained for each animal receiving medical care. The records documenting the care should be available in the housing quarters. When reviewing clinical records, particular attention should be directed to ensure that, for example, a sick animal is assessed by the veterinarian, a treatment plan is in place, and the animal is under veterinary care throughout the entire treatment plan. In all cases, medical records must document the final disposition (i.e., closing out the case) of the animal.

Agriculture Animal Facilities

Inspecting a Farm

An IACUC member inspecting a farm frequently utilizes the following best practices.

The animals.

- When the inspection team visits a farm, it should pay particular attention to the health and well-being of the animals. The inspectors should examine the animals for lesions or any other health-associated problems. For example, swine displaying subtle coughing or sneezing may suggest the presence of a respiratory condition that requires veterinary treatment. Team members may wish to examine the hooves of animals to ensure that they are free of foot conditions such as foot rot.
- Agriculture species such as sheep are considered social animals, and they should be housed in socially compatible groups of two or more unless an exception has been made through an IACUC-approved protocol.
- The housing and feeding areas of the agriculture animals should be clean, dry, and free of excessive fecal matter.

The physical plant. After the health and welfare of the animals have been assessed, the inspection team typically turns its attention toward the physical condition of the facilities. Considering that the facility is a working farm, inspection team members should ensure that physical plant problems that have the potential to compromise the welfare of the animals are identified. For example:

- Sharp edges protruding from feed bunkers, metal buildings, windows, fencing, or stall materials have the potential to injure animals.
- Leaking watering devices could create areas that are excessively wet, which may create health-related concerns for animals.
- Chipping paint in barns may be an issue that should be addressed. The concern becomes elevated when falling paint contaminates animal feed or water supplies.
- The inspection team may want to ensure that gasoline-powered equipment such as skid steers and tractors are not leaking oil or gasoline into areas where animals are being maintained.

Equipment and support structures. The inspection team should look at equipment and other structures that are routinely used by the farm staff to ensure its integrity.

- Transport trailers should be in good shape and routinely sanitized.
- Equipment such as cattle chutes and farrowing crates should be stable and free of sharp edges.
- The inspection team should look at a representative sample of pasture fencing to ensure that it has not been compromised.
- If field shelters are used to provide pastured animals cover from storms, the integrity of those sheds should be validated by the inspection team.

Records review. Inspection team members should take this opportunity to review all relevant records.

- **Veterinary records.** The herd veterinary records should be reviewed by the inspection team.
  - The records should document routine vaccinations and other relevant procedures such as tail docking in sheep, castrations, and needle teeth removal in swine.
  - The veterinary records must also document clinical care from the initial treatment through the final disposition (e.g., healthy or euthanized).
  - If surgically manipulated animals (e.g., cattle with rumen fistulas) are being maintained, records should document that the surgery site is being maintained according to the established veterinary operating procedure.

Field Stations

Occasionally, field stations are established by scientists conducting wildlife research away from campus facilities. In situations such as these the IACUC must establish methods for inspecting the stations. The IACUC frequently works with the PI to conduct the inspection using video tape or some other appropriate technology.

If the IACUC has the opportunity to visit field stations, BP meeting attendees recommended preparing for the inspection by reviewing the protocol and related activities that are occurring at the remote site. For example, if surgeries are being conducted, the IACUC should ensure that the PI has established a specific location for conducting the surgery or collecting tissues.

While conducting the inspection, committee members should discuss and review with the PI the specific animal use procedures occurring, issues associated with the OHSP, and relevant issues specific to the species of interest. In addition, the inspectors should review any relevant research and field notes.

Cage Wash and Support Areas

The cage wash and other support areas should be visited by the inspection team. It must verify that appropriate personnel safety practices have been established. For example, cage wash areas are often loud and appropriate hearing protection equipment should be utilized. In areas where a risk of eye injury exists, safety glasses should be used.

In addition to personnel safety concerns, the IACUC should verify that the cage washer is reaching the temperature required to disinfect the cages. The records must be maintained and document that cages have been effectively sanitized. The temperature-monitoring records for automatic washers should document that wash or rinse water achieves a temperature of 143°F–180°F (the *Guide*, p. 71). If a mechanical cage washer is not being used, the inspection team will wish to verify that a monitoring method is in place that assures the adequate sanitation of supplies. Organizations frequently conduct microbial tests to verify how effectively handwashed items are sanitized.

Procedure and Specialized Areas

Survival and Nonsurvival Surgery Suites

The inspection team should inspect the surgical facilities on a semiannual basis. During the inspection, the team must consider the species of animals that undergo surgery in a specific area. Since surgeries conducted on USDA-covered species must be done in a surgery suite, the AWAR must be used as the standard for the inspection. When inspecting a surgery suite used for nonrodent USDA-covered species, the IACUC inspectors should verify that the facility is dedicated for surgical manipulations only during the time when surgeries are performed. In addition, the facility must be maintained and operated under aseptic conditions. Consequently, the team should review room sanitation records and ensure that all procedures are conducted aseptically. Team members should also ensure that there are practices in place for sterilizing and storing surgical equipment. The inspection team should ensure that there is an area dedicated to pre- and postoperative care and an area for surgeons to prep prior to surgery.

The inspection team should review the surgical records to ensure that they are complete and accurate. The records may be near the animal or could be at the surgeon's office, so multiple stops may be necessary to perform an appropriate inspection. The surgery records should include a description of the procedure being conducted, the type of anesthesia and analgesia being used, and the name of the surgeon. In addition, surgery records may document perioperative physiological parameters. For example, in the case of rodents, basic information such as body temperature, the status of the eyes, and respiration may be considered.

Those attending BP meetings agreed that if a group of rodents undergo the same surgery on the same day, a separate surgery record should be maintained for each animal. However, colleagues from AAALAC International indicated that, in some cases, it is appropriate to maintain group records for mouse surgeries. They indicated that group surgery records should include, at a minimum, the IACUC number and a description of the procedure being conducted.

BP meeting attendees agreed that changes to surgical records should be indicated with a single strike-through and be initialed. Wite-Out® or similar products should not be used and portions of the record should not be scratched out. In all cases, postoperative records of care must be maintained for all animals through recovery.

Specialty Rooms or Labs

In some instances, specialty procedures are performed in areas separate from standard animal support areas (e.g., the PI's laboratory or a specialized equipment area). For example, a PI may have behavior testing equipment or a small-animal MRI unit in his or her laboratory that is utilized in animal experiments. Areas such as this must be visited by the IACUC.

When inspecting PI-managed areas, BP meeting attendees recommended speaking directly with the scientist overseeing the area. During the discussion, the inspection team members should encourage PIs to discuss the details of their experiments rather than using standard phrases such as "I do procedure XYZ exactly as it is described in my protocol."

Euthanasia Rooms

The IACUC should also inspect areas where euthanasia is conducted. In the case of rodent facilities, committee members should inspect gas scavenging systems (if used) and review euthanasia records. The inspection team should question PIs regarding the practices employed. For example, questions such as how long it takes for mice to become unconscious or what type of secondary method of euthanasia is typically used may be asked. To ensure that gas euthanasia chambers are not precharged, IACUC members should make sure that the euthanasia chamber was not filled with CO_2 before the animals were placed inside. Once confirmed, inspectors may inquire as to the flow rate of the CO_2 used to gradually increase the CO_2 levels during euthanasia.

Controlled Drug Storage Areas

Controlled drugs are occasionally used in animal experiments. IACUC administrators reported that their IACUC verifies that the drugs are maintained according to Drug Enforcement Agency (DEA) regulations during facility inspections. At a minimum, committee members ensure that the controlled drugs are secured using a double lock system. The double lock box system is generally defined as a lock box behind a locked door, or the locking of controlled drugs in a refrigerator and then locking the room door. Ultimately, researchers needing to use controlled drugs will need to go through a system of two locks before they are able to acquire the drugs.

In addition to ensuring that the drugs are secured, the inspection team should review the records that document the receipt and use of the controlled substance. The controlled drug records should document the source and when the drugs were received. The records should also provide relevant information of how the substances were used. For example, the record should indicate when and by whom particular substances were removed from inventory and they should indicate how the drugs were used as well as the quantity remaining in inventory. In addition the records should document products that were removed from inventory (i.e., destroyed) upon expiration.

IDENTIFYING, DOCUMENTING, AND CORRECTING DEFICIENCIES

IACUC members should remember that finding what is wrong is not the only reason for conducting the semiannual inspections. During the inspection process it is also important to communicate with PIs and facility managers as to what they are doing right! In addition, committee members should remind the animal users and consistently reinforce that the IACUC is always available to provide them assistance. PIs and managers will feel involved if they are asked in what ways the IACUC can assist them.

In addition to providing encouragement to facility managers and PIs, the committee members are expected to point out deficiencies. However, it is extremely important for this process to be balanced by offering suggestions for corrective the concern. On occasion the IACUC members may discover that some PIs or facility managers have repeated noncompliance (i.e., repeat offenders); this problem may require the IACUC to be more direct and focused with its response.

The IACUC administrator frequently accompanies the inspection team and documents any identified deficiency. If the inspection team discovers a problem, it is classified as either a significant or a minor deficiency. In most cases, organizations define a significant deficiency as one that could endanger the health and welfare of an animal; nevertheless, the IACUC should develop a policy that clearly differentiates a significant from a minor deficiency. When documenting deficiencies, the IACUC administrator notes the problem, the location, the PI, whether the group identified the concern as being a significant or minor problem, and who (e.g., AV, PI, or facility manager) is responsible for correcting the problem. Once the group has completed the inspection for the day, the IACUC administrator frequently summarizes the identified concerns and provides committee members the opportunity to comment on each deficiency.

Once the deficiencies are confirmed by the IACUC chair, the IACUC administrator sends deficiency notices to the responsible parties. The deficiency notice to the relevant individual includes a plan for correcting the problem with a defined time line including a specific required completion date. If a PI does not respond, a second notice is sent, and the concern is referred to the IACUC for corrective action. The plan and time line for correction is typically based on the type of deficiency. For example, a 3-month correction date may be practical for painting the floor of an animal room, but when welfare issues exist, the IACUC may require immediate correction. However, if two IACUC members are present during the inspection, it is appropriate for the inspection team to rectify any issues posing an immediate threat to the health and welfare of an animal. The inspection team would then inform the IACUC of the issue and resolution at the next scheduled meeting.

A correction timetable can be altered by the IACUC, but alterations should occur during a full committee meeting and justification for the change should occur in the minutes, possibly appear as an addendum to the IO report, or be provided as part of a database.

Repeat Deficiencies

On occasion, a deficiency is not corrected in the required time and it is subsequently identified on consecutive inspections (i.e., repeat deficiencies). The repeat deficiencies are documented and treated in the same manner as any other deficiency. However, when repeat deficiencies are noted, many organizations recommend additional training for the PIs as well as follow-up performance inspections (e.g., a surprise visit from an IACUC member, a Compliance Liaison, or the committee chair) to ensure that they do not recur. Deficiency notices for repeat offenders are handled as previously described. However, the IACUC chair may phone the PI's superior or notify him or her in writing of the repeat deficiency. In addition, the PI may be invited to attend an IACUC meeting or serve as a committee member.

CHAPTER 10

Monitoring the Animal Care and Use Program

SCENARIO

GEU has been operating a very large animal care and use program (ACUP) for over 10 years. The organization's IACUC oversees approximately 425 protocols, which are facilitated by approximately 120 research scientists. During the semi-annual facility inspections, the IACUC frequently discovers activities that do not comply with the regulations and policies. For example, during a spring inspection, IACUC members discovered that a scientist conducting rodent surgeries was not maintaining appropriate surgery and postoperative records. The inspections also disclosed that some scientists were beginning procedures on animals before they received IACUC approval and, in some instances, deviations from approved procedures were occurring. Upon completion of a cycle of semiannual inspections, the inspection team referred three potential cases of noncompliance to the IACUC for investigation.

During a full committee meeting of the IACUC, the group resolved the cited concerns but determined that a more focused effort to preventing noncompliance was necessary. The IACUC administrator collated the organization's noncompliance data for a 3-year period in preparation for the meeting. The data indicated that incidences of reportable noncompliance increased each year during the 3-year time period. The IACUC administrator also expressed the concern that the data reflected only those protocols observed during a semiannual inspection. Consequently, the specifics of approximately 80 percent of the protocols had not been monitored during the inspection process. Based on the provided information, the committee concluded that IACUC-approved protocols were not being adequately monitored by the committee, which could constitute an ACUP deficiency. The committee members discussed multiple methods for intensifying the monitoring process.

Although many committee members agreed that a deficiency in the program existed, some felt that the noncompliance concerns were isolated incidents and associated with specific principal investigators (PIs). They believed that the process of reviewing approved protocols during facility inspections was adequate. These members suggested that the inspection team should spend more time with specific PIs during the semiannual inspections and use this opportunity as a means of education.

Other committee members suggested that trends and repeat deficiencies indicated that the inspection process was not effective. They suggested identifying specific individuals who would be tasked to conduct independent regulatory assessments of specific PI research programs. They suggested that the compliance liaisons meet with scientists, discuss their experimental activities, and report the findings to the IACUC during a monthly meeting. This suggestion resulted in a lengthy discussion and focused on whether such a process would imply that the IACUC members did not trust the PIs.

Ultimately, committee members agreed to initiate independent audits for a period of 3 months and to assess the ramifications of the process at the end of the defined time period. The group agreed that if the problem was due to specific PIs and isolated incidents, the random audits should expose no global problems.

During the next 3-month period, multiple concerns were discovered. For example, the assessors noted frequent incidents of untrained personnel working with research animals. The team also identified record-keeping problems, incidents when animal use procedures were modified without first acquiring IACUC approval, and failures to report adverse events with research animals to veterinary staff. During this assessment period, the number of IACUC protocol modifications increased significantly. The committee's interpretation was that the assessors were helping PIs to identify needed corrections and realize the importance of satisfying the regulatory requirements.

Based on the findings of the 3-month assessment, the IACUC supported the idea of extending the program through the remainder of the year to determine whether the program would have any effect on the semiannual inspection findings. By the end of the research year, the IACUC validated that the assessment program resolved the program deficiency of inappropriate IACUC monitoring of approved animal care and use projects.

Regulatory Requirements and Guidance Documents

- **Public Health Service (PHS) policy (Section IV B (1–8), p. 12)** requires IACUCs to oversee the institution's program of animal care and use.
- **Animal Welfare Act regulations (AWARs) (Section 2.31(C))** require IACUCs to oversee the institution's program of animal care and use.

INTRODUCTION

Federal regulations and policies require IACUCs to ensure that an organization's ACUP conforms to applicable regulations. For example, the PHS policy requires the IACUC to conduct semiannual program reviews using the *Guide for the Care and Use of Laboratory Animals* (henceforth referred to as the *Guide*) as the basis for the review. IACUC must review the organization's program at least semiannually using the federal regulations as the standard. The committee must inspect animal housing facilities to ensure that they comply with the standards and IACUC members must ensure that animal care and use procedures are conducted according to the federal standards. To ensure that scientists conduct animal procedures according to the prescribed standards, IACUC members must be aware of how animal care and use procedures are being conducted.

The IACUC must ensure that the federal standards are observed. This process may begin with PIs describing the details of their animal activities in a proposal to the IACUC. The proposal should describe the specific details of animal care and use activities. Once a proposal is submitted, the IACUC members evaluate the details to ensure that it complies with federal standards. For example, the IACUC members will ensure that the proposal includes the details of a surgical procedure (i.e., describes the surgical preparation, the use of pain-relieving medications, how the procedure is conducted, etc.) and that the description of the proposal complies with the federal standards of care and use. Once the IACUC members are confident of the appropriate use of animals for the proposed activity, the committee approves the activity and the research may then be conducted. In order for IACUCs to guarantee continued compliance, organizations develop a method to ensure that the animal care and use activities are being conducted in accordance with the approved IACUC proposal.

MONITORING APPROVED IACUC PROTOCOLS

Best practices (BPs) meeting attendees reported two primary ways organizations ensure that ACUPs are complying with standards and expectations. Although the current terminology, "postapproval monitoring" (PAM), is typically used to describe a specifically defined system for monitoring approved protocols, not all attendees were receptive to the idea of PAM. Some attendees suggested that a formal PAM was an example of regulatory creep while others indicated that they use a PAM to satisfy the regulatory requirements previously referenced.

Some attendees noted a concern with the terminology "postapproval monitoring," believing that the use of the word "monitor" suggested intrusion, "Big Brother," or unnecessary interference. These organizations preferred to use terms like "compliance liaison," "research partner," or "research team member." The intent was to soften the regulatory intrusion of a monitor and encourage a professional relationship such as a liaison, partner, or team member.

Some organizations incorporate the continuing review of approved IACUC protocols while conducting other business associated with their animal care and use program. For example, during facility inspections, many organizations send at least two IACUC members to research locations for the purpose of ensuring that the activities conducted at that specific site conform to requirements. During this visit, representatives from the IACUC talk with PIs to confirm that they are following their approved protocols. The inspection team may ask a PI or his or her laboratory technician to describe how a specific procedure (e.g., euthanasia by cervical dislocation) is conducted. Committee members may ask the PI who typically conducts the procedure and then ask that identified individual to describe the details. During the process, the committee members are evaluating the procedure (e.g., whether what is being described sounds appropriate). The IACUC administrator typically records the details of the conversation in the inspection notes and verifies compliance with the approved protocol once he or she returns to the office.

A second contributing process is the engagement of staff veterinarian observations while conducting routine veterinary visits. During the veterinary visit, the veterinarian can confirm PI observance of humane end points, pain management, and effective monitoring procedures. The veterinarian may discuss approved procedures to confirm that the procedures are being performed as approved or as consistent with the organization's standard operating procedures (SOPs) or policies.

Some IACUC administrators suggested that the best compliance liaisons are those caring for the animals on a daily basis. Animal care technicians have the unique opportunity of daily observation and can assess whether PIs are addressing animal welfare concerns in a timely manner.

Developing a Monitoring Program to Complement the System

IACUC administrators overwhelmingly agreed that organizations are moving toward PAM programs that are intended to complement established ACUPs' monitoring systems. IACUC administrator BP meeting survey data from 2005 and 2012 indicate an increase of PAM-type programs or processes of up to 200 percent. One concern of most attendees was that a PAM program may be considered a quality assurance program. The better reflection of the PAM program is as organizational insurance, which helps to ensure that the organization is meeting its obligations to the animals, the funding agencies, and the public.

Most IACUC administrators agreed that personnel performing these assessments are becoming more common at organizations and that typically these individuals are specific personnel with the single responsibility of monitoring protocols. BP meeting attendees reported that the preferred (i.e., a best practice) approach for monitoring the program would involve specific personnel such as dedicated compliance liaisons, veterinarians, and animal care and research staff. These individuals would use various methods (e.g., a protocol review, a focused procedure review, daily observations, and/or laboratory internal self-assessments) to monitor the program. The monitoring sessions would occur at varying frequencies and might be based upon IACUC-determined criteria such as risk assessment, hazard analysis by procedure, historical trends within a given laboratory, and the species or procedures. Under this umbrella, each organization's process of oversight would look different but all would have the same outcomes goals.

A Simple Method for Conducting Focused Assessments

BP meeting attendees agreed that a simple method would be for PIs to be notified that a compliance liaison would be visiting them to audit their animal research program. Multiple methods for notifying the PI were discussed at past BP meetings. For example, in situations in which a project involved invasive procedures (e.g., tumor development and monitoring studies), organizations included—as part of the project approval letter from the IACUC—the information that the progress of the study would be reviewed by the compliance liaison throughout the year. In other situations, the compliance liaison would directly communicate with the PI and inform him or her that a specific animal use activity approved under a specific project would be audited.

Prior to the audit (generally 30 days but at least 2 weeks before the assessment), the PI is notified in writing as to when the audit will occur. The notice reflects the fact that friendly dialogue is the basis on which the audit will occur. For example, the notice may say, "You and I [in some cases, the biological safety officer or a veterinary technician may attend] may sit down to discuss how your research is progressing. We may discuss specific procedures described in your protocol as well as troubleshoot any problems you may be having." The notice may also indicate approximately how much time will be required for the discussion (e.g., BP meeting attendees indicated that 30–60 minutes was the average).

In addition, the notice would provide the PI with a list of topics that he or she should be prepared to discuss. For example, the PI might be informed that a surgical procedure such as an ovariectomy would be reviewed and that the relevant records would be evaluated. The notice identified other specific review items, such as whether

- The personnel working with animals had been approved to do so by the IACUC
- Daily observations of the animals were occurring and being documented
- Preprocedural activities defined in the IACUC-approved study were being adhered to
- Pain-relieving drugs (e.g., anesthesia/analgesia) were being used appropriately
- Animals were receiving postprocedural care
- The method of euthanasia used was approved by the IACUC

A phone call prior to the actual visit may be helpful to discuss the process and the benefits, answer questions about potential outcomes, and clarify the desired information during the site visit. Some IACUC administrators have found that a written notice with a follow-up phone call that occurs about 10 days before the visit is most helpful.

During the interview process, compliance liaisons should focus on the process improvement. BP meeting attendees who had what they determined as successful oversight programs indicated that terminology is everything! Choosing words carefully can have a significant impact upon successful outcomes. For example, scientists find an interview much more palatable than an audit or monitoring session. The compliance liaisons should do everything in their power to make the experience helpful and friendly. The ultimate goal of the activity is behavior modification and improved performance. The compliance liaison should use the assessment as an opportunity to educate the research staff members. It is important for the liaison to be tentative and never give a definitive answer since the role of the liaison is as the "eyes and ears" of the IACUC; the liaison is not a replacement or substitute for the IACUC and its decision-making process.

It is helpful to take as few notes as possible. People get nervous when the visitor is doing a lot of writing. The notes being taken should be points that will provide specific details of the review and will be critically necessary for the preparation of the postvisit report. For example, notes could include the room used for the procedure, the concentration of the drugs used, the personnel participating, the frequency of monitoring—in other words, the specific information that will create a viable and accurate report. The compliance liaison should inform the research staff before the interview that some notes may be taken but that the notes are for specific details to keep the report accurate and brief. The liaison should facilitate discussion and ask only open-ended questions. Rarely should the liaison ask a "yes" or "no" question. As a general rule, scientists are interested in, and very proud of, their research. Consequently, they enjoy talking about it. Engage, show interest, and discover evidence of good performance; the dysfunctional matters will be readily obvious.

Establishing a Postapproval Monitoring Program

Funding the Program

Perhaps one of the most frequently asked questions at BP meetings is "How do we fund an activity that we recognize is important, valuable, and in the eighth edition of the *Guide*?" In some cases, organizations have assigned new duties to existing staff; in other cases, organizations have hired staff with a specific job description (i.e., dedicated procedural assessor or shared assessor between IACUC, safety, Institutional Biosafety Committee). In yet other cases, organizations have used shared resources—having a single compliance liaison serve multiple units. Attendees representing organizations that have dedicated liaison staff member(s) indicated that their programs are typically funded through the compliance division with funds frequently originating in the Office for the Vice President for Research (or equivalent).

Since monitoring records are frequently maintained using standard programs (e.g., Microsoft Word and Excel), specialized software and computer technology in most cases have not been purchased. Consequently, the principal expense to operate the program is typically the cost associated with the employee (e.g., salary, benefits, computer, etc.). Attendees indicated that when office space is available, the cost of the oversight program is equal to the cost of adding and maintaining a new employee for any other purpose.

While the role of postapproval monitoring is now specifically cited in the eighth edition of the *Guide*, there have not been provisions by many organizations to increase funding support to accomplish the mandate. Since funds are generally tight to begin with, many organizations have chosen to follow the method of monitoring protocols as part of conducting normal business (e.g., IACUC semiannual inspections, veterinary daily care visits, etc.). A few organizations have embarked upon novel methods of assuring compliance by establishing research animal coordinators or research program associates, which is a methodology of engaging the research participants to extend the observations of the IACUC.

Identifying and Training the Compliance Liaisons

BP meeting attendees agreed that the most critical attributes of an effective liaison were the interpersonal skills. Several attendees noted that they could teach their liaison the skills of monitoring effectively but they could not teach anyone to be cordial, communicative, or friendly. Another identified critical characteristic of the liaison is excellent observation skills and the ability to pay attention to detail. One method individuals used to assess the potential liaison was to ask the person what he or she observed from the time of entering the building until arriving at the supervisor's office. The best liaisons effectively observed their surroundings and noted things that they had not been told to observe. These skills would be very helpful in a research laboratory to build a complete picture of the situation where research is being conducted.

Although organizations reported that education and experience were secondary to communication and observation skills, many agreed that veterinary and laboratory technicians make the best compliance liaisons as their professional training included such areas as animal well-being and many procedures approved in most protocols (e.g., blood collection, monitoring frequencies, medication management and administration, etc.). Researchers and scientists also make good compliance liaisons but they would need to shelve their interest in the science of the activities and focus upon the animals' well-being and the procedures being performed. Those in attendance agreed that if an individual had good observation and communication skills, the chances were that he or she could become an exceptional compliance liaison.

To avoid auditor burnout, IACUC administrators suggested structuring the liaison's job description to include other responsibilities. At some organizations the compliance liaison also served as a trainer; in this role, liaisons spent a portion of their time conducting orientations for new IACUC members and research staff as well as serving as guest lecturers in classrooms to discuss animal-related topics. Organizations should be careful to avoid creating a culture of "animal police" in the minds of the research staff. Liaisons should never see themselves in this role. Rather, they should view themselves as a partner in the research enterprise, serving as the extension of the IACUC, fulfilling the organization's commitment to quality animal care and use, providing confidence that animals are being used as described and in humane and ethical manners, and using those observations as education content issues for the research and potential research staff on campus and beyond.

Liaisons frequently receive extensive training prior to conducting postapproval monitoring sessions on proper personal and professional conduct during an assessment. The compliance liaison should never be confrontational; he or she should not change, interfere with, or try to guide the research activities being performed. In addition, it is critical for the compliance liaison to only observe the activities and not to provide commentary during the performance of any procedures—in other words, "be a fly on the wall." In order to conduct a thorough and accurate assessment, it is critical that liaisons are familiar with and have a comprehensive understanding of general IACUC and PI-specific SOPs, forms, and policies. They should have a clear understanding of the animal welfare regulations and policies. It may be helpful for the IACUC or the compliance supervisor to craft exercises that teach liaisons the specifics of performing an assessment. For example, one common practice noted was for a newly hired liaison to review a project and then conduct a "mock" PAM session with perhaps his or her supervisor, another senior liaison, or an IACUC member. The IACUC must have some method to assure that what the liaison is observing and documenting is consistent with expectations of IACUC oversight. One best practice involved having newly hired liaisons review protocols (as an additional reviewer) to sharpen their skills of process, and then begin with assessing breeding or teaching protocols (i.e., less invasive or complicated activities) and slowly move into more invasive protocols and procedures.

Reporting Lines and the Compliance Liaison's Relationship with the IACUC

BP meeting attendees noted that the liaison staff members typically have two reporting lines. The first is an operational reporting line to the IACUC and the second is an administrative or duty reporting line.

The most common administrative reporting line was to the director or supervisor of the organization's ACUP compliance division. Attendees agreed that the only administrative reporting prohibition was that liaisons should not administratively report to the attending veterinarian (AV) or the director of animal care. Since these individuals have a primary role in the operation of animal care, this reporting structure would be a clear conflict of interest. In fact, since liaisons assess IACUC approved protocols (e.g., experiments) and AV-approved SOPs (e.g., laboratory animal husbandry), the liaison reporting administratively to either the AV or animal care director would require that the liaison assess the quality of the program that his or her supervisor administered. To reiterate this concern, liaisons should interview and observe veterinarians conducting procedures to ensure that they are conducting their activities as described in the approved SOPs; therefore, the compliance liaisons cannot report administratively to the veterinarians. IACUC administrators attending past BP meetings indicated that reporting structures in which liaisons report to the AV or animal resource program director would be equivalent to not complying with the regulatory requirement, which states that IACUC members may not engage in deliberations regarding projects in which they are involved. Organizations must consider potential conflicts of interest when determining the reporting lines of their compliance liaisons.

Liaisons also have collegial reporting lines to the IACUC. For most organizations, compliance liaisons are not IACUC members. The reasoning behind this approach was to clearly separate the liaison's activities from the committee's official actions. There was a concern that the liaison might assume more authority than might be appropriate if he or she were an IACUC member or that researchers may perceive liaisons as having more authority than is appropriate. Liaisons must never become a surrogate or replacement for the IACUC and keeping them as ad hoc consultants to the committee will assure that relationship. Serving as nonmembers, liaisons are able to help PIs address minor concerns and partner with the researcher in crafting a probable response to the IACUC without PIs getting the perception that the issue is settled by their interaction with the liaison. The nonmember role also provides the liaison some leeway and allows the development of a professional and mutually respectful relationship with the PI.

Policy and Standard Operating Procedure Development

Organizations developing compliance assessment programs should also develop clear and concise procedures and policies and make them available to everyone. Program documents should be "soft" but clear. For example, rather than using terminology such as "audits" and "inspections," verbiage such as "reviews," "discussions," or "assessments" can be used. The documents should indicate, for example, that liaisons will assist the researcher with meeting the regulatory expectations, the IACUC with preparing for semiannual inspections, and the PI in preparing protocol amendments.

Conducting the Visit

Prior to conducting the PAM visits, an organization must decide how frequently assessments will be conducted and what protocols will be reviewed. BP meeting attendees offered different suggestions for timing audits and determining exactly what types of protocols should be reviewed.

Some institutional representatives at past BP meetings indicated that they review all of their protocols annually to ensure that PIs have the opportunity to benefit from the review. Conversely, others indicated that they conduct random audits with emphasis placed on projects that involve, for example, distressful or painful procedures being conducted on species of animals overseen by the USDA. Others in attendance indicated that they review each project at least once in a 3-year period. Still others noted that each researcher received an assessment every year but, in many cases, researchers maintaining several protocols did not have every protocol reviewed every year.

One common concern BP meeting attendees expressed was the need to review the entire approved proposal or just specific procedure(s) within it. Attendees from smaller organizations generally noted that their liaisons reviewed the entire protocol, while liaisons at larger organizations tended to review procedures within the protocol rather than the entire document. This procedural difference is likely the result of available personnel time and the number of protocols approved by the IACUC.

Conducting the Interview

Prior to meeting with the researcher, the liaison should decide what will be reviewed. In some cases, a liaison may decide to review only the surgical procedure listed on the protocol and follow the observation by discussing the procedure with the surgeon. The liaison should review every aspect of the protocol that will be discussed during the assessment, with emphasis on the particular component of the project that will be observed (in this case, the surgery). For example, the liaison may review the details of a surgical procedure listed in an IACUC-approved protocol; in addition, he or she may assess all of the relevant IACUC-approved policies and guidelines as well as any PI-specific SOPs. BP meeting attendees indicated that preparing for the assessment session typically takes much longer than conducting the actual review.

As part of the review process, the liaison should talk with the researcher to explain the difference between the types of concerns that must be addressed by the IACUC and those that can be immediately addressed by the researcher at the time of the visit. The liaison should explain to the researcher that if an issue requiring a report to the IACUC is identified, he or she, as the liaison, will serve as the PI's advocate and help bring the issue before the committee and ultimately resolve it. In all cases, the liaison should assure the PI that issues of noncompliance can be rectified and that he or she will help the research program become and remain compliant. The liaison should encourage self-reporting and emphasize that the IACUC would rather hear about potential problems from the research team rather than from other sources.

BP meeting attendees stressed the importance of the timeliness of processing the review outcomes (i.e., quickly compare gathered information to approved protocols). Timely responsiveness to the research team is a significant component of partnering with them to facilitate effective shifts in behavior and effective animal care and use improvement. Immediacy of site visit processing is also important because the marginal notes taken during the review may lose relevancy as memories fade. For example, the timely review of minor notes taken at the time of the interview may prompt the reviewer to recall important details that he or she may not be able to recall a week after the visit. It is best to process the review, develop conclusions, and provide the report as soon after the assessment as possible.

Resolving Discrepancies

Liaisons should make every effort to immediately resolve issues. It makes no sense for a research partner to allow a noncompliance event to continue. In some cases in which animal welfare or care issues exist, the liaison may need to immediately contact the AV, IACUC chair, compliance director, and/or institutional official (IO). Animal welfare concerns require immediate and appropriate veterinary care.

In other circumstances, the liaison may discover issues that might be significant and must be addressed. For example, if a liaison discovers animals are not receiving analgesia as described in the approved protocol, a significant (i.e., an incident that has the potential to affect the health and welfare of the animal) violation of the animal welfare regulations may have occurred. If research animals have potentially been experiencing pain or distress due to a noncompliant pain management plan, this is a significant issue affecting animal welfare and the contact tree noted before is appropriate.

However, if the animal was effectively medicated and did not experience any pain or distress, there remains the issue of deviation from the IACUC-approved protocol pain management plan. In some cases, the liaison might inform the researcher that IACUC-approved analgesia must be used in the future and that the IACUC must be informed of the protocol noncompliance. This is a case where a researcher self-report would be favorably received by the IACUC. The liaison may explain to the PI that the IACUC must be informed and that he or she will assist in preparing correspondence to the committee. In addition, the liaison may discuss with the researcher the issues associated with the observation: *what* the problem is, *why* it is a problem, and *what* the IACUC generally would expect the method of correction to be. The liaison may include an explanation of the sequence of events that will likely occur. For example, the liaison may explain that when a noncompliant event associated with PHS-funded activities is identified, a prompt report must be made to the Office of Laboratory Animal Welfare. The IACUC will consider the report, may investigate the allegation by speaking to the parties involved, and will require the PI to develop a resolution that will ensure that the problem will not continue or repeat. The liaison may choose to discuss with the PI the fact that a final written report will be prepared to summarize the events of the issue and that this report will be made available to, for example, the IACUC, research team, or IO.

At times, the liaison may identify less acute noncompliance. For example, the liaison may discover that trained animal handlers are working with animals and are not listed on an approved animal care and use protocol. He or she may find that the skills of the individuals are appropriate and the procedures were performed expertly, so the only problem is that the IACUC has not had the opportunity to validate the qualifications as required by regulatory provisions. At some organizations, IACUCs have authorized the liaison to immediately resolve this concern by facilitating an amendment to add the personnel to the protocol; this could be accomplished by using an administrative review process. Consequently, some liaisons are able to immediately assist the researcher, prevent unnecessary delay in research, and assure a fully compliant atmosphere—the evidence of a partnership between the researcher and the program.

POSTASSESSMENT ACTIVITIES (FOLLOW-UPS)

The most important action is to commend the PIs for what they have done right. BP meeting attendees noted that the best commendation seems to be written notice because many PIs proudly display these notices in their labs.

In the event that problems are discovered, the organization should have established a process to ensure that the concern is reported to the appropriate institutional officials. For example, the SOP should identify which findings must be reported to the IACUC for investigation. While the IACUC should be notified of all assessments (remember that the compliance liaison is the eyes and ears of the committee and cannot serve separately from them), the manner of notification may differ depending upon the IACUC's predetermined policy. For example, the policy may first identify what issues (e.g., the personnel amendment) are managed on behalf of the committee by the liaison and later reported to the committee. It may specifically identify examples of animal welfare concerns, instruction on how to report them to the AV for management, and how the issue is brought before the committee for discussion. The policy may also identify those issues that must be reported prior to being managed so that the IACUC can determine the appropriate continuing process of investigation, review, and corrective actions.

Sanctions

BP meeting attendees agreed that methodologies to deter repeat problems should be employed. They also agreed that protocol suspensions should be an absolute last resort; a suspension would delay research and could lead to returning federal funds or requiring the unnecessary euthanasia of animals. IACUC administrators indicated that the best way to view sanctions was to consider those that would correct the issue within the specific laboratory and also consider if the issue would benefit from a more global measure. Such measures could be crafting a new program policy, including the issue in the continuing education program, or transforming the concern into IACUC or PI training. IACUC administrators agreed that although all issues were caused by an individual, almost all issues had a programmatic component that allowed, fostered, and even encouraged the poor performance. By using noncompliance events as indicators of programmatic health and function, the program could benefit by tracking performance and noncompliance trends, and other researchers could be protected by a well thought out policy, procedure plan, or SOP.

Is Your PAM Working?

BP meeting attendees reported that one of the simplest health monitoring tools for the program was the number of protocol amendments processed by the IACUC office. Some attendees indicated that when their PAM was initiated, they received a significant number of modification requests; after about a year and a half, modifications decreased. This suggested to them that their researchers were doing a better job of writing accurate and complete protocols and keeping their proposals current. Attendees indicated that PAM reports confirmed this measurement to be accurate since fewer incidents of noncompliance were identified as the number of modifications decreased. In most instances, IACUC administrators indicated that, after a couple of years, researchers started to participate more fully in the program of institutional self-assessment. Organizations have seen an increase in self-reports and requests from researchers for assistance in corrective actions.

Certain organizations noted that their IACUC had established a strategic plan to facilitate institutional compliance. Such plans often included preparing trend reports that identified the numbers and types of adverse events; the frequency, timing, and recurrence of the events; and how the research community was engaged to facilitate the building of a compliant culture at the organization. In addition, IACUCs have identified objective measures of successful trending to evaluate the success of their PAM programs—for example, a decrease in the overall number of perceived noncompliant reportable events, a decrease in the occurrence of significant adverse events, an increasing number of researchers self-reporting and implementing self-corrective measures, and fluctuations in the number of protocol submissions (generally an increase upon program initiation with the number of amendments per month stabilizing after 1 or 2 years). In addition, the IACUC begins to see the quality of submitted protocols increase. In other words, as researchers become educated as relates to identifying the important details, they begin to ensure that those issues are included in their submissions.

On occasion, select organizations have incorporated the most common deficiencies identified through the year by liaisons into their research community's annual or ongoing training. In these situations, organizations may have tracked the adverse events by category (many use the chapters or section headings in the *Guide*). They then craft annual training questions around the top seven to ten noncompliant issues that occurred during the past year. This process has proven to be a highly effective method for identifying and developing training as it focuses limited training time on the specific areas of performance failures. Organizations engaging in this process reported that they routinely saw a dramatic decrease in the number of the same types of significant identified adverse events in the subsequent year and that they believed that bringing the issue to the research community in a focused and pro-active approach was another measure of good partnership in the process of animal care and use.

CHAPTER 11

Facilitating Communication

SCENARIO

GEU operates a diverse and decentralized animal care and use program. The animal facilities are located across campus with two outlying facilities being ten miles apart. The animal care and use program is well staffed. Personnel from the animal resource program and the research administrative office have specifically defined responsibilities. In addition, staff members from Environmental Health and Safety (EHS), Occupational Medicine (OM), Public Relations (PR), and the Office of the Vice President for Research have an active role in the animal care and use program.

GEU's sheep farm is remote and primarily used to teach ovine management and production. During the lambing season many students are present for parturition classes. In fact, the facility manager and numerous students deliver lambs as part of their classroom activities. Approximately 2 weeks after working at the lambing facility, a student started to experience flu-like symptoms. As a result, she met with the OM physician for consultation and examination. The university physician performed several evaluations and ultimately diagnosed the student with Q-fever.

The doctor discussed the illness with the student and her parents, including the treatment and potential complications. The illness progressed and as a result the student was unable to attend classes during the spring semester, which had the potential of negatively impacting her scholarship.

Proactively, the student's parents contacted university officials to ensure that her illness would not affect the status of her scholarship. They also asked that university officials take action to ensure that a similar situation would not occur to other university students. Her parents were able to resolve the concerns relating to the scholarship, but they also had their story published in the local newspaper and broadcasted on the world news. The story included the following statements:

- GEU conducts dangerous research using animals.
- GEU does not consider the safety of students when working with research animals.
- Students conducting research can get extremely sick from handling experimental animals.
- GEU has no practices in place to protect students from animal-related diseases.

Once the story was released, the CEO was overwhelmed with phone calls and e-mails. As the CEO was not familiar with the intricacies or ramifications associated with the animal care and use program, she referred all of the questions to the institutional officer (IO), the vice president for research. The IO immediately consulted the attending veterinarian (AV), IACUC administrator, and the EHS and PR directors and found all to be unaware of the incident. Until the IACUC administrator contacted the university physician, those offices managing the animal care and use program had no record of the incident. In the event that GEU had effective communication practices, this incident may have had an entirely different outcome. For example; the OM physician could have informed the IACUC administrator of the Q-fever exposure. The IACUC administrator could have coordinated a meeting with the physician, AV, EHS personnel, the IO, the IACUC chair, and the PR director. The group could have collectively discussed all relative concerns and may have taken the following actions (as examples):

- The IACUC administrator could have reviewed the IACUC-approved protocol for the lambing facility to confirm that appropriate precautions were in place to protect animal handlers from Q-fever.
- The AV could have performed clinical evaluations to identify or refute the presence of Q-fever in the university's sheep herd.
- EHS staff members could have visited the sheep facility and observed those delivering lambs to ensure that appropriate and required safety practices were being engaged.

Once these actions were taken, the PR director, and perhaps the AV could have met with the student and her parents to discuss the details of Q-fever. During their meeting, GEU representatives could have provided a pamphlet or some other form of written materials to the student and her family explaining the illness and related issues. They could have discussed how the disease is transmitted from sheep to people and answered the questions posed by the family. For example, they could have explained that the sheep in question were not research animals and that the acquired illness was due to a disease naturally carried by sheep. They may have explained that an exposure such as this one is more likely to occur at a private farm than in the university setting since GEU's animals are routinely monitored for Q-fever. In addition, the university representatives could have discussed the fact that all animal handlers are trained to recognize symptoms of the disease and, in fact, that is the reason the illness did not progress in their daughter; she knew to contact the university physician and inform him that she worked on a farm delivering lambs. The veterinarian may have informed her parents that EHS staff members equip the lambing facility with personal protective equipment (e.g., safety glasses, respirators, latex gloves, and coveralls) to help to ensure that an exposure does not occur. Upon completion of the conversation, university officials would have then been available to answer any additional questions relative to the incident. Upon completing the conversation, the PR representative could have asked the family for permission to use their daughter's name in a brief press release discussing the details of the exposure.

Had the university established a strong communication program, which would have resulted in a timely response as described in this scenario, there would have most certainly been a different outcome: The parents would have been better educated on the issues relative to Q-fever, and there would have been a significantly decreased likelihood of the matter being addressed by the world news. On the other hand, if the incident were reported by the local and world news, the report could have included the details presented by the university officials to the parents.

EFFECTIVE COMMUNICATION

Many situations require both internal and external forms of communication. In addition to information on zoonotic diseases, situations relating to natural disasters and animal activists activities also require effective forms of communication. Organizations operating successful and complex animal care and use programs involve multiple departments when facilitating various components of the program. For example, to ensure that occupational health and safety (OHS) requirements are satisfied, the offices of EHS and OM participate in OHS program activities. For example, EHS conducts risk assessments of animal housing and use areas, and OM performs personal health assessments on animal users. In addition to regularly disseminating information to the community, the PR office may also prepare press releases, for example, of animal activist activities occurring on campus. Additionally, police services, the local FBI, and IACUC administrative and veterinary staff members have critical functions within the program.

In order to facilitate an efficient and successful program, organizations should develop a communication plan that incorporates input from individuals representing all departments that have an interest in the operation of the animal care and use program. The overall goal is to ensure that entities within the organization are working together.

Communications among Groups Administering the Program

Many organizations develop a process for communicating important information between its different departments. Best practices (BP) meeting attendees frequently noted that the IACUC administrator serves as the central point of contact and disseminates information throughout the leaders of the organization's animal care and use program. For example, in the event that vandalism occurs at a university animal facility that results in a call to police services, the police officer immediately notifies the IACUC administrator with specific information that is to be shared with the rest of the program leadership.

Depending on the degree of concern, the IACUC administrator may schedule an immediate meeting for those on the animal program contact list, or schedule a meeting between only those with relevance as it relates to the issue. The key is for those administering the program to have a successful and effective communication process that is described, for example, in a standard operating procedure. The written policy will ensure that a consistent and efficient process is followed by the organization.

Communicating with the IO

The standards governing animal use programs indicate that the CEO of an organization or his or her designee (the institutional official) bears ultimate responsibility for the program. Consequently, an effective method for communicating with the IO should be established.

During IACUC administrators' BP meetings, attendees generally reported that their IO was open, available, and communicative. Attending IOs indicated that they take their role very seriously. At one such meeting, the attendees invited an IO to participate and offer suggestions to enhance communication. The IO offered the following tips for facilitating communication with the IO. IACUC administrators need to realize that IOs have multiple interests competing for their time. He or she should prepare statements and keep to the facts that are necessary to make an informed decision. The information should focus on the core of the issue and be presented succinctly.

In addition, IACUC administrators should understand that the IO frequently has no money in the budget to solve new problems. Consequently, the IO needs suggestions on how to resolve the issue with particular emphasis placed on those that are no- or low-cost options. IACUC administrators should recognize that just because there is no money does not mean that IOs do not care. For example, IOs take significant issues of noncompliance very seriously because noncompliant practices or public perception of poor practices can cause damage to an organization's reputation. In addition, IOs understand that noncompliance is not to be treated lightly.

When meeting with IOs, IACUC administrators need to prepare a list of options to address discussion points. There should be three or four possibilities prepared, with one or more recommended as the preferred solution. A clear, logical, effective, and—as much as possible—painless solution that resolves the problem and prevents recurrence is the preferred choice!

IOs are busy, and they appreciate having reasonable and effective solutions recommended. Therefore, when communicating with the IO, the IACUC administrator should begin the conversation with a brief, concise description of the problem. Before bringing any concerns to the IO, the IACUC administrator should verify the facts. To reiterate, the IACUC administrator should bring not only concerns but also solutions to the IO. Once the concern is accurately summarized and described, the IACUC administrator should recommend two or three cost-effective (or no-cost) solutions to solving the problem. In the discussion with the IO, as many details as necessary should be provided to allow the IO the ability to make an informed decision.

IACUC administrators should keep in mind that IOs have little free time, so once the administrator has accurately described the concern and options, he or she should stop talking. If the meeting diverges into complex issues that an administrator is not prepared to discuss, he or she should offer to look into the matter and schedule another meeting in the future. To help facilitate conversation outside the context of

problems, IACUC administrators may choose to ask the IO to periodically participate in IACUC meetings or inspections. Try to schedule regular meetings between
the IACUC administrator, the AV and the IO to discuss programmatic issues.

Communicating with the AV and Veterinary Staff

In addition to communicating with the IO, the IACUC administrator must also
establish an effective method for communicating with the AV and the clinical staff.
IACUC administrators with effective communication practices discussed the important points of communication during past BP meetings.

IACUC administrators indicated that before any meaningful communications
can occur; the IACUC administrator and the AV should first identify those responsibilities that will be addressed by the veterinarian and those that will be addressed
by the IACUC administrator. While BP meeting attendees noted during discussions
that having the IACUC administrator report to the AV may not be a best practice,
in those situations where such an arrangement exists, the AV will make the final
decision. An emerging common practice is for organizations to establish a separate
compliance office, which includes the IACUC administrator. In situations such as
this one, defining responsibilities can be critical.

Those BP meeting attendees agreed that IACUC administrators should handle all matters relating to research compliance. For example, under the authority
of the IACUC, IACUC administrators should be authorized to initiate investigations into alleged cases of noncompliance. They should apply the federal expectations and report actions to the IACUC. The IACUC will deliberate the outcomes
and determine which are reportable to the Office of Laboratory Animal Welfare
(OLAW), US Department of Agriculture (USDA), or Association for Assessment
and Accreditation of Laboratory Animal Care (AAALAC). The IACUC administrator will draft the reports to the appropriate organizations and acquire the IO's
signature before distribution.

IACUC administrators will frequently follow cases of noncompliance to ensure
that imposed sanctions are satisfied and that recurrences are minimal. The IACUC
administrator often meets with the veterinary medical officer (VMO) during USDA
inspections to coordinate the record review process. He or she will also communicate with AAALAC International, when applicable, to manage AAALAC reac-
creditation visits.

Once the delineation of duties is clear, the next step is to develop means for
open discussion. With the understanding that both the IACUC administrator and the
AV have an interest in ensuring that the program satisfies the needs of the animals
and remains compliant with the regulations, both parties should meet to discuss the
associated activities.

Some IACUC administrators schedule monthly meetings with the veterinary
staff to develop and/or review IACUC meeting agendas. Issues of concern identified
by the AV and IACUC administrator are routinely scheduled for an IACUC discussion during upcoming meetings. For example, the IACUC administrator and the AV
may agree that there are concerns regarding postapproval monitoring. The issues can

be presented to the IACUC with both the IACUC administrator and the AV agreeing on a set of proposed solutions for IACUC consideration. The advantage of this activity is that the AV and the IACUC administrator are dedicated program staff who can meet, discuss, and explore options to programmatic concerns. In addition, this process does not consume volunteer time of the IACUC members until the issue is distilled and the required actions are clearly defined.

Communicating with Principal Investigators

IACUC administrators must also develop effective methods for communicating with PIs. Many IACUC administrators meet with PIs on a regular basis to have informal conversations relating to the IACUC process. During this time, they frequently identify the PIs' frustrations and irritations relating to the process. The IACUC administrator then works to resolve those issues identified by the PIs, and in doing so the administrator gains creditability and establishes a good working relationship with the research community.

Some BP meeting attendees reported having regular "questions and answers" sessions for groups of PIs. The IACUC administrator may conduct these sessions over lunch in a town meeting atmosphere. Another method effectively used by IACUC administrators is to have a specific e-mail box to accept questions from PIs. This process gives PIs an opportunity to ask specific questions, which may evolve into a one-on-one meeting or an in-depth IACUC discussion.

Communicating with the IACUC

The IACUC administrator must develop effective measures for communicating with the committee. He or she frequently serves the IACUC in an advisory role relating to research compliance, so finding a way to effectively disseminate information to the IACUC is a critically significant tool for successful IACUC administrators. For example, they frequently conduct a training session during each IACUC meeting to keep members current on programmatic issues. In addition, when developing the meeting agenda, an IACUC administrator's report is frequently used to update committee members on administrative issues that were addressed since the last meeting. Many administrators frequently disseminate information to the IACUC by e-mail; the development of an IACUC member LISTSERV may facilitate conversation on relevant matters when preparing for monthly meetings.

CHAPTER 12

Field Studies

SCENARIO

GEU boasts a comprehensive animal research program that includes scientists actively engaged in field research that involves, for example, wild rodents. In fact, Dr. Smith's research program focuses on vole speciation (i.e., the identification of new species of voles in the Americas). As part of her research, Dr. Smith travels to remote locations throughout North, South, and Central America to trap voles. Once the animals are trapped, tissue samples are collected and sent to the laboratory for genetic analysis. On occasion, Dr. Smith will euthanize and retain an animal cadaver as a reference specimen (i.e., voucher specimen).

GEU's IACUC conducted a thorough review of Dr. Smith's research activities, which ultimately led to approval. The IACUC generated an approval memo to Dr. Smith that included a statement indicating that no work should be conducted until required federal and state permits were acquired. The memo also reminded this principal investigator (PI) that the activities associated with her protocol must comply with state and federal regulations governing activities involving wild animals.

Dr. Smith's research included trapping animals in locations that were a few miles south of the US border in Mexico. In this particular location, Dr. Smith captured four animals that displayed unusual markings that she believed may be indicative of a subspecies of vole. Dr. Smith decided to euthanize two specimens and transport them to her lab as voucher specimens for in-depth genetic analysis.

The animals were euthanized in accordance with her approved IACUC protocol and packed in ice for transportation back to her GEU laboratory. At the US and Mexican border, Dr. Smith met with customs officials and explained that she was a scientist researching vole speciation and was transporting specimens back to her US lab for evaluation. Upon inspection of Dr. Smith's samples, custom officers noted that the euthanized animals were an endangered species of vole. The officers reminded Dr. Smith that international laws protect endangered species and that importing this species of vole required a specific permit—in particular, a Convention on International Trade in Endangered Species (CITES) permit. Dr. Smith was asked to produce her CITES permit for importing her animals.

She indicated that she had intended to secure the permit but had not yet done it because this opportunity was time sensitive and she did not want to miss it. Consequently, she had neglected to acquire the appropriate permits to conduct the research. Since Dr. Smith did not have the appropriate permits to euthanize and collect endangered species, she was arrested for failure to adhere to international wildlife regulations.

By the end of the day, the situation had escalated and the incident was reported as a local news headline. The incident was soon broadcasted by a local news affiliate in the GEU community, which prompted protests by animal protection agencies. As a result of the protests, federal officials from the US Fish and Wildlife Service (USFWS) conducted an audit of Dr. Smith's research activities. Officials soon discovered that Dr. Smith was also conducting field studies in the United States without the required federal and state permits.

The incident tarnished GEU's animal research program and, as a result, the organization's Public Relations Office prepared a statement in an attempt to mitigate the situation. The statement discussed the fact that GEU had an IACUC that ensured that all research animals were handled and treated humanely. It continued that the organization complied with all federal and state regulations governing experimental activities conducted on animals, including euthanasia. The statement indicated that GEU understood that international, federal, and state permits were issued to the scientist rather than the organization. Therefore, GEU had expected Dr. Smith to acquire the appropriate permits before conducting her field activities. The statement concluded with the assurance that "appropriate actions will be taken by the university to ensure that this problem does not recur."

In this particular situation, the IACUC's decision to remind the PI, but leave completion of permitting up to the researcher, may not have been the best choice. Although federal regulations governing research activities do not require IACUCs to verify that researchers acquire appropriate permits before conducting field research, had the IACUC required such confirmation, the incident with the customs officers and the associated ramifications may not have occurred.

Regulatory Requirements and Resources

- **Animal Welfare Act regulations (AWAR) (Section 1.1)** define a field study as "any study conducted on free-living wild animals in their natural habitat, which does not involve an invasive procedure, and which does not harm or materially alter the behavior of the animals under study."
- **AWAR (Section 2.31[d][1])** indicate that field studies as defined in Section 1.1 do not require IACUC review.
- **Public Health Service (PHS) policy** does not differentiate between vertebrate animals. Consequently, providing that an organization's assurance covers all species equally, the IACUC must review and approve animal care and use activities involving both free-living and laboratory animals as well as any other vertebrate animals used for research, teaching, or testing.

- **Title 50 code of federal regulations (Sections 1–100)** governs activities that involve wild species of animals. The federal mandates are enforced through the USFWS.
- **Wildlife protocol review form** is a template form that can be found on the Office of Laboratory Animal Welfare (OLAW) website. Although using the form is not required, it is available for use by any organization.

IACUC REVIEW OF FIELD STUDIES

While research projects can vary significantly, in certain situations, research conducted on free-ranging vertebrate animals is covered under animal welfare regulations and standards. For example, using *Peromyscus* sp. (deer mouse) will require a protocol since warm-blooded wild vertebrates are a US Department of Agriculture (USDA)-covered species. Using reptiles for field studies will not require a protocol under USDA regulations since reptiles are cold-blooded animals. However, if reptile work is funded by the National Institutes of Health (NIH), then a protocol will be required as PHS policy covers both warm- and cold-blooded vertebrates.

Sometimes it is not clear when a protocol is required by federal regulations and when a protocol is necessary for local institutional reasons. For example, one PI may be gluing a transmitter to the backs of rattlesnakes to track their migratory movements, while another ecologist may be observing different species of fish caught by vacationers in a given area. Best practices (BP) meeting attendees agreed that a scientist handling or conducting any procedure on wild species requires IACUC approval, but the basis for a protocol may not be driven from the federal perspective. In the examples noted, a protocol was required, but such a position exceeds the USDA requirement for a study to be "conducted on free-living wild animals in their natural habitat, which does not involve an invasive procedure, and which does not harm or materially alter the behavior of the animals under study." Although some organizations may consider gluing the transmitter to the back of a snake as an alteration of its environment, almost all organizations would consider the observation of fish being caught by local anglers as neither harming nor materially altering the behavior of the fish.

In other situations, BP meeting attendees were less consistent with their institutional opinions—for example, deciding whether a protocol would be needed for a study that involved placing animals' favorite foods in an area such that the foods would draw them out of their natural environment to better facilitate observation of their behavioral patterns, sitting in a field of nesting birds with binoculars trying to identify new species, or sitting on the tundra observing the migration patterns of caribou. Most attendees indicated that their organization still required a protocol, if not for the USDA, then for the integrity and oversight requirements of the organization. In other words, BP meeting attendees believed that IACUC-approved protocols provided a degree of protection to the researcher and the organization from allegations of animal abuse or misuse even when a protocol is not required by federal overseers.

A subset of BP meeting attendees reported that their organizations did not require a protocol if there were no animal handling or manipulation procedures. Consequently, observational studies or documentation of animal activities did not require a protocol. A few organizations noted that while their organization would not require a protocol in these circumstances, the organization would require a memo from the researcher to the IACUC outlining the activities to be performed. This memo served two purposes: (1) to confirm that a protocol is not required, and (2) to alert the organization that work was being performed that did not require a protocol but might require an international, federal, or state permit and the health protection of students and researchers.

Attendees noted that an increasing number of journals were requiring confirmation that the IACUC had reviewed and approved all reported animal-related manuscripts. Therefore, many organizations were providing a protocol review for any animal-related activities if there was an intent or desire for the activity to be published at a later date. One best practice suggestion was to review the *Guidance for the Description of Animal Research in Scientific Publications*, published by the Institute for Laboratory Animal Research (ILAR), and confirm that the organization's research community was aware of developing guidelines for publication.

Field Studies: What Information Does the IACUC Need?

Organizations conducting a significant amount of field research frequently had a wildlife field biologist on their IACUC to provide technical expertise during protocol reviews. The biologist frequently developed questions for the IACUC to ask of the PI relating to field studies. Organizations heavily engaged in field studies often develop an addendum to their core protocol form to cover field activities.

BP meeting attendees generally addressed the issues of "protocol" or "no protocol" by answering the following question: "Will animals be harmed or their behavior materially altered as a result of the research?" Considering that any intervention with a wild animal will materially alter its behavior, a "yes" response to this question would require the IACUC to oversee the activity, while a "no" response would not require IACUC oversight.

Upon confirmation that IACUC oversight is required, committees usually ask a number of follow-up questions to gather the information they need to appropriately evaluate the proposal. In addition to the routine questions that describe laboratory animal care and use activities (e.g., "Will you be conducting survival surgery? If yes, describe."), additional questions were frequently asked of those conducting field research—for example:

"Will animals be trapped in any manner? If yes, describe."
"If animals are trapped and euthanasia is not planned, how will injured animals be handled?"
"Will animals be euthanized? If yes, describe the method that will be used."
"Will wild animals be housed in the field for periods greater than 12 hours? If yes, please describe the housing conditions."

"Will surgical procedures be conducted in the field? If yes, describe the area in which the surgery will be conducted, and how you will practice asepsis."

"Are you working with endangered or threatened species of animals?"

Protocol Review and IACUC Oversight

Once the IACUC has gathered the relevant information in the submission form, members consider the proposal as they would any other IACUC submission. However, there may be unique situations that must be considered when work is conducted with wild species. For example, if the PI plans to conduct field surgeries or house wild animals for extended periods of time, the committee must develop appropriate means of oversight. When field surgery is conducted, the PI usually describes the surgery and recovery as part of the application's surgical section.

The question remains as to how the IACUC can inspect and validate the field surgery area. Representative IACUC administrators offered a number of ways to validate the appropriateness of field surgery locations. One way was that the organization would send two IACUC members (e.g., the AV and a scientific member with experience in field biology) to the surgery site to observe the initial surgeries. This process allowed the committee to verify and document that the surgeries are being conducted as aseptically as possible given the field conditions. An alternative method discussed, especially for those conducting surgical procedures in remote areas, involved the use of video technology. In certain circumstances, the PI would take still digital photographs of his or her surgery area and e-mail them to the IACUC administrator, who would forward them to the IACUC for comments. In addition, some have used video to allow the IACUC to view a live broadcast by the PI. In this particular situation, the IACUC is able to ask specific questions of the PI. In both instances, the IACUC is able to evaluate the appropriateness of the surgical area.

If animals are held in the field housing areas, new issues can arise. How can the IACUC inspect the housing facilities prior to housing the animals and conduct continuous inspections on a semiannual basis? BP meeting attendees offered a couple of methodologies that could be employed. The most obvious method is to have two representatives from the IACUC visit the field housing location. The physical inspection scenario is the option preferred by the USDA, but other methods have been proven to be successful. For example, video transmission is a process proven to be effective. In this particular scenario, the PI would transmit video of the housing facility to an IACUC meeting where committee members would have an opportunity to ask questions while the inspection was conducted. In addition to asking questions, committee members were also able to direct the videographers to certain areas of the housing site—for example, close-ups of feeding and watering systems, cage construction, and bedding. In many locations, a video wireless signal was not possible, so while it was not a preferred option, having the researcher take still photographs and mail or e-mail them to the IACUC administrator for sharing with the IACUC was also a feasible option.

Field Housing

Since field housing (e.g., cages, perimeter fencing) contains free living wild animals that would otherwise be thriving in their natural habitat, those housing areas must be semiannually inspected by the IACUC. BP meeting attendees discussed various field housing methodologies and IACUC oversight. For example, consider a researcher who plans to use wire to isolate fish to a specific section of a stream: Is this considered housing since the fish are still able to get their own feed, thus maintaining their own sustainability? BP meeting attendees offered different approaches for overseeing these fish. Some organizations followed a very conservative approach and enlisted at least two IACUC members to inspect the housing area. In other situations, attendees indicated that IACUC members would use video to conduct the inspections while others simply asked the PIs very specific questions about the housing area as part of the protocol review. In almost all of the circumstances, the concerns were the same: animal well-being and care (e.g., assuring that water levels would not diminish and eventually leave the fish out of water). Attendees also reported that their organizations required assurances that the confinement of the fish would not lead to an increased risk of predation and diminished food supplies.

In a second example, a researcher planned to confine wild vertebrate mammals by fencing off large acreage where the animals could fulfill most to all of their behavioral needs except that they were restricted to the fenced areas. BP meeting attendees reported oversight practices and IACUC concerns similar to those for the fish example previously discussed. In some cases, IACUC members visited the confined areas, while at other organizations they viewed videos or simply asked specific questions. The primary concerns were, again, related to the animals' ability to acquire adequate amounts of food and water, protection from the elements, and predation. An additional example involved a researcher wishing to establish a mesocosm to study the interaction of stream residents including fish, amphibians, and small mammals. BP meeting attendees noted that a mesocosm is considered to be no different from any other style of animal housing found in a vivarium. In other words, a standard operating procedure must be developed that describes the care of the animals, mesocosm change frequencies, feeding and watering practices, daily animal checks, and veterinary oversight. Most organizations have sent IACUC members to the fields to inspect the mesocosm, while others have effectively used video or still photographs to accomplish the activities of oversight and procedure monitoring.

Vertebrate Animals as Bait in Field Studies

Live animals are occasionally used as bait or feed for research animals. Live animals that will be used for bait or as food must be maintained in accordance with the provisions of the *Guide for the Care and Use of Laboratory Animals* before their use as bait or animal feed.

A researcher proposes using pigeons or small rodents for catching raptors, minnows for catching water snakes, and rabbits for catching coyotes. In all instances, BP meeting attendees agreed that the "bait animals" should be included in the submission to the IACUC and that the IACUC should consider the distress associated with the predator interaction. All organizational representatives indicated that they require the researcher to justify the need to use live vertebrate animals as bait in the protocol as opposed to recently euthanized animals or inanimate bait. In some cases, the IACUC administrators reported that their IACUC required a pilot study using recently killed animals for bait or the use of inanimate bait to prove that the process was ineffective prior to authorizing the use of live animal bait stations.

Accounting for Animals Used in Invasive Research Field Studies

When research includes procedures that are conducted on animals as part of a field study, a specific number of animals must be requested and justified. As with any other project involving animals, the three Rs (replacement, reduction, refinement) must be observed. When considering the number of animals needed to conduct the study, the minimal number of animals needed to produce statistically valid results (i.e., reduction) should be requested and justified in the IACUC submission. The number requested should include sufficient numbers to produce statistically valid results. For example, if a scientist plans to monitor a bear's physiology during hibernation, he should identify how many bears are needed to produce statistically valid results. In his protocol, he should include language such as "data must be collected from twenty bears based on a statistical test to produce valid results." He should indicate the statistical test used and explain why it was chosen.

Accounting for Animals Used in Noninvasive Research Field Studies

Perhaps not as obvious is research that involves noninvasive studies. For example, let us consider a situation in which a scientist is interested in learning what percentage of a population of bats roosting in a condemned factory are endangered Indiana bats. In summary, his project involves setting up mist nets to trap bats as they are leaving their roosts. He plans to catch the bats and, after removing them from the nets, identify the species, record weights, measure, and ultimately tag the animals before releasing them. The scientist has indicated that he wants to trap as many of the three thousand plus bats in the colony as he can to effectively determine the percentage of Indiana bats using the factory for housing.

BP meeting attendees agreed that when a project does not necessarily involve an experimental design (i.e., the lack of experimental groups and no control groups), statistical validation may not be possible while the results of the study may remain valid. In this particular example, the attendees agreed that the scientist justified the need to capture as many of the bats in the colony as possible and, therefore, most attendees concurred that this project would be approved to include an unlimited number of bats in the specific condemned factory area with the caveat that the PI report the number of bats captured and involved in the project at least annually. The animal numbers report was necessary to accomplish the annual reporting requirements of the USDA and the Association for Assessment and Accreditation of Laboratory Animal Care (if the organization is accredited). If the study were PHS funded in accordance with the policy, the PI would need to provide an approximate number of animals to be used.

A second example involved the state fish commission. The fish commission granted a scientist a permit to determine the number of exotic fish (e.g., tilapia) populating an area of the river having elevated water temperatures due to industry discharge. The scientist's project summary discussed the fact that tilapia typically inhabit the warmer rivers of Asia and Africa but, on occasion, some of these fish have been released into colder US waters. The premise was that some of these fish have relocated to areas of the river where water temperatures stay consistently warm due to industrial discharge into the waters. In addition, they thrive and reproduce in these waters. The scientist planned to electrostun the fish in industrial areas of the river to determine if tilapia were present and, if so, determine the density of tilapia per section of the river. BP meeting attendees expressed concern that electrostunning would affect all species of fish in the area, not just the tilapia. The experimental design was limited to identifying warm-water species of fish in cold-water environments. If the study were PHS funded, the project would also need to include an estimation of animal use. While the protocol might require adjustment to include anticipated species other than tilapia, the project was approved with assurance that the participants were trained on the use of the electrostunning device and that the state fish commission permit included more than just tilapia. Even so, the researcher had to report to the IACUC all numbers of fishes affected by the electrostunning, including nontarget species.

OHSP and Field Research

Since the IACUC should ensure that the animal activities do not put researchers at an increased risk for injury or disease, organizations must ensure that they have measures to cover field researchers. Investigators conducting wildlife studies must enroll in the organization's occupational health and safety program (OHSP). Consequently, organizations that regularly support field research frequently enhance their OHSP to include information relevant to wild species. Many organizations supporting wildlife studies conduct risk identifications and assessments relating to wild species and their habitats. For example, the risk assessment officer at an organization may identify a number of unique zoonotic diseases associated with wild animal research and supplement training programs to include information on zoonotic diseases that are often specific to wild species. Organizations working with field mice may include a training module on hantavirus, those working with raccoons or skunks may develop a rabies module, and those working with wild rodents may develop a plague module.

In addition to the training modules, organizations may require researchers to receive certain prophylactic immunizations (e.g., working with rabies carriers such as bats requires a current protective rabies titer). In addition, medical personnel may conduct specific training courses to teach scientists about tick-borne diseases such as Lyme disease. In almost all cases, organizations conducting field research supplement their OHSP to ensure that researchers are protected from the related hazards. Many institutions reference the "Safety Guidelines for Field Researchers" produced by Arizona State University.

Permit Requirements

Prior to conducting field research, researchers must acquire the appropriate permits. Many organizations place the responsibility of acquiring the necessary permits for field work on the researcher. BP meeting attendees generally indicated that, while the IACUC advises the researcher of requirements for permits, it does note in the protocol approval letter that it is the responsibility of the scientist to acquire the required permits for the field work (e.g., a bird-banding permit before initiating the research). A best practice includes language such as, "IACUC approval of the research does not eliminate the need for scientists to acquire licenses and permits to conduct field research."

If wild animals are to be transported, most organizations perceive the risk to the organization as greater and generally ensure that appropriate permits are in place. When researchers plan to transport cadavers or live animals from organization to organization or over state/national borders, the IACUC often requires copies of CITES permits prior to approving the research.

What Should IACUC Administrators Know about Permitting Requirements?

BP meeting attendees generally believed that it was essential to have a good working knowledge of permitting and oversight requirements associated with field studies, especially as it involves free-ranging wild species.

Providing Guidance to the PI

Investigators should assure the IACUC that collection of species will comply with federal, state, and international requirements. In addition to the animal welfare regulations, there are international, federal, and state laws that must be satisfied when conducting research using wild vertebrate species. The IACUC administrator frequently serves in an advisory role to the principal investigator. At many organizations, the IACUC administrator was requested to assist the researcher with satisfying permitting requirements.

State Regulations

In some manner, each state oversees scientific activities involving wildlife. The only common theme between the state wildlife agencies is that they have and enforce regulations that protect wild species of animals. Some organizations delegate the responsibility of ensuring that principal investigators comply with state wildlife laws to the IACUC administrator. A valuable resource when seeking guidance on wildlife law for IACUC administrators is the Center for Wildlife Law, which was established at the Institute of Public Law, University of New Mexico School of Law. Its mission is to educate the public on wildlife federal and international laws. One of the primary guidance documents offered by the center is the *State Wildlife Laws Handbook*, which is available through its website. The center also offers a guidance document entitled *Federal Wildlife Laws Handbook*. In addition to the handbook, the center publishes the *Wildlife Law News Quarterly*. The articles in this magazine frequently discuss legal concerns associated with wildlife law. Although the center's publications do not specifically discuss regulatory requirements associated with wildlife research, they do provide references to the applicable regulations.

An additional valuable resource for IACUC administrators is state wildlife agencies' websites. Although these websites often focus on recreational activities such as camping, fishing, and hunting, many also include information relating to education and outreach. The sites reference state statutes and include contact information for state employees that can often provide essentials to IACUC administrators. Since the permitting requirements are specific to state laws, the IACUC administrator frequently will contact state officials or research websites to ensure that appropriate permits are secured before the IACUC approves a field study. Consequently, due to the diversity of state regulations when principal investigators conduct field studies, the IACUC administrator will need to review and determine the permitting requirements for each state in which the PI plans to conduct research.

Federal Regulations

The USFWS issues permits to those planning to conduct research involving wild species in the United States. The service also regulates and oversees activities involving wild animals. The IACUC administrator can locate the wildlife regulations as well as guidance on permitting requirements in Sections 1–100 of the 50 CFR. To be able to assist principal investigators with permitting requirements, the IACUC administrator should have a working knowledge of wildlife regulatory requirements.

CHAPTER 13

Personnel Qualifications and Training Programs

SCENARIO

GEU IACUC administrators believe that they have a very robust training program. They have a very effective process for training new IACUC members to review protocols and oversee the organization's animal care and use program (ACUP). Once an individual is appointed to the GEU IACUC, the IACUC administrator schedules individualized training with him or her. The IACUC administrator provides new appointees electronic copies of all the relevant documents, including the *Guide for the Care and Use of Laboratory Animals* (henceforth referred to as the *Guide*), Public Health Service (PHS) policy, the *Institutional Animal Care and Use Committee Guidebook*, and a copy of the Animal Welfare Act regulations (AWARs).

The appointee then meets with the IACUC chair to discuss committee member responsibilities, as documented in PHS policy IV.B.1–8, such as reviewing animal activities and semiannual program reviews and inspections. The GEU IACUC chair assigns a veteran committee member to serve as the new member's mentor for the next year. During this time, the appointee reviews the same protocols as the seasoned IACUC member, and the two collectively discuss each protocol until their reviews become generally consistent.

During semiannual inspections, new committee appointees shadow voting members. The voting members identify potential concerns during the facility inspection and discuss each with the new committee members. For example, if the site team members notice rust on a laboratory animal room door, they explain to their new IACUC colleagues that all surfaces in the animal rooms must be impervious and easily sanitized. The veteran committee members also explain to their new colleagues that since the room door is not in direct contact with the animals, the deficiency does not create a welfare concern for the animals and therefore would not be considered a significant deficiency. The committee members also explain to the trainees that the problem must be corrected since the regulation requires animal rooms to be properly sanitized. The inspection team explains to the trainees that the committee will list the problem as a minor deficiency and require a corrective action by the next semiannual cycle.

After reviewing the GEU ACUP during an Association for Assessment and Accreditation of Laboratory Animal Care (AAALAC) site visit, the site visitors commended GEU for the method they used to train new committee members. However, after thoroughly discussing the training program with GEU IACUC administrators, they learned that the organization's training program did not extend much beyond new committee member training. Consequently, the AAALAC indicated that the organization would need to expand its training program to include all members of the ACUP before AAALAC accreditation could be granted. For example, GEU's training program should also include training modules for animal care technicians and researchers and their staffs, as well as students and others having contact with GEU's animals.

Regulatory Resources

The overall theme of the federal standards is that only trained personnel can participate in animal care and use activities.

- The *Guide* and the *Guide for the Care and Use of Agricultural Animals in Research and Teaching* (henceforth referred to as the *Ag Guide*) require organizations to appropriately train individuals to satisfy the expectations of their positions. For example, each IACUC member must be trained to oversee the organization's ACUP, and animal care technicians must be trained to care for the species of animals they oversee. Both guides stipulate that only qualified individuals can conduct animal activities (e.g., experimental or animal care procedures). In addition, both guides emphasize that animal care staff members should participate in continuing education activities.
- **PHS policy** does not discuss specific training requirements but requires that personnel be trained. For example, the PHS policy (I.A.1) states that an assured organization must use the *Guide* as the basis for its ACUP and that the *Guide* thoroughly describes training expectations. The *Guide* also indicates the IACUC must verify that researchers are qualified to conduct animal care and use procedures. The PHS policy (IV.C.1.f.) also states that "personnel conducting procedures on the species being maintained or studied will be appropriately qualified and trained in those procedures." The PHS policy (IV.A.1.g) also requires organizations to establish training programs for animal users.
- **AWAR (Section 2.32)** discuss personnel qualifications and indicates that personnel involved in animal care and use activities must be qualified to perform their duties. In addition, according to AWAR (Section 2.32, a), it is the organization's responsibility to ensure that animal users are adequately trained. AWAR also indicate that IACUC members must be qualified to serve on the committee, which would imply that training is also required for committee members.

ESTABLISHING A TRAINING PROGRAM

Establishing a training program is a regulatory requirement. Scientists, instructors, and clinicians must be qualified to work with animals and those qualifications may be provided or confirmed through a training program. Under regulatory expectations, everyone engaging in the program must be qualified and, when necessary, be trained.

While the regulatory standards governing animal use activities differ slightly, they all indicate that individuals engaged in an organization's ACUP must be adequately trained to satisfy their roles. In addition, the regulations state that it is the responsibility of the research organization to provide resources and fiscal means for establishing and maintaining an effective training program. Consequently, the Office of Laboratory Animal Welfare (OLAW) and the US Department of Agriculture (USDA) agree that individuals involved in an organization's program must be trained and qualified to work with research animals prior to initiating their responsibilities. As a result, whether individuals are reviewing and approving protocols, caring for animals, or conducting a survival surgery, they must be qualified or appropriately trained to perform their specific duties.

COMMON TRAINING EMPLOYED BY ORGANIZATIONS USING ANIMALS

There are certain types of training that everyone participating in animal care, animal use, or animal oversight will require. This should be considered "core" training and may be obtained from a commercial service or a locally produced module. Common or core training programs routinely include animal user and occupational health and safety training programs.

Training for Individuals Handling Animals

Best practices (BPs) meeting attendees endorsed several training initiatives for different organization populations. Frequently, a tiered training approach, which captured varying populations at each organization, was employed.

For example, most organizations have an initial training for investigators, research associates, and students. The research staff training typically covers a wide range of activities (e.g., regulations, animal use activities, and reporting animal welfare concerns) and requires initial and ongoing training. The IACUC must ensure that animal users are appropriately qualified to work with their species of interest and to conduct the proposed procedures listed on their IACUC proposal. The goal of all training is to assure appropriate qualifications and effective skills of those working with animals.

A critical component of the user training program is to first identify the correct instructor. In certain situations, the principal investigator (PI) may be the most proficient at conducting specific procedure training. For example, a PI would be skilled at conducting a surgery that has been part his or her research for a number of years. Consequently, training to conduct this surgical procedure should be done by the PI. Conversely, a PI may require training on conducting tail vein injections into mice and acquire that training through a veterinarian-conducted workshop.

To expand on the tiered training approach, organizations may develop various phases of training. For example, phase 1 training may constitute a number of online tutorials covering regulations, ethics, alternatives, IACUC functions, pain and distress, surgery, euthanasia, and adequate veterinary care. This level of training is typically required of all animal users, including the PI, postdoctoral fellows, research fellows, residents, students, and research technicians. In other words, anyone working with animals or having an oversight role for animal use requires this type of training. Phase 1 training could also include an organization's annual refresher training (discussed in more detail in the section on refresher training). Typically, satisfactory completion of the phase 1 training is confirmed by the student successfully passing a brief online quiz. Successful completion of the phase 1 online training and quiz can serve as a core stipulation for IACUC approval of an animal care and use protocol. Completion of training can be documented by retaining verification that the relevant quizzes were passed by the trainee.

Phase 1 training. The phase 1 training program typically initiates the process of ensuring that members of the ACUP team receive all of the required standard training. Standards have identified the expectation that organizations familiarize animal users on federal laws and policies. Many organizations choose to develop training modules with the intent of educating animal users on the importance of and complying with federal regulations. This training module is frequently delivered to animal users through online training modules or printed meeting packets. In both cases, the module allows the users to do the training at their convenience and remains available to the users.

The phase 1 training modules frequently include information on the topics relating to regulatory compliance. The training may first cover the relevant regulatory agencies and accrediting organizations. It often includes a brief discussion on each of the regulating agencies and the AAALAC. For example, the PHS policy governs activities funded by PHS agencies and is administered by OLAW. The AWAR must be applied any time covered species (e.g., dogs, cats, nonhuman primates) are employed in animal activities, and the act is enforced by the Animal and Plant Health Inspection Service (APHIS). If relevant, the fact that AAALAC International is not a regulating agency may be discussed during the training. The instructor may inform the trainees that program accreditation by AAALAC is strictly voluntary and consists of simply inviting a group of the organization's peers to assess their program of animal care and make recommendations based on the program evaluation.

Training may be used to ensure that PIs are informed of specific mandated regulatory requirements that they must satisfy. For example, PIs are trained that all governing documents require an IACUC to review and approve animal use activities before they are initiated by the PI. They are taught that they should not conduct any animal use activities that are not first approved by the IACUC. PIs are informed that the consequences for not complying with the mandate to acquire prior IACUC approvals could be to return any funding that was spent on nonapproved activities. Additionally, PIs may be informed that any adverse events occurring in research animals must be reported to the IACUC and the attending veterinarian (AV). They will also be educated that all active protocols must not expire as long as the research is being conducted. For example, the trainer must discuss the urgency of maintaining IACUC approval through the life of the grant. The discussion would typically include the fact that if a PI allows an approved project to expire, no National Institutes of Health (NIH) funds can be spent on animal care and use until the project is reviewed and approved by the IACUC. A primary goal of this training session is to teach animal users that the IACUC is federally required to oversee animal use activities and that open communication between the PI and committee is vital.

Phase 1 training may also elaborate on the ramifications of not complying with the guiding standards. Organizations routinely develop sessions to explain that not satisfying the regulatory expectations will negatively affect the researcher and the organization's program of animal care. For example, if a research project is funded by the NIH, and the PI deviates from the PHS policy, the funding could be jeopardized. In extreme situations, such as conducting procedures that have not been approved by the IACUC, the result may be the loss of the granted funds.

Phase 2 training. Phase 2 training typically consists of species-specific training. Participants may view commercially available video tapes, slides, or other multimedia material as a first layer of education regarding basic biology, animal husbandry, handling, and research techniques pertaining to the species to be used. This training is encouraged for all protocol participants but is typically required in cases where the IACUC determines that individuals do not have an adequate skill set to properly achieve the expected level of animal care or use. Alternatively, the IACUC may require such training as corrective action for a committed instance of noncompliance.

Phase 3 training. Phase 3 traditionally consists of hands-on activities. Students may receive training on euthanasia techniques or surgical skills, or be educated on appropriate techniques for blood collection. Hands-on training may be provided by the veterinary staff, the PI, or other specialized staff.

A common practice is for the organization's AV to establish a training protocol. The protocol authorizes the veterinarians to train animal users on all IACUC-approved procedures. This type of protocol typically covers standard animal care and use activities such as collecting blood, performing surgical procedures, performing injections, and euthanizing animals.

Organizations commonly use two different processes for conducting these training sessions. In some cases, the AV will visit the PI's lab and conduct personalized training sessions with relevant staff members. Conversely, the AV may establish a training laboratory in the core vivarium and conduct regularly scheduled training sessions open to all interested animal caretakers and users.

In certain situations, the PI trains personnel. The PI may add the trainees to the protocol and include specific provisions for training in the protocol. The PI may choose to have untrained technicians repeatedly observe a specific procedure (e.g., a specialized surgical technique). The PI may choose to have the trainees participate in components of the procedure and then gradually allow them to assume responsibility for the procedure.

Training husbandry staff. Most organizations require the Department of Veterinary Resources to provide training for animal care staff and veterinary technicians. In addition, promotions for husbandry staff and veterinary technicians are often predicated on increasing levels of certification or training. Many organizations provide American Association for Laboratory Animal Science (AALAS) training programs internally for staff members pursuing AALAS certification. For example, a training program for a technician caring for a colony of mice may include multiple components:

- **The mouse environment.** This section of the training may discuss the appropriate environmental parameters for mice. It may include a discussion as to why it is important to maintain a relative humidity between 30 and 70 percent and room temperature of 68°F–79°F for laboratory mice. In addition, the lesson may include a discussion on the difference between static and ventilated caging. The trainer may want to discuss the possibility of ammonia gases building up more quickly in static cages and the potential need to change bedding in static cages more frequently. The trainer may also decide to include a section on room illumination and discuss not only the need to maintain appropriate light cycles for mice, but also how light levels exceeding thirty foot candles can result in phototoxic retinopathy for certain albino mice. Additional topics that the educator may choose to cover in the mouse environment section can include the importance of noise control, appropriate air quality, the adequate use of environmental enrichment devices, and the importance of social housing of mice.
- **Space requirements for mice.** The mouse training program may also discuss understanding appropriate cage size requirements. Since the floor space per animal is based on each animal's weight, the trainer may choose to show the technicians mice of varying weights during this discussion. The trainer may also wish to familiarize the technician with "Table 3.2: Recommended Minimum Space for Commonly Used Laboratory Rodents Housed in Groups" on p. 57 of the *Guide* and explain the importance of knowing where to find the appropriate resources. The trainer may decide to discuss such items as the maximum number of mice that can occupy a standard breeding cage and when litters should be weaned.
- **Appropriate husbandry for mice.** As part of the husbandry training for mice, the trainer may choose to discuss not only the need for providing mice food ad labium, but also information on proper feed storage and handling. This training may include discussions on how to determine if feed has expired, the appropriate environmental storage parameters, and how to ensure that diets are not contaminated. The session may also include discussions about the appropriate bedding for mice. This training may discuss how certain types of environmental enrichment can lead to health problems (e.g., conjunctivitis in nude mice). In addition, the trainer may decide to discuss the pros and cons of various types of bedding (e.g., how aromatic wood shavings, such as cedar, can promote animal welfare) and also take this opportunity to discuss cage and bedding changes. The trainer may choose to discuss an institutional operating procedure used to ensure that bedding changes occur on a regular basis and discuss the parameters that must be monitored (e.g., excessive amounts of urine, feces, or odors) to ensure adherence to the appropriate change intervals.
- **Mouse health and well-being.** The technician should also be trained to recognize diseases commonly acquired by laboratory mice and the symptoms associated with pain and distress. For this particular session, the trainer may choose to have a veterinarian facilitate the discussion. The veterinarian may discuss symptoms of

infectious diseases and also include topics such as tumor growth, barbering, and other common problems that may be encountered by those caring for the mice. The veterinarian may also identify and discuss characteristics typical of distressed mice such as hunching, rough coats, and lethargy. During this discussion, the trainer may also use the opportunity to explain how the technician can report what is believed to be incidents of inappropriate animal use to either the veterinarian or the IACUC.

- **Mice and special circumstances.** The trainer may also choose to discuss special circumstances that may require additional care or attention. For example, a technician caring for a colony of diabetic mice may need to change their cages more frequently than those of normal mice since they tend to urinate more often. The trainer may choose to include topics such as handling biohazardous cages, the use of nontraditional housing (e.g., hanging wire cages), and special handling required for immunocompromised animals.

Similar training programs should be developed and facilitated by the organization for technicians caring for other species of animals. For example, organizations maintaining colonies of rabbits should have a similar training program specific to rabbits that is used to educate staff caring for rabbits. Some organizations have chosen to use online computer-based systems to educate their staff members. For example, organizations frequently train husbandry staff using either the AALAS Learning Library or Collaborative Institutional Training Initiative (CITI) training modules. Both commercially available training programs include a series of training modules on numerous species-specific modules. Students are able to learn how to specifically care for rabbits, guinea pigs, and rats. In addition to using the commercially available online training modules, some organizations have developed and use their own in-house species-specific training modules. Many sites frequently use on-the-job training practices for husbandry staff facilitated by a veterinarian or, perhaps, a senior technician. In situations such as this, institutional veterinarians frequently spend time in the vivarium and provide personalized training to animal care technicians.

Training students. In an academic setting, students may handle and work with vertebrate animals as part of their course curriculum. Since students may only be working with animals for a short period of time (e.g., 2–3 weeks of a 12-week semester) during a school year, organizations have developed some unique methods for training students. One common method is to have the course instructor train the students as part of the classroom activities. In a situation such as this, the PI frequently includes a copy of the class syllabus with the IACUC protocol submission. The IACUC members look for the syllabus to include, for example, a section on zoonotic diseases or perhaps the safe handling of animals. Alternatively, the PI may choose to submit a copy of the class plan covering the training initiatives. In both cases, the PI will lecture the class on risks associated with vertebrate animals and methods to minimize these risks. The training program should include specific information about the species the students will be handling as well as any relevant safety procedures. In certain instances and to document that each student received the training, faculty members require students to sign an attendance roster on the day the training occurs.

Training research animal coordinators (RACs). Some organizations have developed enhanced training programs for research staff, who are sometimes referred to as research animal coordinators or laboratory animal coordinators. These programs seek to enhance training within the research laboratory for specially selected individuals. For active laboratories or large laboratories with several staff members, having an on-site trained RAC may be value added to the research team, the IACUC, and the animals. Training for the RAC can be crafted to meet the individual needs of the organization; however, the training is generally similar to that of IACUC members, although it is frequently focused on the issues of significance from the laboratory perspective (e.g., protocol writing, internal lab audits, identifying, correcting, and reporting potential noncompliances).

OHSP Training

A second critical core training conducted by most organizations using animals is occupational health and safety program (OHSP) training. Anyone working with or around animals must be aware of the risks associated with animal work. The risk of using any animal species can increase dependent upon the test agents used or procedures being performed upon that animal. For example, mice or rat behavior work is unlikely to result in a zoonotic infection but may cause allergies due to highly allergenic rodent proteins found on the rodent pelt or in saliva and urine. Those risks may not be significant for most persons, but could be a serious risk requiring preventive measures if the animal handler or researcher already has documented allergies. On the other hand, working with primates carries specific human health risks (e.g., B-virus, shigellosis, bites). Compromising the immunity of those animals in a research environment may allow suppressed disease to become active and cause increased risk to humans in the area. OHSP training must recognize that there are risks associated with animals that may be modified (either greater or lesser) when stressing the animals (e.g., test agents, medications, research activities). That same risk will not be identical to all groups of people. OHSP training should recognize core risk and individual risk and provide measures to protect each individual, identify signs of the condition, and discuss measures to take corrective action if necessary.

Every institutional training program should have an OHSP component. Organizations use various methods to provide OHSP training. Some organizations develop web-based training locally. Others establish didactic training for high-risk groups that can be accessed by students at their leisure. Still others choose to subscribe to nationally available resources such as the AALAS Learning Library or CITI training modules.

OHSP training programs frequently include a section on zoonotic diseases. A zoonosis is any disease that may be transmitted between animals and humans. Since animals can carry a host of agents that can cause disease in humans, training programs should include specific zoonosis training. Frequently, this training is incorporated into global OHS training (if the risks of zoonosis are consistent across campus) or provided as segmental training (if the risks of zoonosis vary across campus and across species on campus). Zoonosis training may include a section that describes personal protective equipment (PPE) that can be used to protect an individual from infection. For example, an OHSP training program may include training modules for individuals working with sheep (example 1) and rodents (example 2).

Training Module Example 1

The causative agent of Q-fever: *Coxiella burnetii*

How can I encounter Q-fever? You can be exposed to Q-fever by delivering lambs from ewes carrying the causative agent.

What happens if I get Q-fever? Most people do not experience or immediately develop symptoms of the disease. In certain situations, individuals may develop a high fever with muscle pain (flu-like symptoms). Some will develop an uncontrollable cough and have shortness of breath; generally, this will occur within 2–3 weeks of exposure. Individuals with cardiac concerns may be at higher risk of disease.

How can I protect myself from acquiring Q-fever? To help ensure that you are not infected with *Coxiella*, use appropriate protective equipment (such as laboratory overalls) and a respirator (you must be medically cleared for respirator use). It is also important to practice good personal hygiene by not eating or drinking in the lambing area and regularly washing your hands with hot soapy water. Always wash your hands when leaving the research area.

What should I do if I believe that I have Q-fever? If you believe that you may have acquired Q-fever, immediately seek medical attention at the medical clinic and inform the clinician that you work with sheep and may have been exposed to Q-fever.

Training Module Example 2

Allergies: Developing allergies to laboratory animals is a common health risk for animal workers. Allergies occur when your body's immune system reacts to an external agent such as animal dander. The most common symptoms of an allergic reaction include watery eyes, sneezing, nasal congestion, and hives. If you experience these symptoms, especially soon after entering an animal room, you should immediately contact and consult a medical clinician. If allergies are left untreated, they could develop into more serious health concerns (such as asthma).

Chemical and physical hazards: The OHS training program should also include relevant information as it relates to physical or chemical hazards. Animal researchers, on occasion, use hazardous chemicals, physical hazards, or radio-isotopes (radiological hazards) as part of their experimental design. One organization's best practice is to label cages that contain hazards. Included in the room is a best practice for handling and disposing of the hazards. For example, if a carcinogen were being used to induce tumors in mice, then all of the cages containing the carcinogen would be labeled with a specific card indicating "Carcinogen Present." A handling notice would be posted on the animal room door explaining that dirty bedding should only be handled in a chemical fume hood. The notice would include instructions on how to prepare the bedding for disposal.

Guidelines and policies on personal hygiene: The training program may include specifications for mitigating risk. Organizations should have developed guidance documents on personal hygiene focusing on minimizing risks associated with the activity. For example, the organization's policy may be a prohibition of eating, drinking, applying cosmetics, or using tobacco products in any animal housing or support areas. The organization may also mandate that at the beginning of each day, workers change into the provided laboratory clothing and, at the end of the work day, the organization can require that workers change back into personal clothes prior to departing from the facility. A shower should be available and may be required in certain circumstances (e.g., when leaving a biocontainment facility). Processes to familiarize animal users on these methods to protect themselves from identified hazards are frequently included in the training.

The OHS training program may be expanded to include additional modules based on the risks at each organization. Organizations working with macaques may choose to develop a specific training program on herpes B virus infections and those using poisonous reptiles may develop a program on safe handling of these animals.

Refresher Training (Also Called Annual or On-Going Training)

A refresher training module is considered to be core training by some organizations. It is important for animal care and use operations to provide ongoing training. Some organizations create a tailored annual training module at the end of each calendar year. The goal of the customized training is to highlight problems from the previous years and focus on processes to minimize or eliminate them during the upcoming year. One method of crafting effective training is to use the current year's compliance reports as the core for the year-end annual training module (e.g., the top five to seven noncompliance trends from the previous year). Using this approach, each year's refresher module is different since compliance trends shift over time. The refresher training may also include a new regulations and programs update (e.g., emergency veterinary pager number). Some organizations use an "Annual Progress Report" submission that is specific to each protocol to verify that all research staff completed their required training. In addition, the IACUC may identify additional training needs for research staff based upon reported concerns relating to animal care or use activities and the presence of suboptimal performance.

IACUC Member Training

Since the IACUC is the central body of the animal care and use program, organizations must develop a thorough and efficient training program for both new and veteran committee members.

New member training. The axiom that only qualified personnel should be working with animals in concept also applies to IACUC members: Only qualified personnel should be reviewing, approving, and overseeing animal care and use activities. New IACUC member training is fairly common and provided by most organizations. The individuals conducting the training vary from organization to organization. Even though some organizations employ training coordinators, the tendency is for IACUC administrators, the IACUC chair, and/or attending veterinarians (or a combination thereof) to conduct the new member training and orientation. According to BP meeting attendees, most organizations initiate new member training through a one-on-one meeting between the new member and the IACUC administrator or chair. In situations when multiple new members are appointed at the same time, training is generally collectively conducted.

Organizations have established different methods for conducting the training. In some situations, new members first complete prescribed web-based training (e.g., AALAS Learning Library, CITI), which is followed by discussion of the modules. One of the more common methods used to educate new IACUC members is for trainers, generally the IACUC administrator, to develop a PowerPoint presentation that covers the critical topics at that organization. The trainers then use the slides to cover all of the relevant topics in a systematic fashion.

Prior to initiating the training, new members are frequently provided reference materials. Some organizations provide committee members printed copies of the regulatory documents, while others may provide an electronic copy (e.g., CD, thumb drive, or server access) of each relevant document. The materials provided by most organizations include a copy of the Health Research Extension Act of 1985, PHS policy, the Animal Welfare Act (AWA), AWAR, the *Guide*, the *Ag Guide*, *Institutional Animal Care and Use Committee Guidebook*, *American Veterinary Medical Association (AVMA) Guidelines on Euthanasia*, and relevant institutional policies.

Formal education programs. Formal education programs have been established by some organizations to educate their committee members. The constraint associated with formal programs is often associated with budgetary restrictions. Although each type of program has its benefits, the cost to send staff can vary significantly. For example, some conferences would cost an organization several hundreds of dollars (i.e., travel, lodging, registration) while the cost of others may be limited to a nominal fee and travel and lodging expenses.

Some formal educational opportunities for IACUC members are activities that were specifically developed to educate ACUP staff members. IACUC 101 is a formal activity that provides basic information about ACUPs. Many organizations ask new IACUC members to attend the IACUC 101 course to become educated in the history and the basic principles of ACUPs.

In addition, many organizations use OLAW, Foundation for Biomedical Research (FBR), Public Responsibility in Medicine and Research (PRIM&R), AAALAC, and National Association for Biomedical Research (NABR) webinars to provide specific training information. IACUC administrators reported gathering IACUC members, with a light lunch or refreshments, into a central location where the webinar was viewed and discussed. OLAW webinar topics of particular value to IACUC members have included "IACUC Responsibilities beyond Protocol Review," "Facilities Inspections," "Writing a Good Assurance," "Occupational Health and Safety Programs," and "Grants Policy and Congruence." These webinars have been recorded and are available to view on the OLAW website. Organizations can register on the OLAW website to participate in new webinars in real time.

An annual conference offered by PRIM&R for ACUP staff members is also an excellent opportunity for formal training. This particular program is typically attended by over five hundred animal program staff members. The program format is often classroom style and breakout sessions with many outstanding lectures. PRIM&R attendees frequently listen to experts and attend specific-topic working group sessions.

In addition, IACUC administrators' BP meetings provide opportunities for committee members to meet many of their peers and to discuss in detail specifics about particular topics such as committee member training programs. These BP meetings are limited to approximately sixty participants. The program format involves a 10-minute background presentation followed by 45–60 minutes of interactive discussion by attendees. The agenda topics for the BP meetings are selected by the attendees; this differs from most other meetings where the topics are selected by a committee or leadership team. This venue gives attendees the opportunities to ask about and discuss specific issues they encounter at their home organizations.

"HomeGrown" Training

In addition to utilizing commercially available formal training, organizations have developed their own in-house training programs. The advantage to homegrown training is that it can be customized to the organization's research portfolio environment and institutional policies. Successful training methods and programs are highlighted in the following discussion.

A Web-Based Training Module

While an organization's options for web module training are extensive, many organizations choose modules relevant to committee oversight and protocol review. Modules on the following topics have been proven to establish strong foundations for new IACUC members:

"Euthanasia of Research Animals"
"AVMA Guidelines"
"Occupational Health and Safety in the Care and Use of Research Animals"
"Animal Welfare Act Regulations"
"Public Health Service Policy on Humane Care and Use of Laboratory Animals"
Guide for the Care and Use of Laboratory Animals, eighth edition (2011)
"Ethical Decision Making in Animal Research"
"Working with the IACUC"
"Semiannual Facility Inspection"
"Essentials for IACUC Members"
"Post-Approval Monitoring"

These modules (or similar ones) are available on the AALAS Learning Library or through CITI. Each module will require 30–45 minutes to complete, so the time is not significant, but the foundation provided by a common resource for all IACUC members is important.

Face-to-Face Orientation

In addition to online training, the IACUC administrator often provides individualized training. During this time, the two members may discuss multiple topics. BP meeting attendees offered topics that have been effective at their organizations. For example, one of the primary components of many new IACUC member orientation programs is a review of the federal laws and policies governing animal care and use. The laws and policies section of the training frequently covers the primary regulatory documents, which include PHS policy, AWAR, the *Guide*, and, when applicable, the *Ag Guide*.

PHS policy. The discussion about the PHS policy typically begins with a brief history on the policy. The trainer often discusses how the policy was established and updated after an incident of research animal neglect that was discovered in a group of monkeys in Silver Springs, Maryland. The discussion may also include the notion that the PHS policy ensures that government funds are spent only on projects in which animals are humanely treated. The conversation typically includes such statements as that the policy is enforced under the Health Research Extension Act and that OLAW, NIH, administers the PHS policy on behalf of the NIH director.

An additional topic covered during the PHS policy discussion often relates to the applicability of the policy. For example, the trainer often uses this opportunity to inform the trainees that the PHS policy must be applied to every project funded by a PHS agency (e.g., NIH, the Centers for Disease Control and Prevention [CDC], and the Food and Drug Administration [FDA]), and the National Science Foundation (NSF).

Trainers usually discuss the organization's animal welfare assurance and the impact it can have on the program. For example, if an organization indicates that its assurance covers all vertebrate animal projects, irrespective of the funding, then the PHS policy must be applied to all projects rather than only those funded through PHS organizations. Training also often includes a discussion that outlines the consequences of not complying with the PHS policy.

Animal Welfare Act regulations. The AWAR training session typically begins with a brief history on how and why the act was established. The discussion focuses on how vertebrate animals for research purposes became a prime commodity in the mid-1960s and that pets were being stolen and sold into research. The history discussion typically concludes with a summary of how the act has been revised in more recent years to include research institutes and to focus on selected species (e.g., covered species). A primary point made by the trainers is that the AWAR do not apply to *mus* or *rattus* bred for research, animals used in food and fiber studies (i.e., farm animals' use in food production research), birds, or cold-blooded vertebrates. Trainers frequently remind new IACUC members that the AWAR are federal law and if violated by an organization, fines can be imposed for each violation.

Guide for Care and Use of Laboratory Animals. The *Guide* training session is typically highlighted by informing new members that the eighth edition is a primary standard by which laboratory ACUPs are evaluated by the AAALAC. The PHS policy requires that institutions base their programs of animal care and use on the *Guide*. The discussion frequently centers on the five chapters of the *Guide*: (1) "Key Concepts"; (2) "Animal Care and Use Program"; (3) "Environment, Housing, and Management"; (4) "Veterinary Care"; and (5) "Physical Plant."

Trainers generally advise new members that each chapter of the *Guide* can be used to reference specific requirements within the animal program. For example, the chapter on veterinary care can be used to identify the required components of a veterinary care program.

The training may also include how the *Guide* can be used by the IACUC member and how it is used by the AAALAC. The trainee is advised that the AAALAC will use the *Guide* as the standard to evaluate the organization's animal care and use program during accreditation visits.

Guide for Care and Use of Agricultural Animals in Research and Teaching. When an organization maintains working farms, it often applies the *Ag Guide*. Trainees are instructed that the *Ag Guide* must be applied when evaluating programs that include maintaining agricultural animals such as cattle, sheep, swine, and chickens. Much like the *Guide*, the *Ag Guide* training session typically focuses on the chapters of the standard. The first five chapters concentrate on programmatic issues such as policies and veterinary care and animal handling; however, this document also includes specific chapters that detail husbandry practices for common species of agricultural animals (i.e., cattle, horses, poultry, sheep, goats, and swine).

Other Reference Resources. Several other documents exist for specific species or specific styles of research. Training will follow a similar path as described above, but the focus would be to the specific species or activity under consideration. Examples include: Guidelines for the Euthanasia of Animals (2013); Guidelines for Behavioral Research Using Animals; Guidelines for Fish in Research; Guidelines for Field Researchers; Guidelines for Live Amphibians & Reptiles in the Field & Lab Research; Guidelines for the Use of Wild Mammals in Research; Guidelines for Use of Wild Birds in Research; Guidelines for Training in Surgical Research with Animals; or Guidelines for Removal of Blood.

Protocol Review Training

In addition to understanding and being familiar with the regulatory documents, IACUC members must also know how to review an animal care and use proposal. Consequently, new IACUC member training programs frequently include a protocol review component. Organizations have employed a variety of methods for training IACUC member candidates:

- **Mentor programs.** Some organizations assign IACUC member candidates to "shadow" veteran members. A new member is partnered with a senior IACUC member who has a thorough understanding of protocol review. The protocols are distributed to the IACUC member and his or her mentee. The member conducts the protocol review and the aspiring member performs a mirror review. Once both individuals have completed the review, the veteran IACUC member and his or her protégé compare the two reviews for consistency. The veteran IACUC member provides explanatory advice and helps the new member to develop a consistent review methodology and to identify protocol inconsistencies.

- **Auditing and debriefing.** Many organizations have the new member attend an IACUC meeting as an observer after he or she has completed training (web and didactic). During the meeting, the new member absorbs the process and meeting flow, noting issues that seem odd or unusual. At the conclusion of this first IACUC meeting, the new member meets with the IACUC chair and/or IACUC administrator to debrief. This is an opportunity for the new member to ask questions about processes that seemed odd or an opportunity for administration to determine if additional training is necessary. If all appears to be in order, the new member will be receiving documents for review and be expected to participate in the discussions and deliberations of subsequent committee meetings.

Continuing Education for IACUC Members

The regulatory standards require organizations to provide IACUC members with the necessary resources to be an engaged committee member. Consequently, many organizations have developed continuing education programs to ensure that their committee members are current on regulatory standards and ethical expectations. Organizations have developed various methods to provide continuing education to their committee members. BP meeting attendees noted several ways in which their committee members remained current with the greater animal care and use community:

- **Education sessions during IACUC meetings.** Many BP meeting attendees indicated that they conduct brief education sessions during IACUC meetings. In a situation such as this one, IACUC administrators frequently allotted approximately 15 minutes during each meeting to discuss relevant topics. In preparation for an upcoming meeting, the IACUC administrator identified a topic for discussion and the relevant topic facilitator. In one example, the IACUC administrator chose to discuss "Conducting Literature Searches for Alternatives to Painful Procedures" during a meeting. He asked a librarian to be a guest facilitator and focused on the type of key words that should be present in an alternatives review for a specific topic, the databases that should be utilized, and the filtering options. The discussion provided gentle reminders to the IACUC members on markers of an appropriate search. Once the session was completed, committee members were given the opportunity to ask questions of the guest facilitator and to reflect on the discussed topic.
- **Online resources and periodical publications.** IACUC administrators at many organizations indicated that they frequently share online resources with their IACUC members as a form of continuing education. For example, IACUC members may be asked to subscribe to OLAW's LISTSERV or RSS feed, which will help keep them current on issues relative to the PHS policy and related documents. In some situations, the IACUC administrator directly e-mails relevant information such as AAALAC International's newsletter, *Connections*, to committee members.

Additional resources that have been used to provide continuing education to IACUC members include articles from *Lab Animal*, *ALN Magazine*, the *AALAS in Action* newsletter, and e-clips.

Additional Training Programs Frequently Developed for Animal Users

BP meeting attendees have identified some additional training components, listed here, that have served to improve their ACUPs:

- **PI-facilitated training.** On occasion, the PI conducts the training. For example, a PI may have been conducting a unique surgical procedure for years as part of his research. Based on his experience with the surgery, the PI is undoubtedly the resident expert. Consequently, each time the PI adds new personnel to his research proposal, he conducts the surgery training. In situations such as this one, the PI should be maintaining training records that are usually kept in his lab. During semiannual IACUC inspections, committee members frequently review the records for completeness and clarity.
- **Facilities orientation.** In addition to animal users going through all three phases of training, they also frequently receive a facility orientation. Many organizations' program participants are required to attend a facility orientation prior to obtaining access rights to animal care or housing facilities. This orientation session provides participants with information and materials outlining topics such as basic care services, animal order information, disposal of laboratory animals, security, facility operation, the use of PPE, and facility traffic flow patterns.
- **Corrective action training.** Some organizations use commercial programs such as the AALAS Learning Library and CITI for corrective action training as required by the IACUC for noncompliance retraining. Corrective action training most often concludes with a face-to-face interview between the "learner" and the IACUC chair, the IACUC, and/or the compliance director/administrator. The follow-up sessions are generally used to reiterate the significance of the item discussed in the training module and affirm the IACUC's position that continued noncompliant behavior will not be tolerated.
- **The IACUC and program oversight training.** In addition to species-specific training, the training team may choose to include a training session discussing the IACUC and its role in the ACUP. The conversation may include a discussion of the IACUC's routine visits to animal facilities to ensure that the animals are receiving appropriate care. It may also focus on the need for technicians to help contribute to quality ACUP by being the eyes and ears of the IACUC and encouraging them to bring any concerns they may have to the IACUC.
- **Training record maintenance.** The *Guide* indicates that "all program personnel training should be documented." Consequently, organizations typically develop methods to document that individuals associated with the ACUP have received appropriate training. The methods used by organizations may vary according to their specific processes.

Many organizations use web-based systems to document training. In some situations, once web-based training is completed by a trainee, his or her transcript is automatically updated. In other situations, the IACUC administrators may update an Excel spreadsheet that is manually maintained. Additional methods have been employed, including documenting training on the approved protocol or in meeting minutes, and using course attendance records.

Some organizations employ training coordinators who conduct or facilitate all of the required training. In situations such as this one, the coordinator typically maintains the training records in a central database. The training coordinator is usually able to produce a transcript of training activities for each individual involved in animal care and use.

Evaluating the Effectiveness of Training Programs

Although the regulations require organizations to establish training programs for those facilitating the ACUP, both the AWAR and the PHS policy state that it is the responsibility of the IACUC to determine if the training is effective. IACUCs satisfy this requirement using multiple methods. For example, during semiannual inspections, IACUC members talk to animal users to ensure familiarity with the critical components of the program. An IACUC member may ask a laboratory technician what he would do if he noticed an animal welfare concern. If the technician responds by summarizing the policy for reporting animal welfare concerns to the IACUC member, the committee uses this incident to document effective training.

In addition, organizations frequently use incidents of noncompliance to evaluate and refine their training programs. For example, if PIs allow new personnel to work with animals before they are added to the protocol, training may need to be modified to include more information on adding personnel to protocols. BP attendees noted that often monitoring the noncompliances that occur within a program provides good clues on when individualized training is necessary and where global training should be modified. Animal care and use programs are dynamic, and training must also be dynamic to meet the needs of an every changing landscape.

Final New Member Actions before New Members Go to Work

Many organizations also provide specific training and ask committee members to sign a statement to acknowledge their concurrence with a process and participation in the training. For example, a frequent concern is to identify a process to notify the committee member of the confidentiality expectation. A training session may discuss the need to maintain confidentiality of those issues discussed during full committee meetings. Generally, this training is conducted during the face-to-face session so that discussions pertaining to IACUC member expectations, performance, and protections can also be reviewed. Many organizations have new committee members sign a confidentiality statement to document that this training has occurred.

CHAPTER 14

Tracking Animal Use on Protocols

SCENARIO

The GEU IACUC reviewed a protocol involving primate work at a nearby organization. GEU did not have the equipment necessary to complete the research project, so a researcher at Great Western University (GWU) who had the equipment agreed to collaborate with the GEU researcher. The IACUCs at both organizations approved the work for nine animals and the researchers began the activity. At the end of the year it was time to file the US Department of Agriculture (USDA) annual report, and the IACUC administrator for GEU collected the animal use numbers for all protocols at the organization. The primates used at the collaborator's organization were not reported on GEU's USDA annual report because the animal work was performed at GWU. Approximately 6 months later, the USDA inspector visited GEU and, while reviewing the protocol, discovered that the primates were not listed on the annual report. While the GEU IACUC administrator argued that the work was done at a different organization, the USDA inspector issued GEU a citation for under-reporting animal use. The citation was based upon animal ownership, which had been retained by GEU.

Regulatory Requirements and Guidance Documents

Neither federal regulations nor policies require organizations to develop procedures to account for the number of animals used under each specific protocol. However,

- **Animal Welfare Act regulations (AWAR 2.35[b] [8] and 2.36[b] [5])** require organizations to maintain records on animal use. The records should include the number of animals experiencing unrelieved, painful procedures as well as the number experiencing painful procedures that were alleviated using pain-relieving drugs. The total number of animals in each category must be reported to the USDA in an annual report.
- **AWAR (2.31[e] [1] and 2.31[2] [2]) and Public Health Service (PHS) policy (IV D [1] [b])** require principal investigators (PIs) to justify the number of animals needed to conduct a research study. The IACUC should ensure that the justification

is documented in the protocol and that the minimum number of animals necessary to conduct the experiments is used. When feasible, the justification should be based on statistical calculations.

To satisfy the regulations, most organizations have developed methodologies to track the number of animals used in each IACUC-approved protocol. Best practices (BP) meeting attendees agreed that any time a procedure is conducted on an animal, measures to count those animals must be implemented. However, challenges remain when determining what an animal is and under what conditions animals should be tracked, counted, and reported to the USDA, the National Institutes of Health (NIH), or the Association for Assessment and Accreditation of Laboratory Animal Care (AAALAC).

ANIMAL USE

BP meeting attendees confirmed that their organizations have developed methods for tracking the number of animals used under each protocol. However, before developing their accounting systems, most organizations struggled with defining "animal use." Two BP commonly applied involved counting animals against a research protocol either when they are transferred to that project from another approved protocol or when they are purchased for that protocol. Let us say that a PI is conducting research that involves rabbits and requisitions five animals through the animal resource group. Those animals are debited against that specific protocol on the date of the requisition. Although accounting against research protocols when animals are purchased presented minimal problems, other situations often create greater challenges.

Identifying Use When Transferring Animals between Research Proposals

On occasion, research animals are transferred from one protocol to another. When an animal is used on two separate protocols, it must be accounted on each protocol. For instance, when a protocol is approved, the use of a specific number of animals is justified and approved for that project. Consequently, irrelevant of the source, only a specific number of animals can be used on a specific project. If a rabbit is used under project number 1234 and that rabbit is later transferred to project 5678 for use, then that rabbit is counted as being "used" on both protocols. Using the same scenario, project 1234 is a USDA category C project (blood collection only), while project 5678 is a USDA category D project (subcutaneous device implant). When that specific animal is reported on the USDA annual report, it is only counted once on the report, in the highest pain category—in this case, category D (i.e., pain appropriately relieved).

Agricultural Production Animals

Many organizations (e.g., land grant universities) raise agricultural animals in a production setting. These animals (e.g., cattle, sheep, horses, and swine) not only are used for research and teaching but also are frequently used to supplement finances. For example, beef cattle, dairy cattle, lambs, and swine are processed into food products. In addition to agricultural production, many of these agricultural animals are held for research or teaching activities. In the event that an agricultural animal is transferred to a protocol for the purpose of conducting research or teaching students, those specific animals should be accounted for under the research or teaching project. Perhaps a pig is removed from a production herd for use in biomedical research. That specific pig is counted as a research animal against that specific project, while the others in the herd remain production animals. Only the research use of the pig is reported in the USDA annual report.

Rodent Breeding Colonies

Research staff members conducting projects that involve laboratory rodents, such as mice (*Mus*) and rats (*Rattus*), frequently maintain colonies because they have unique needs (e.g., strains that are not available commercially). Accounting for breeding rodents can create unique challenges. Should preweaned mice be counted as used by research facilities? Should an organization count culled mice as being used?

The USDA utilizes a pain classification system (USDA pain cat B, C, D, and E) for animals used in research, teaching, or testing. It is only required for USDA-covered species. This practice is commonly used throughout the research community. Considering the common practice, should breeding mice that undergo a procedure (e.g., ova implantation) be counted as used in USDA category D or category B?

BP meeting attendees agreed that when mice undergo any procedure, they should be counted as used by the organization (in our example they would be counted as USDA category D due to the surgery for ova implantation). Several other procedures, while minimal or routine, require animals to be shifted from a breeding category (USDA category B) to a use category (USDA categories C, D, or E). Let us say that mice are euthanized as part of the culling process (USDA category C) or that they undergo tail snips for genotyping (USDA category D); they would be considered used for research. BP meeting attendees noted that the only time that mice would be counted as USDA category B animals would be if they were born and died without any use (death unrelated to research or procedure use). In essence, this means that most organizations that are breeding, genotyping, or culling mice would rarely have animals in USDA category B. (Note: USDA annual reports do not include mice or rats bred for research, so this discussion involves accounting for animals under a common process but not reported to the USDA).

BP meeting attendees discussed whether fetuses used to establish tissue cultures should be accounted for by an organization. They agreed that when PIs euthanize pregnant dams to collect fetuses for the purpose of establishing tissue cultures, they need only count the female mice used and not the collected fetuses. However, if the fetuses were the test subject and, upon euthanasia of the dam, the fetuses were removed for tissue assessment, then the accounting process would be slightly different. While only the dam would be reported to the agencies as an "animal used," the number of fetuses required must be justified to the IACUC as a basis for requiring a set number of dams to acquire the necessary number of fetuses.

As a general rule, BP meeting attendees did not count animals born until a predetermined time period after birth. Organizations determined different set points, but many used the "first cage change" as the first accounting of animals against a project or protocol. The justification was that some animals would die shortly after birth for any number of natural reasons (e.g., mutation, lethal gene, etc.) and that to count animals very early could disturb the dam, causing her to cannibalize, for example, the healthy pups. Therefore, it was assumed that animals surviving until the first cage change would continue to thrive and therefore would be accounted animals.

Counting Hatched Species

BP meeting attendees identified another challenge in counting animals that come from eggs or are born in community tanks or containers. The challenge forced organizations into defining an arbitrary time point that could be used for defining when the embryo moves from embryo to animal state. For example, fish or amphibians may be counted when there is no longer a yolk sac, and birds or some reptiles at 3 days after hatching. In cases where research is performed on animals not yet counted under the organization's set points, those animals are counted as being used in a research project because the numbers must be justified on the animal proposal to the IACUC. As a best practice, counting animals at an early age requires a logical thought process on what is practical, will not interfere with normal animal well-being and care, and will exercise good judgment. BP meeting attendees did not identify any one clear set point for all species.

Methodologies for Tracking the Number of Animals Used on Each Protocol

Organizations have devised multiple processes for tracking animal use on specific protocols. In some instances, organizations ask PIs to track and regularly report their animal use to the IACUC administrative office; this is most commonly used when the PI has a breeding colony. In other situations, organizations associate animal use with husbandry per diems and track use with the help of sophisticated equipment and census programs including bar coding. On occasion, organizations develop processes to track animal purchases and only purchase animals based on the number of approved animals listed on the IACUC-approved protocols. Some organizations manage and maintain rodent breeding colonies and transfer animals to research projects when requisitions are made by the PI.

Electronic Reporting

Organizations have developed or employed electronic management systems to monitor animal use. In situations such as this, animal facility managers frequently place bar codes on animal cages or cage cards. The bar code is associated with a particular animal (or cage of animals) and a specific protocol. Each time new animals are added to a protocol, they or their cage is assigned a bar code and tracked. The codes are scanned regularly (e.g., daily or weekly) and the information is used to charge per diems and to subtract used animals from the approved total on a protocol. The electronic system accounts for animal use and PIs are notified when their protocol has utilized nearly all of their approved animals.

Using bar coding or cage counts for reporting the number of animals used on each protocol can be problematic. Consider that an electronic process (bar coded or cage counts) to count animals assumes a set number of animals in each cage. If an organization determines that its average cage density is three or four animals, then a census by bar code may over-report a researcher keeping only two or three mice in a cage and the system may under-report a researcher keeping four or five mice in the cage. IACUC administrators need to be aware that a system report of a researcher overusing animals may be a system-generated error. All animal use numbers should be verified by the researcher prior to submitting final counts to federal oversight agencies.

PI Reporting

In some circumstances, PIs report animal use to the IACUC office during annual review submissions. The use is recorded and tabulated by administration and the PIs are notified when they begin to approach their approved number limit.

De Novo Reviews and Animal Numbers

The de novo review required by the PHS occurs 3 years after a research project has been approved. During the review, the IACUC approves the number of animals that can be used for a research activity. IACUC members should ensure that used animals are not carried over into the de novo submission after a 3-year period, unless they remain on study. Animal use must be justified in the 3-year de novo review only for the remaining portion of the project. For example, if a research project involves 5 years of experiments and one hundred animals are used each year, at the end of 3 years only two hundred animals remain. Consequently, the de novo submission should only include the remaining two hundred animals and should not be a repeat of the previously approved protocol. Only in the event that this cannot be done should animals be euthanized before the IACUC protocol expires.

CHAPTER 15

Veterinary Care Programs

SCENARIO

GEU veterinarians provide care for over 50,000 animals including dogs and cats; biomedical sheep and swine; a small group of rhesus monkeys; some rabbits, ferrets, and guinea pigs; and about 35,000 laboratory mice and 15,000 rats. Their veterinary care program is complex and includes all of the necessary components to be compliant with the regulating documents.

The GEU IACUC conducts program reviews semiannually, at which time members also review the veterinary care program. During the review process, IACUC members had few questions regarding those components of the program in which they were actively involved. For example, committee members, being aware of the required annual reports sent to the Office of Laboratory Animal Welfare (OLAW), US Department of Agriculture (USDA), and Association for Assessment and Accreditation of Laboratory Animal Care (AAALAC), had no questions about reporting requirements. They understood the process of semiannual inspections since they participated in them biannually; they were aware of the protocol review process, and committee members understood the intricacies of the organization's occupational health and safety program (OHSP). However, the committee members were not familiar with the components of the veterinary care program. Consequently, it was the understanding of IACUC members that since animals were not getting sick, the veterinary program was satisfying the primary objective of ensuring animal welfare. Therefore, IACUC members concurred that the veterinary staff were adhering to the regulating guidance, and that the veterinary care program was compliant.

Since the OLAW program review checklist includes all of the components of the program of animal care and use that IACUC members must assess semiannually, the GEU IACUC frequently uses it as a guidance document to ensure that it evaluates all of the relevant components of the program. For example, the checklist identifies the protocol review process, the OHSP, and the unit emergency disaster plan as program components that must be assessed. It also includes a detailed list on the mechanisms of the veterinary care program that must be reviewed during the semiannual program reviews. In this particular scenario, traditionally when the GEU IACUC gets to the section on veterinary care, the IACUC chair would defer the review to the attending veterinarian (AV). The IACUC chair would ask the AV if the program of veterinary care complied with all of those issues (i.e., collectively rather than going through each point) listed on the checklist. The AV would typically touch on a couple of minor points, but eventually confirm that all was in order and compliant. Once the confirmation was made, the committee would move to the next section of the program for review and consideration.

Over a 4-month period following the program review, a number of veterinary issues were investigated by the IACUC. The initial concern involved a researcher who was conducting a project using mice. A critical component of the project was to conduct tail tip amputations for multiple blood collections, which were scheduled to occur over a period of 12 hours. In accordance with the protocol, the blood collection procedure was conducted on the same mouse bimonthly. During a facility inspection, IACUC members identified a number of mice with severely necrotic tails. Their tails were discolored and half the length of those of normal mice. The committee members questioned the researcher, who indicated that the veterinarians had provided training on the proper technique for conducting the amputations and collecting the blood. The principal investigator (PI) also explained that the veterinary technicians monitored the animals on a regular basis and to date they had not identified any adverse events and had not treated any animals for problems relating to tail amputations.

Later a second issue was identified and investigated by the IACUC. It involved Dr. Smith, a researcher who was maintaining a unique line of transgenic mice. In this particular situation, veterinary staff notified the researcher that his once clean line of mice was now infected with parvovirus. The researcher was concerned and surprised since parvovirus had not been found in his biosecure facility in over 15 years. During an IACUC meeting, the AV explained that a shipment of mice for a new researcher had been received a few months before the parvovirus was identified. The AV explained that these animals were immediately moved to the new researcher's housing facility, which was adjacent to the facility housing the transgenic strain; the new shipment of animals was later confirmed to be carriers of the disease and they were the source of the Dr. Smith's problem.

During an inspection later that month, the veterinary medical officer (VMO) reviewed multiple veterinary records. The VMO identified a number of additional treatment records that did not list the final disposition of animals under treatment. For example, one of the records indicated that a rhesus monkey was being treated for a bite wound it experienced from a cage mate. The record indicated that the monkey was sedated, the wound was cleaned and stitched, and a course of antibiotic was initiated to prevent infection. During the inspection, the VMO noted that the incident had occurred at least 50 days prior to the visit and, although the animal appeared to be healthy, the records were incomplete since they documented no follow-ups after the initial treatment.

During a follow-up visit, the USDA VMO identified two dogs that were being singly housed. He noted that according to the organization's enrichment policy, singly housed animals would either receive toys as forms of enrichment or they would be walked and played with by staff daily. The VMO observed that the dogs had no enrichment devices available in their cages. After discussing the matter with the animal care technicians, he also learned that time restraints had resulted in the animals not receiving specialized attention (i.e., walks or hands-on time with the technicians) for the last few weeks.

After the USDA inspection, the IACUC chair asked the IACUC administrator to summarize the incidents of the past 4 months and schedule time to discuss the concerns during the next regularly scheduled IACUC meeting.

During the discussion, the IACUC considered all of the points summarized by the IACUC administrator. Committee members discussed the fact that necrosis in the tails of mice is a chronic problem (i.e., it does not happen overnight). Their discussions ultimately led to the opinion that the necrosis was as a result of insufficient oversight by the veterinary staff members as well as the PI.

Committee members also agreed that the parvovirus infection may have been avoided if the organization's "New Animal Quarantine" policy had been followed. The committee members agreed that if the incoming mice had been quarantined and evaluated according to their policy, the researcher's valuable line of transgenic animals would probably not have been exposed to the parvovirus.

Finally, the committee considered the findings of the VMO during the USDA annual inspections. IACUC members were particularly concerned since, during a very recent semiannual program review, they had agreed that appropriate clinical records were being maintained and the enrichment policy was being followed. Committee members were very concerned that the USDA findings and subsequent citation for inadequate veterinary records (primates) and inappropriate use of enrichment (dogs) contradicted their findings.

As a result of the committee's discussions, the IACUC concluded that the veterinary care program might not be as stellar as members had thought and that a thorough assessment of it was warranted. Since all members of the veterinary team were also IACUC members, the committee decided to enlist the services of an ad hoc reviewer (i.e., local veterinarian) to conduct the assessment. As a result of the review, the program was found to be deficient in multiple areas. Accordingly, the GEU IACUC developed a plan of action and correction timetables to eliminate the deficiencies. The plan included a process to ensure that the veterinary care program remained compliant, which involved hiring an outside veterinarian to serve as an ad hoc consultant to the IACUC during semiannual program reviews and inspections.

Regulatory Guidance

- **Public Health Service (PHS) policy (Section IV [C] [1] [e], p. 14)** indicates that adequate veterinary care will be provided to the animals.
- **Animal Welfare Act regulations (AWAR) (Section 2.33[b])** require organizations to develop a program of veterinary care that includes (1) appropriate facilities and equipment; (2) appropriate methods to prevent, control, diagnose, and treat diseases and injury and the availability of emergency care on weekends and holidays; (3) daily observation of animals; (4) guidance provided to PIs on handling, immobilization, anesthesia, analgesia, tranquilization, and euthanasia; and (5) animals receiving adequate pre- and postprocedural care.

- **The *Guide for the Care and Use of Laboratory Animals* (Chapter 4; henceforth referred to as the *Guide*)** outlines all of the expectations relevant to the veterinary care program. For example, it discusses the details of preventative medicine, clinical care and management, surgery, pain and distress, anesthesia and analgesia, and euthanasia.

THE VETERINARY CARE PROGRAM

An organization's program of veterinary care must satisfy the requirements identified in the regulations and must be overseen by the IACUC.

The veterinary care program must include professional veterinary care staff. A board-certified veterinarian with training or experience in laboratory animal medicine and responsibility for activities involving animals at the organization is preferred. In other words, the institution should have a staff veterinarian, or one should have been contracted to oversee the unit's veterinary care program for the research animals. The veterinarian must have unencumbered access to all of the research and teaching animals and the authority to euthanize or medicate any animal at the institution to prevent unnecessary pain, distress, or suffering.

The organization should institute methods for procuring laboratory animals. The procurement process should include methodologies that help to ensure that sick or infectious animals are not brought into the vivarium. The organization may want to develop processes to ensure that animals are only acquired from reputable vendors. For example, an organization may institute a policy in which USDA-covered species are only purchased from USDA-registered vendors. As part of the procurement process, the AV may require copies of health reports for all animals brought into the facility. For some species, such as mice and rats, the AV may also require serology tests to be conducted on all animals prior to entering into the organization's vivarium to validate that they are free of infectious agents. In addition to receiving individual health records for some animals, the AV may also choose to review colony records for species such as mice and rats and to look for specific health trends.

The organization's veterinary care program should include quarantine and isolation procedures, as well as stabilization periods for imported research animals. At some organizations the quarantine process is often initiated with the veterinarian or another trained individual conducting initial examinations on all animals that are part of a shipment. For example, the AV may check to ensure that the shipping cage is not damaged, that the animals being received are exactly what was ordered (i.e., the correct strain of mice), and that the general health of the animals is good (i.e., active and without coughing or sneezing). The organization may decide to quarantine any animals that are not from an approved vendor and that have not been cleared for use as a result of their initial health assessment.

The primary goal of the organization's quarantine program is to prevent animals from entering the general population in the vivarium until they are cleared (i.e., free of infectious disease) to be used in research. As with animals being used for research, those held in quarantine must have housing accommodations that are appropriate (i.e., adhere to the *Guide* expectations) and specific for the species needs. The quarantine time periods experienced by animals may differ from species to species and may be impacted by the procedures that will be conducted on them. For example, animals that will undergo survival surgery procedures may require a longer quarantine and acclimation period than those that are being euthanized for tissue collection or undergoing a nonsurvival procedure. In certain situations—for example, quarantined mice—the animals may be quarantined and sentinel animals will be used to help validate their clean health status. A sentinel animal is one that is typically housed in the dirty bedding of other animals. After an adequate exposure period, blood is collected from the sentinel animal for serological testing. The animal is often euthanized and necropsied to ensure that it is disease free. The common practice is for disease-free sentinels to equate to disease-free research animals.

It is recommended that the program include provisions for species separation. To minimize disease transmission and anxiety that may be experienced by housing incompatible species together, best practices (BP) meeting attendees noted that their organizations develop policies requiring different species to be physically separated (e.g., through a physical room or cubicle separation).

Portions of the veterinary care program can be executed by the AV as well as individuals that were trained by the veterinarians. For example, the AV may train research staff to recognize sick animals. This training may facilitate the PI's ability to observe animals daily for abnormal physical signs or behavioral patterns. The training may also include, for example, instruction on how to notify the AV when an adverse situation is identified. In the event that any abnormal issues are observed, the AV should be notified and should respond accordingly.

Processes should be in place to ensure that sick animals receive appropriate veterinary care or, when necessary, are humanely euthanized. For example, the PI or perhaps an animal care technician may notify a veterinary technician of a particular animal health-related issue. The type of notification may depend on the severity of the situation, but could be through a phone call or a posting on the animal room door. The call or posting typically results in veterinary action, which may include treatment and establishing a medical record or, in severe cases, euthanasia.

EVALUATING THE VETERINARY CARE PROGRAM

The veterinary care program must be reviewed for compliance with the regulatory standards at least annually—and, in most cases, semiannually—by the IACUC. Since the veterinary program is managed by the veterinary staff, organizations frequently consider it a conflict of interest for the AV to conduct or be part of the review of the program that they manage, including participating in the IACUC decision-making regarding adverse events which occur within the veterinary care system. Organizations conduct the reviews in multiple ways. Some of the methods provided by BP meeting attendees are summarized here.

- **During facility inspections.** Some organizations choose to access portions of their veterinary care program during facility inspections. In this situation, IACUC members review relevant veterinary records (e.g., animal medical records, daily veterinary sick-call records, treatment records, etc.) during their semiannual assessments. For example, the daily animal observation logs should show that the animals were viewed each day. Since the observation records are typically current on routine work days (i.e., Monday–Friday), it is often beneficial to look back through the records and confirm that technicians are looking at the animals on weekends and holidays. IACUC members may also want to check duty logs to ensure, for example, that floors, walls, and relevant equipment are being sanitized according to the program standards. Committee members may want to ensure that the cage washing records validate that the cages are being sanitized during each run. In the event that standard operating procedures are in place to support the veterinary care program, IACUC members may want to review them during the inspections.
- **During the semiannual program review.** Another opportunity for the IACUC to assess the organization's program of veterinary care is during the semiannual program review. Some organizations create a subcommittee of the IACUC to conduct a focused assessment of the program of veterinary care. The subcommittee may include IACUC members, a veterinary consultant, and the IACUC administrator. The subcommittee may discuss all of the components of the program and provide the IACUC with a recommendation. In addition, subcommittee members may use the OLAW program review checklist as a guidance document to ensure that they consider each component of the veterinary care program. For example, the subcommittee may discuss the veterinary triage process, how guidance on surgical procedures is provided to scientists, and the animal procurement process.

Clinical Records

The committee members should also check clinical care records. For committee members to thoroughly assess clinical records, it is helpful for them to understand a method that veterinarians frequently use to organize medical records. Traditionally referred to by the acronym SOAP (subjective, objective, assessment, and plan), this method is typically used by veterinarians to cover the components that should be evident in the medical records. All of the records should include treatment dates, a general diagnosis based on observation, perhaps a specific diagnosis based on an examination and tests, and finally a plan of treatment. In addition, the record should document the details of all specific treatments (e.g., a daily dose of gentamicin was provided) and the final disposition of the animal (e.g., the animal recovered and was returned to the herd, or treatment was ineffective and the animal was appropriately euthanized using Euthasol®).

Using Research Animals in More Than One Project

The veterinary care program should also have a process that determines whether a research animal can be used in more than one research project. For example, a researcher may use a dog in a series of behavior studies. Once the behavior work is completed, the dog may be available to another researcher to use for additional research activities. Conversely, a researcher may involve a dog in a project that includes survival surgery. Once the project is completed and the dog recovers from the surgery, the organization may decide that the animal cannot be used on other projects involving painful and distressing procedures.

Consequently, an IACUC and the staff veterinarians should develop an animal reuse policy. This policy should identify when an animal is eligible for reuse. It may say, for example, that only animals used in noninvasive procedures are eligible to be used in additional research projects. In the event that an organization develops this type of policy, the veterinary records should identify the types of procedures conducted on each specific animal to ensure that animals involved in category D or E procedures (i.e., a painful or distressful procedure) are not used in an additional invasive research project.

Controlled Drug Records

The eighth edition of the *Guide* places the onus of reviewing drug records and storage procedures on the organization. With regard to controlled drugs that are used in animal research activities, organizations have delegated IACUC members responsible for auditing controlled drugs and related records. In particular, any time controlled drugs are being used in animal research activities, the IACUC members are responsible for the oversight. The OLAW checklist guidance document includes information regarding monitoring controlled substance use. IACUC administrators must be aware that while there is a single federal position on controlled substance use, regulatory requirements are very different from state to state and that the federal agents generally defer policy and process to the state agents.

As a result of the delegated oversight, during semiannual inspections, IACUC members should ensure that the veterinary care program includes and effectively uses methods for maintaining accurate records and secure drug storage.

For committee members, inspecting the pharmacy is usually a two-step process. The initial step is to discuss with the facility manager or the veterinarian the process for handling expired drugs. The process typically includes a method for physically separating expired drugs from those that can be used for research or veterinary purposes. In addition the practice should include processes for the final disposition of expired drugs. For example, in certain situations, depending on the state requirements, the expired drugs must be sent back to the distributer for disposal (i.e., a reverse distributor). In other situations, with a witness present, expired drugs may be injected into the carcass of a deceased animal and the carcass then incinerated, or the substance can be independently incinerated. Once this process is defined, the inspector randomly samples the pharmaceutics to ensure that no expired drugs are part of the inventory that is available for use.

Committee members also review how controlled drugs are stored and the inventory recorded. Controlled substances must be stored using a two-lock system. BP meeting attendees offered a couple of common methods as examples. A common procedure was to use a drug lockbox. In this particular situation, if the lockbox is considered one of the two locks, it must be secured in a manner that will not allow someone to pick it up and walk out of the room with it (i.e., bolted to the wall or to a heavy piece of furniture). The second lock in this scenario would be the room door lock.

Another popular method discussed by attendees included placing a metal storage box into a lockable filing cabinet with the second lock being the room door. In addition to being secured, controlled substances should be secured only with other controlled substances. In other words, a lockable general chemical storage facility should not include an inventory of routinely used chemicals mixed with controlled narcotics.

The second part of the inspection process involves checking the records for completeness. A controlled drug log typically includes, at a minimum, a log entry date, the quantity of drug received and from where, the quantity of drug removed and by whom, and the total quantity available in storage. When inspecting controlled drug records, committee members routinely select a drug—for example, ketamine—and count the number of vials. They then compare the number with the log book entries.

Pharmaceutical versus Reagent Grade Drugs

An organization's program should also include a policy for the use of reagent grade drugs for animal research. During past BP meetings, attendees discussed with representatives from the USDA OLAW, and AALAC that drugs must be used when available, unless justified for scientific or clinical reasons. OLAW has published specific guidance on the use of pharmaceutical grade versus reagent grade materials to ensure that IACUC administrators are well informed on this topic.

BP meeting attendees agreed that regardless of whether a reagent is the focus of the experiment or simply a tool to help the experimental model be successful (e.g., doxycycline to turn on a gene in a transgenic mouse), a pharmaceutical grade reagent is required unless there are clearly justifiable reasons why a lesser grade will be used. If a reagent grade doxycycline is used to turn on a gene in a transgenic mouse, the impurities present in the reagent chemical may have an impact on the experiment and perhaps the well-being of the animals. For example, if one of the impurities is an endotoxin, depending on the levels of endotoxin in the product, it could induce an immune response in the animal or cause illness—both outcomes potentially modifying research data and the usefulness of the study.

However, there is no prohibition of using nonpharmaceutics in research. When such use is contemplated, the PI must provide to the IACUC a convincing argument that the reagent grade use is critical to the science and is justified from a purely scientific perspective. For example, if a researcher was interested in determining the health effects of impurities in nonpharmaceutical grade medications, he or she could justify the use of reagent grade drugs as the only means to fully ascertain the effects upon the animal.

CHAPTER 16

Whistle-Blower Policy

SCENARIO

During a visit to GEU's Cancer Institute, Dr. Smith, a visiting scientist, notices several rodents with ulcerative tumors. Since the policy at his organization is to euthanize animals before tumors ulcerate, he is concerned that the animals are experiencing unnecessary pain and distress.

Given that Dr. Smith is a guest, he is apprehensive about reporting his concern to GEU's IACUC and decides that he will make the report anonymously at a later time. Since Dr. Smith is an IACUC member at his home organization, he is aware that organizations must develop a process for individuals to report animal welfare concerns to the IACUC for investigation. As he continues his facility tour, he looks for a notice that describes how to report animal welfare concerns. During his tour of the vivarium, Dr. Smith is unable to locate the whistle-blower policy. After returning home, he is still concerned that the animals require attention. He decides to search GEU's website for a whistle-blower policy, but still cannot locate a process for reporting animal welfare concerns. Consequently, Dr. Smith decides that because the IACUC is overseeing the research, the ulcerative tumors must be an approved part of the protocol.

The standards governing the use of vertebrate animals for research, teaching, or testing require organizations to establish a method for individuals to report animal welfare concerns. For the process to be effective, it must be promoted to facility personnel and made accessible (e.g., posted notices or on the website) to anyone visiting the animal facilities. In this particular case, a perceived animal welfare concern was noted and went unreported because a clear process for making the report was not available.

Regulatory Requirements and Resources

- **Animal Welfare Act regulations (AWAR) (Section 2.32 [c] [4]) and *Guide for the Care and Use of Laboratory Animals* (henceforth referred to as the *Guide*)** require animal care and use programs to establish methods for reporting concerns associated with the care and treatment of animals. The regulatory requirements also require organizations to train personnel to recognize deficiencies in animal care and for the IACUC to review and, if warranted, investigate reports. In addition, the reporting process must include provisions for protecting whistle-blowers from discrimination or any reprisals resulting from the report.

THE POLICY

IACUCs are required to oversee and monitor animal use activities. The primary goal of the committee is to ensure that the animals receive the appropriate care and treatment as outlined in the federal standards. One requirement and a process for ensuring oversight is for each organization to institute a whistle-blower policy. The purpose of the policy is to establish a method for individuals to report perceived animal welfare concerns. The policy at a minimum should include:

1. A statement confirming the organization's commitment to the ethical care and treatment of vertebrate animals
2. A process for reporting animal-related concerns that includes the appropriate office and contact information for reporting concerns (i.e., an office address, telephone number, fax number, or an e-mail address)
3. Documentation that the IACUC will investigate the reports when warranted
4. A statement indicating that reports can be made anonymously and that federal law (i.e., AWAR) prohibits discrimination against persons bringing forth legitimate concerns

Publicize the Policy

A key component of an effective policy is to make it accessible to everyone. Some methods that are effectively used to publicize the policy include:

1. Posting the written policy in animal housing and procedure areas
2. Posting the policy in teaching laboratories where animals are used for educational purposes
3. Posting the policy on the organization's public website
4. Publishing the policy semiannually in newspapers or newsletters (i.e., university student or local newspaper) to inform facility personnel and the public
5. Conducting seminars for staff, students, and faculty
6. Conducting educational sessions to inform the public

Sample Policy 1

Reporting Noncompliance or the Misuse of Animals

The IACUC has the responsibility for ensuring that all animals used in research, education, or testing activities at GEU are treated humanely and in accordance with all federal, state, and local policies and regulations.

These activities are coordinated through the IACUC administrative office (IAO), which is located at 25 Main Street. Concerns or questions related to projects involving animals conducted at or under the auspices of the Great Eastern University can be brought to the attention of the director of the IACUC Administrative Office (telephone: [12345]; fax: [6789]; e-mail: IAO@GEU.edu).

The matter will be referred to the IACUC chairperson, attending veterinarian, and the IACUC for investigation. Reported concerns will be handled confidentially. Federal law prohibits discrimination against persons that bring forth legitimate concerns for investigation.

Sample Policy 1: Advantages and Disadvantages

Advantages
1. The relevant information is a concise policy that is readily available and can be effortlessly read.
2. A brief policy can be transcribed onto a five- by eight-inch index card, laminated, and distributed throughout the organization. For example, the IACUC administrator can distribute the cards throughout the organization while conducting the semiannual facility inspections.
3. The laminated policy cards can be effortlessly posted throughout the animal housing facilities.
4. A brief policy can be easily published in local newspapers and relevant publications.

Disadvantages
1. Although sample policy 1 highlights the main points that should be included in a whistle-blower policy, it does not clearly define the intent of the policy. Individuals reading the policy would not necessarily understand its applicability. For example, a concerned party from outside the organization may not be aware that the policy also applies to him or her. The person may not understand that it should also be applied anytime that he or she perceives there is a concern with animals—not just when a policy or regulation is violated.
2. Sample policy 1 requires the organization to incorporate into its education program information on the applicability and the purpose of the whistle-blower policy. For example, the organization education session will need to discuss the applicability of the policy and who can make reports.

Sample Policy 2

GEU's Whistle-Blower Policy

The IACUC will investigate concerns relating to the humane care and treatment of university animals. The concerns can include, for example, alleged instances of animal cruelty, neglect, or abuse. In addition, violations of government standards (e.g., Animal Welfare Act regulations, the *Public Health Service Policy on the Humane Care and Use of Laboratory Animals*, the *Guide for Care and Use of Laboratory Animals*) should be reported to the IACUC. The IACUC should also be notified if deviation from approved research protocols occurs, or if vertebrate animals are being utilized without prior IACUC approval.

This policy applies to all animals housed at GEU that will be used for activities (e.g., research, teaching, demonstration, or testing) conducted under the auspices of the university. For example, concerns relating to animals housed in university laboratories, wild, or agricultural animal facilities can be reported to the IACUC for investigation.

The concerns can be reported by anyone witnessing what he or she perceives as being a questionable activity involving animals. For example, reports can be made by university faculty, staff, and students. In addition, concerned citizens and bystanders can also report concerns to the IACUC.

At the request of the petitioner, only the committee chairperson and the director of research compliance will know the identity of the concerned party. When a report is referred to the IACUC for investigation, information relative to the identity of the complainant will be redacted from the documentation.

Reporting a Concern

If you are concerned about the welfare of a university-owned animal or believe that an animal is being mistreated, attempt to discuss the concern with the involved individuals (e.g., principal investigator [PI], facility supervisor, or animal care technician). If you are uncomfortable approaching the person or for any reason do not wish to speak to the people involved with the activity, your concern should be directed to the institution for investigation. Please contact either the IACUC chairperson (e-mail and phone number), or the director of research compliance (e-mail and phone number) to discuss specific concerns. If you wish to file a formal complaint, the following process should be observed.

Formal complaints should be made in writing, addressed to the director of research compliance, and sent to the office of the vice president for research, 1600 Pennsylvania Avenue, President's Place, MD 12345.

The notice must include:

1. The complainant's name and contact information
2. The specific date, time, and location when the problem was witnessed
3. A complete description of the incident
4. The signature of the concerned party

GEU conforms to the Animal Welfare Act, Section 2.32, (c), (4). Consequently, no facility employee, committee member, or laboratory personnel shall be discriminated against or subject to reprisal for reporting violations of any regulation or standards under the AWAR.

Sample Policy 2: Advantages and Disadvantages

Advantages
1. The principal advantage of sample policy 2 is that it serves as a stand-alone document, which also educates the reader. Once an individual reviews the policy, he or she understands that the policy applies to all university-owned

animals, covers any adverse event involving animals, and is applicable to everyone. The policy thoroughly discusses the types of activities that should be reported and covers all the relevant points recommended in the standards.

Disadvantages
1. Although sample policy 2 is a very thorough document, there are some disadvantages associated with such a detailed whistle-blower policy. A complex policy cannot be read in a brief glance. Consequently, Dr. Smith, in our scenario, may not feel comfortable stopping his tour to study the policy.
2. The complex policy is not easily posted in animal use areas. A one- to two-page document such as this one would need to be posted on a bulletin board or maintained in a technician's break room, for example. In situations such as this, the policy would not be as accessible to potentially interested parties.

INVESTIGATING WHISTLE-BLOWER POLICY REPORTS

The whistle-blower policies facilitate the opportunity for both organization employees and private citizens to report perceived animal welfare concerns. Consequently, there is the potential to receive a variety of reports that involve apparent concerns relating to the care and treatment of animals. For example, the organization may receive a report from an internal staff member regarding concerns relating to tumor burden in mice that are involved in cancer research. Conversely, the organization could receive a report when horses are being wintered in pastures that do not appear to have shelters that enable them to escape the harsh environment.

Although both reports are equally important, in some cases the concern can be immediately resolved. For example, the individual reporting the concern about whether the horses had shelter from the winter weather did not realize that the barn is about five hundred yards on the other side of the hill, and that the horses have chosen not to use the available housing arrangements. Alternatively, the report relating to tumor burden in mice may be substantial since it was made by an animal care technician that had been caring for these animals for over 10 years.

Preferring to err on the side of caution, some organizations encourage reporting. These organizations have developed a process for screening reports and only direct substantiated reports to the IACUC for investigation.

Prescreening Reports

Many organizations have reports prescreened by either a veterinarian, the IACUC administrator, or the director of research compliance before submitting reports to the IACUC for investigation. In many cases, this party is able to resolve the concern. The types of concerns that are typically resolved before they are referred to the IACUC include, for example, (1) animals that have escaped from their confinements, (2) agricultural animals that have not been provided appropriate winter housing, or

(3) a particular animal that has an apparent injury or appears to be ill. In each of these examples, the veterinarian, for example, is able to explain that animals are either being returned to their housing facilities or are being treated for their injuries by a veterinarian. When situations are resolved by a simple explanation, the activity and the resolution are reported to the IACUC during monthly meetings.

Some reports, especially those made by trained animal users (e.g., animal care technicians and research staff utilizing animals) can be significant. For example, after a PI collects tail biopsies for genotyping, an animal care technician may notice that a large percentage of the cages are stained with excessive amounts of blood. As a result, the technician may report the excessive amounts of blood and indicate that the PI is either taking too much of the tail for biopsy or not stopping the bleeding prior to placing the animal back in its cage. In situations such as this one, the concern is referred to the IACUC for investigation.

IACUC Investigation of Reports

When a report is referred to the IACUC for investigation, the following process is often followed:

1. If a report is formal, the IACUC administrator will redact the complainant's personal information and provide the report to the IACUC for discussion at the next full committee meeting. Similarly, if the report is verbal, the administrative staff member will prepare a summary of the allegation, which will be reviewed and discussed at the next regularly scheduled IACUC meeting. In the event that a significant concern warrants immediate action to protect the welfare of animals, an emergency meeting is convened at the earliest date possible. If this meeting cannot occur within a reasonable time frame, the attending veterinarian may stop the research activities until the issue is resolved.
2. During the meeting, the involved party (e.g., PI, research technician) will be given the opportunity to meet and discuss the concern with the IACUC. At this time, the PI, for example, can provide the details of the incident and be provided the opportunity to resolve any misunderstandings and work together to find solutions to the alleged concern.
3. After discussing the concern and considering input from the relevant sources, the IACUC may take action or request additional information prior to addressing the concern. The IACUC, for example, may consider one of the following actions as a resolution for the concern:
   a. The complaint may be dismissed.
   b. The project, or a portion of the project, may be suspended.
   c. Animals may be referred for additional veterinary care.
   d. A submission (e.g., new or a modification to an existing protocol) to the IACUC may be required.
   e. The IACUC may ask for an increase in oversight.

The ultimate goal of the whistle-blower policy is to establish a mechanism for interested parties to report animal welfare concerns to the organization for investigation. Multiple institutions shared their methods of satisfying this goal. The following checklist outlines the recommendations of the meeting participants and satisfies the federal standards governing the care and use of animals.

Whistle-Blower Policy (Reporting Animal Welfare Concerns) Checklist

1. Does the policy include the following?
   a. From a public relations standpoint, it is beneficial to indicate that the institution is committed to the humane care and treatment of its animals.
   b. A process for reporting animal welfare concerns identifies:
      i. A responsible individual or position (e.g., attending veterinarian)
      ii. A phone number and/or e-mail address and/or
      iii. A mailing address
   c. Notice that reports can be made anonymously and the confidentiality of the complainant will be maintained
   d. A statement indicating that all reported concerns will be considered and, if warranted, investigated by the Animal Care and Use Committee (IACUC)
   e. A statement indicating that concerned parties making reports will not be discriminated against and will be protected from reprisals
2. Is the policy easily found by concerned individuals? It should be posted in prominent places. Examples include:
   a. Throughout the animal housing areas including barns
   b. In animal procedure rooms
   c. In teaching labs where animals are used
   d. The institution's website
   e. On occasion, published in departmental newsletters and student and local newspapers
3. Does the process include central points of contact that are empowered to prescreen reports and, when warranted, refer them to the IACUC for investigation?
   a. The central points of contact should be individuals in senior management (e.g., the director of research compliance, institutional official, IACUC chair, attending veterinarian) trained to recognize the circumstances that would require IACUC investigation.
   b. The process includes a means for informing the IACUC of reported issues that did not require committee investigation. For example, the IACUC administrator reports addressed concerns to the IACUC during monthly meetings.
4. Does the process include measures to protect the identity of the individual that filed the report?
   a. The name of the concerned party and his or her personal identifiers are redacted from the report prior to providing it to the IACUC for investigation.
   b. The details of the oral reports are summarized in writing by the central point of contact. The report does not include any information about the complainant when it is provided to the IACUC for investigation.
5. Were the report and resolution process appropriately documented?
   a. In almost all cases, report activities are documented in the meeting minutes. For example, when a report is resolved by the point of contact, the IACUC administrator summarizes the report and discusses the actions taken during a

full committee meeting. The relevant points of that discussion are documented in the meeting minutes.

b. The same is true for reports that are investigated by the IACUC. The issue is provided to the IACUC in the form of a formal report. The committee discusses the concern during a monthly meeting and takes the appropriate actions to resolve the issue. The relevant points about the investigation are discussed and captured in the meeting minutes.

c. Reports of the IACUC's findings are mailed to the complainant.

CHAPTER 17

Occupational Health and Safety Program

SCENARIO

At GEU, an animal care technician was bitten by a rat while performing routine cage change operations. The bite was severe and required medical attention. The technician was concerned that reporting this bite could result in counseling or retraining. He chose not to report the incident, but rather to remove his gloves, wash his hands, apply a bandage, don new gloves, and continue conducting his duties. What concerns do you have about this scenario?

An organization's occupational health and safety program (OHSP) should identify and mitigate risks inherent with animal care and use. A properly constituted and engaged OHSP will establish policies that facilitate a safe working environment and train personnel to follow them, encourage communication when incidents occur, and foster an institutional partnership for the good of the worker and the research. In this case, the worker did not feel part of the OHSP; he was confident that he would be blamed for the incident and not receive support; also, he was not adequately educated on proper cleansing and disinfecting procedures.

Regulatory References and Guidance Documents

- **Title 29 of the Code of Federal Regulations (CFR):** Federal law mandates that organizations provide a safe and healthy working environment for employees. A compilation of these laws is listed in the CFR, Title 29 (29 CFR). These regulations are enforced by the Occupational Safety and Health Administration (OSHA); they are traditionally known as the OSHA laws, which are the overarching rules that form the primary elements of health and safety programs.

  The OSHA regulations are applicable to all work environments. Accordingly, the 29 CFR can be used as a guidance document when developing the OHSP component of an animal care and use program (ACUP). OSHA regulations include, for example, information about emergency evacuation plans, noise exposure levels, fire protection programs, working with hazardous materials, and using personal protective equipment (PPE).

  Safety programs, developed in accordance with OSHA mandates, must take into consideration any workplace hazards defined in the OSHA laws. These laws do not place special emphasis on animal-related hazards. Federal directives governing the use of animals in research, teaching, and testing provide guidance on developing and operating an OHSP that specifically focuses on hazards intrinsic to working with animals. Consequently, the animal care and use OHSP is usually part of an organization's overall health and safety program.

- **Public Health Service (PHS) policy:** Must be followed by organizations receiving support from agencies within the Public Health Service. These agencies are the (1) National Institutes of Health (NIH), (2) Food and Drug Administration (FDA), (3) Substance Abuse and Mental Health Services Administration, (4) Agency for Healthcare Research and Quality, (5) Agency for Toxic Substances and Disease Registry, (6) Centers for Disease Control and Prevention (CDC), (7) Health Resources and Services Administration, and (8) Indian Health Service with only NIH, FDA, and CDC supporting animal activities.

  The PHS policy requires organizations to base their animal care and use programs on the *Guide for the Care and Use of Laboratory Animals* (henceforth referred to as the *Guide*). The *Guide* mandates that an OHSP must be part of the ACUP and refers to the National Research Council (NRC) publication *Occupational Health and Safety in the Care and Use of Animals* as a resource for establishing and maintaining a comprehensive OHSP.

- *Guide for the Care and Use of Agricultural Animals in Research and Teaching* **(henceforth referred to as the *Ag Guide*):** Although the PHS policy does not specifically reference the *Ag Guide*, organizations establishing programs that involve agricultural animals should also consider the *Ag Guide* standards. The *Ag Guide* mandates that ACUPs must include an OHSP and also refer to the 1997 NRC reference.

- **Animal Welfare Act regulations (AWAR):** are federal laws enforced by the US Department of Agriculture (USDA). They must be applied by any organization using USDA-covered animals for research, teaching, testing, experimentation, exhibition, or as a pet. USDA-covered animals include all warm-blooded animals with the exception of birds, rats of the genus *Rattus* (bred for research), mice of the genus *Mus* (bred for research), and horses not used for research purposes. In contrast, traditional agricultural animals (e.g., cattle, poultry, and swine) used for food and fiber or in research intended to improve the quality of food and fiber are not covered by the AWAR. The AWAR do not discuss and consequently do not require organizations to establish an OHSP. In theory, if an organization only used USDA-covered species and was not receiving support from a PHS agency, an OHSP specific to the ACUP would not be required.

- **Additional relevant guidance:** Protocols may involve hazards that pose unique risks to personnel and the environment. In certain circumstances, provisions described in additional federal regulations and policies will need to be implemented. Title 40 of the CFR will provide guidance on how to handle hazardous waste, while Title 10 of the CFR will discuss how to protect personnel from ionizing radiation. In addition, the NIH's *Recombinant DNA Guidelines* provide the requirements for personnel working with genetically modified animals, and the CDC's *Biosafety in Microbiological and Biomedical Laboratories* will provide guidance on protecting personnel from biological agents that will cause human disease.

THE OHSP AND ACCREDITATION

The Association for Assessment and Accreditation of Laboratory Animal Care (AAALAC) International accreditation certifies an organization's program when it meets or exceeds the regulations and policies governing the use of animals for research, teaching, or testing. During program assessments, AAALAC site visitors evaluate an organization's program. In the event that an organization is not required to adhere to the PHS policy and AWAR, AAALAC site visitors will base their assessment on the standards outlined in the *Guide* and/or the *Ag Guide*, depending upon the species utilized at the organization. AAALAC site visitors evaluate the effectiveness of an organization's OHSP.

The IACUC Administrator's Role

Since fiscally supporting and operating an OHSP is the responsibility of the organization, the IACUC administrator may contact the institutional official (IO) to discuss the federal requirements. Prior to the meeting, the IACUC administrator may want to prepare a list of personnel who are relevant to the development of the OHSP. The list should be institutionally specific, but at a minimum include representatives from Health Services, Environmental Health and Safety, the Animal Resource Program, Physical Plant, and IACUC Administration. In addition to the list of prospective program developers, the IACUC administrator should prepare a brief background that explains the requirements for and significance of the OHSP.

The initial discussion with the IO may cover topics such as program development, fiscal support, and long-term staff support, but ultimately the IACUC administrator's goal should be to ask the IO to consider supporting the development of the program by assigning relevant staff members to participate in its development.

Developing the OHSP

At minimum, the OHSP should include processes for analyzing potential hazards, assessing risks, and determining effective mitigation measures. It should include required training activities, which are based on identified risks. The training should teach personnel to recognize the risks and provide measures for protection. The organization should have a process to ensure that all individuals at risk of developing an animal-acquired injury or zoonotic disease have been given the opportunity to enroll in the OHSP. The requirement is for organizations to provide individuals the opportunity to enroll in the program. However, allowing individuals who are not enrolled in the program to participate in animal activities can be a significant liability for the organization. While the organization cannot force anyone to enroll in the OHSP, it is also not obligated to give unprotected personnel access to higher risk activities or risk-associated spaces such as animal use facilities and laboratories.

The OHSP should also include methods and processes employed to protect personnel from hazards inherent and specific to protocols. Finally, a complete program includes administrative support to thoroughly document activities related to the OHSP.

Hazard Identification and Analysis

Since hazards associated with animal care and use can compromise the health and well-being of individuals working in or around the animal use or housing areas, it is essential for organizations to identify and evaluate these risks. The concerns may differ among organizations, but the primary types of hazards that have been identified are those related to the environment, research, animals, and the individual's personal health disposition.

In certain situations, the hazards may not be apparent. Consequently, the *Guide* emphasizes the importance of selecting appropriately trained individuals to identify the programmatic hazards. For example, organizations have involved certified industrial hygienists, veterinarians, physicians, and IACUC administrative staff members in the hazard identification process.

Identifying Environmental Hazards

Environmental hazards are traditionally those associated with the ACUP physical plant. For example, equipment such as cage washers, sterilizers, compressed gas cylinders, silos, tractors, manure pits, and walk-in freezers pose unique risks to personnel. In addition, stored supplies blocking emergency fire exits and fish facilities without appropriate ground fault receptacles are environmental hazards that have the potential to harm personnel. The following two methods have been effectively employed by best practices (BP) meeting attendees to identify environmental hazards.

Common Practice 1

Before scheduling an environment assessment, prepare a list of facilities that will be visited during the audit. This list may be found in an organization's PHS assurance and, if applicable, the AAALAC accredited program description. For example, an accredited program description typically includes a table identifying the building name and total square feet of animal housing and support areas. In most cases, a similar supplement is part of an organization's OLAW-approved PHS assurance. For the purpose of this assessment, it may be necessary to expand the list of general facilities to include specific areas as the following two examples illustrate.

> **Example 1.** Some specific assessment areas at a dairy farm may include (1) research barns; (2) general housing areas and facilities; (3) surgery suites and procedure rooms; (4) feed processing, bagging, and food storage areas; (5) milking parlors; (6) equipment and hand tool storage areas; (7) personnel lockers and break and support rooms; and (8) manure and waste disposal areas.

Example 2. Specific assessment areas for a general laboratory animal vivarium may include (1) housing facilities; (2) cage wash (dirty and clean) areas; (3) research facilities involving hazardous materials (e.g., infectious agents, carcinogens, and radioactive isotopes); (4) animal care technicians' changing and shower facilities; (5) storage areas (e.g., walk-in freezers, animal feed, and equipment); and (6) receiving docks.

The assessment team in both cases—the laboratory animal vivarium and the dairy farm—should include professionals with specific expertise. In addition to the IACUC administrator who typically coordinates and documents the activity, the team must include an individual trained to identify workplace hazards (e.g., industrial hygienist or another environmental health and safety specialist). Since the facility manager knows the intricacies of the animal use and housing areas, he or she may be ideal to lead the facility tour. A number of organizations have also found it beneficial to strengthen their team by including, for example, physicians, veterinarians, and physical plant staff members to participate in the facility review.

Physicians from the Occupational Medicine (OM) office, for example, were able to identify the types of injuries routinely treated in OM. Consequently, these specific injuries and their causes were a focal point for the review committee. In addition, veterinarians focused on risks (e.g., zoonotic diseases, bites, and other physical injuries) specific to the animals. The physical plant staff members with expertise in electrical systems, equipment safety, and/or facility maintenance contributed their unique knowledge to the group. They were able to concentrate on electrical issues as well as issues such as the safe use of farm machinery or cage washers.

An essential component of developing a strong and effective OHSP is identifying and understanding the hazards associated with the organization's specific ACUP. Since the ultimate goal is to develop processes to protect and minimize personnel's exposure to risks, the OHSP will be built on the information gathered during the hazard identification process.

Ensure that adequate time is allotted to conduct a thorough assessment of each facility and that the review is conducted at an optimal time. In some instances, the assessment must be conducted at a specific time of the day for it to be useful. For example, a dairy facility's milking parlor should be evaluated during the milking process. The review committee should observe how, for example, the cattle are moved through the parlor, the milking equipment is used, the facility is cleaned and sanitized, and the milk is processed. Similarly, the cage wash area in a laboratory animal vivarium should be assessed when it is in use.

Since many organizations have multiple areas supporting the animal care and use program, conducting thorough hazard assessments may lead to a large amount of data to analyze and update. For these reasons, organizations have developed efficient data acquisition and organization methods for the findings. In addition to the creation of traditional data, photography, video recording, and audio recordings can be used for documentation.

Although many organizations follow similar assessment processes, unique findings among organizations have been documented. However, a majority of the identified risks appear to be consistent between organizations. For example, the types of hazards typically identified at agriculture animal facilities (e.g., cattle, horses, sheep, swine, poultry, deer, and goats) include those associated with the species of animal (e.g., zoonotic diseases, bites, scratches, and trauma injuries), enclosed spaces (e.g., silos and manure pits), farm equipment (e.g., tractors, power takeoffs, feed grinders, mixers, and livestock chutes), ergonomics (e.g., lifting feed and equipment), the physical environment (e.g., housekeeping issues that create trip hazards or block emergency exits, electrical hazards created by wet environments or electric fences, and fall hazards from barn storage lofts), and biological and chemical hazards (e.g., herbicides, pesticides, and inhalation of molds).

Laboratory animal facility hazards are also similar among organizations. For example, the types of hazards typically identified are associated with the species of animal (e.g., zoonotic diseases, bites, and scratches), equipment (e.g., walk-in freezers, cage washers, and autoclaves), ergonomics (e.g., moving compressed gas cylinders and repetitive tasks), the physical environment (e.g., housekeeping issues that create trip hazards or block emergency exits, electrical hazards created by wet environments, noise levels detrimental to hearing, eye hazards), and chemical hazards (e.g., cleaning and research reagents).

Once the hazard identification process is completed, the IACUC administrator typically collates the information into a report that identifies risks and mitigation strategies for protecting personnel from the hazards.

Common Practice 2

A second popular practice is for the hazard assessment to be conducted using the OSHA standards as guidance to minimize workplace risk. Since the OSHA standards do not differentiate between the overall workplace and the animal facilities, the hazard assessment is conducted throughout the entire establishment without uniquely identifying the ACUP areas.

The organization often delegates the responsibility of administering the OSHA standards to, for example, the director of Environmental Health and Safety (EHS), who frequently coordinates the hazard assessment process. The assessment team may be a group of subcommittees comprising health and safety professionals with expertise in, for example, biological, chemical, and fire safety; ergonomics; waste management; and industrial hygiene. Subcommittees usually conduct independent assessments and focus on hazards associated with their specific area of expertise. For example, the biological safety officer will assess only those areas of the organization where biological hazards are utilized. The overall assessment is complete once each subcommittee has completed its review.

An essential component of developing a strong and effective OHSP is for the assessment team to completely apply the OSHA regulations. The standards were specifically developed and should be fully implemented to ensure that every employee is provided a safer and healthier workplace. For example, the OSHA standards indicate that exposure to average sound levels of, or exceeding, eighty-five decibels over an 8-hour period will impair an individual's ability to hear. Consequently, an industrial hygienist would determine how long individuals are exposed to what noise level—for example, assessing the noise level in the cage washing area using a decibel meter. The data would then be analyzed, using the specific guidance available in the OSHA standards, to determine whether the sound levels in a specific work environment are hazardous.

Once the hazard identification process is completed, the documentation is usually collated and maintained by the EHS as part of the organization's comprehensive occupational health and safety program. The relevant reports are made available, and specific questions are addressed by health and safety personnel.

Ongoing Hazard Identification and Assessment

Although hazards inherent to the environment may not vary (e.g., farm- or vivarium-related hazards), the hazards related to research projects may be initially diverse and subject to change over time. Consequently, it is important to develop a program that includes an ongoing hazard identification component. For example, a new scientist may propose conducting research that involves a viral pathogen never before used at the organization. The program should include measures to identify and assess the new hazard. IACUC administrators attending past BP meetings identified two effective methods that are commonly employed during the new hazards identification process.

Common Practice 1

The first and perhaps one of the most common practices is to utilize the experience of the organization's Institutional Biosafety Committee (IBC). In some cases, principal investigators (PIs) are asked to indicate when they are using hazards directly on their animal care and use protocol. Once a PI identifies, a hazard he or she is informed that the project must be submitted to and reviewed and approved by the IBC before the research can begin. During the IBC review, the committee identifies and assesses the hazard and then employs—or informs PIs how to employ—best practices to protect the staff, the environment, and the organization from related hazards. The IACUC submission is approved by the committee only after the biosafety review has been conducted and the biosafety submission is approved. In addition, some organizations require animal rooms containing biohazards (i.e., bacterial, viral, or chemical) to be identified with signage that explains the nature of the hazard and how individuals can protect themselves.

Common Practice 2

The second process routinely practiced involves equipping the IACUC with staff members that can assess new hazards as part of the protocol review process. In a situation such as this one, an organization's committee roster may include, for example, a biological safety officer, an industrial hygienist, and/or a medical doctor. When a submission is provided for assessment, relevant committee members identify the risk as well as appropriate personal protective equipment and best practices to contain the reagent of concern and protect the organization's personnel from associated hazards. An additional advantage of having a safety professional serve as an IACUC member is that it allows the organization to use the person in a dual role. For example, the safety professional who is also an IACUC member can participate in facility inspections. In this case, the IACUC voting member who is also serving on the safety committee can inspect the animal facilities from an animal care and use prospective and at the same time conduct an environment risk assessment. The hazards can then be quantified, and the assessment can be documented in the institution's semiannual report to the IO.

Quantifying Identified Environmental Hazards

To determine the degree of risk that each hazard poses to personnel working in animal support areas, the hazards must be quantified. Those who specialize in safety and safety monitoring (e.g., certified industrial hygienists) have specific expertise in quantifying the degree of risk that a hazard poses to workers (i.e., animal users).

However, to provide some guidance on quantifying hazards, BP meeting attendees, including attending safety specialists, offered advice on risk identification and quantification. For example, when considering the risks associated with biological and chemical hazards, OHSP developers should reference available documents. One example is a chemical material safety data sheet (MSDS), which clearly identifies the potential risks and methods (i.e., personal protective equipment and training) that individuals can use to mitigate the risks. Assessors should consider that infectious biologics will cause illness, and that mitigation processes such as the use of biological containment cabinets, lab coats, a mask and gloves can minimize the risks.

In addition, OSHA regulations quantify some risks. For example, they identify sound levels that now require hearing protection and discuss when safety glasses and steel-toe shoes should be used to protect workers from eye or foot injuries. The OSHA regulations also discuss confined spaces (e.g., silos and manure pits) and provide requirements to ensure safe conditions.

PERSONAL DISPOSITIONS (PERSONAL RISK ASSESSMENTS)

Animal handlers and users may experience inherent risks due to their personal health profile. For this reason, the organization's program must include a component to evaluate, quantify, and, when necessary, mitigate any risks that an individual may experience.

The most common practice identified by BP meeting attendees involved a process occurring between the animal user and his or her doctor. Since an individual's personal health disposition is confidential between the patient and doctor, the process of conducting personal risk assessments should be coordinated through the organization's office of OM or a contracted medical professional.

The assessment is typically initiated when a participating individual completes a health history questionnaire. This document includes inquiries about an individual's personal health and contains questions such as

Do you suffer from allergies?
Are you immunocompromised?
Do you have high blood pressure or cardiac-related problems?
Is your tetanus vaccination current?

Once the questionnaire has been completed, it is assessed by medical staff and may necessitate a meeting with the physician. In the event that the physician identifies health concerns that elevate the individual's risks associated with animal research, he or she may not clear the individual to work with animals. Alternatively, the physician may require an individual with known allergies to wear a respirator when working with or around animals.

Some organizations required their OHSP physician to clear individuals before allowing them access to animal facilities. The clearance process was frequently associated with the protocol review process. In other words, a protocol submitted to the IACUC for review was not approved by the IACUC until all those listed as personnel on the study received medical clearance from their physicians to work with animals. In some cases, if medical clearance could not be obtained, individuals were not permitted to work directly with animals and may have been reassigned duties that did not require animal exposure, such as in vitro work.

During the health assessment process, a physician may consider not only an individual's personal health status, but also risks inherent to an individual's life stage. To clarify, a pregnant worker may have elevated risks as compared to a male college student. An immunocompromised person may have elevated risk compared to a veterinarian who receives regular prophylactic vaccinations.

Consider a scenario that involves individuals working in an old, drafty barn that utilizes pregnant sheep for fetus development studies. For the purpose of the example, let us imagine that the PI's submission to the IACUC includes a diverse list of research personnel. The PI and co-PI have extensive experience working with sheep; in fact, they grew up raising sheep. However, the first person named on the protocol personnel list is a recent graduate and a newly wed female interested in becoming a first-time mother and the second, although healthy, received a donor kidney during his childhood and is taking tissue antirejection drugs. A third is a freshmen with absolutely no sheep-handling experience that has no known adverse medical conditions, and the fourth is a trained shepherd who recently recovered from heart-related problems that led to permanent heart valve damage.

In this example, the risk of Q-fever is inherent for all since they would be working with pregnant sheep. In addition, the personal health status of each individual involved in research on the PI's application must be considered. Upon assessment, the physician may choose to clear only the freshman, a healthy, immune-competent male, to work with the sheep, even though he has absolutely no experience working with the animals. The risks of the other individuals may be above the organization's acceptable threshold. Specifically, a Q-fever infection for the immunocompromised male and the individual with heart valve problems could result in their deaths; a Q-fever infection in a pregnant woman could result in devastating problems for her unborn child.

In the event that all of these individuals are cleared to work with sheep, each may be required to take different protective measures. But, again, some may simply not qualify for work in a given area based upon their own personal risk profiles.

Risk Protection and Mitigation

The program should include training that informs staff members of the identified risks and how they may protect themselves. In some cases, training is conducted by AVs and OM staff. BP meeting attendees identified a number of successful training methodologies.

Organizations across the country distribute training brochures to relevant personnel (i.e., maintenance staff and those having incidental contact with animals), while others effectively use web-based training on PPE and zoonotic diseases. In certain situations, specific, personalized individual training sessions (e.g., training for primate users) are developed for animal users.

Program Participation

As previously indicated, the OHSP must be available to everyone in contact with the facilities (e.g., animal housing, husbandry support, storage, and procedure areas) that are part of the ACUP. Consequently, the program must cover all researchers, technicians, volunteers, and students conducting or supporting animal care and use activities. It must also include faculty, staff, students, and teaching assistants that utilize animals in course-related activities. Additionally, the program should cover anyone else who has only incidental contact with the organization's animals or animal facilities (e.g., housekeeping, physical plant staff, and program support staff). The costs associated with establishing and operating the OHSP must be assumed by the organization (i.e., those required to participate in the program cannot be charged for the service).

What Constitutes Participation in the Occupational Health and Safety Program?

The degree of participation in the program may depend on the individual's personal risks or the degree of his or her animal exposure. Some BP meeting attendees suggested that program involvement could be based on an individual's risk of experiencing health problems as a result of contact with animals or animal tissue. In other cases, IACUC administrators agreed that airborne disease may have an impact on individual OHSP participation. Therefore, BP meeting attendees agreed that 100 percent participation should occur, but the level of participation may vary based on individual circumstances.

Workers performing janitorial duties in an animal facility would experience minimal risk (generally, from fomite transmission or airborne transmission—e.g., allergens); therefore, their program involvement may only be receiving a training brochure during new employee orientation. In a situation such as this, some organizations determine the risk level associated with the position and treat everyone filling that position equally. However, during the initial point of contact (i.e., when the training brochures are provided), everyone who will be exposed to animals is informed of unique circumstances (i.e., allergy issues).

A second instance relates to students who utilize animals in the course of their studies. Students involved in classes that utilize animals for instruction may be at risk only from certain species (e.g., rabbit fur, rodent allergens) or certain products (e.g., formalin); therefore, they might only receive OHSP-specific training from the instructor of the class.

Those having more and direct contact with animals may perhaps be required to complete all of the required training and acquire medical clearance prior to working with animals. In other words, individuals listed on IACUC research protocols may need to complete OHSP online training and hands-on species-specific training, as well as complete a personal risk assessment form prior to meeting with the OM physician.

Organizations may identify hazards of particular concern and require individuals to complete species-specific and specialized training. For instance, individuals working with nonhuman primates may be required to complete herpes B virus training and undergo a thorough medical evaluation.

An alternative to degrees of OHSP enrollment has been practiced by others and involved requiring everyone in contact with animals (including incidental contact—researchers, custodial staff, and IACUC members) to be fully enrolled in the program. They identified full enrollment as completing a personal risk assessment, medical evaluation and clearance from a physician, and being involved in extensive training sessions.

CHAPTER 18

Emergency Disaster Plans

SCENARIO

A major research institute in southern Texas, Tri Bio, was preparing to contend with a category 2 hurricane. A number of mice that were a critical component of research activities supported through millions of National Institutes of Health (NIH) research dollars were housed in the vivarium basement. The tower vivarium facility was used to maintain foundation stocks of these extremely valuable strains of transgenic mice, which were housed nowhere else in the world. Without these mice, multiple research activities could not be conducted and millions of dollars would need to be returned to the granting agency.

Since Tri Bio had developed and instituted a comprehensive emergency disaster plan (EDP) and had successfully managed the ramifications of category 2 hurricanes in the past, the institute was confident that its million dollar resources—the transgenic mice—were secure. Tri Bio's EDP was extensive and included, emergency communication procedures, provisions for getting animal care staff to the facility, acquiring backup generators in the event of a power outage, and relocating research animals.

The storm intensified to a category 4 storm as it approached the coast and the impact on Tri Bio and the surrounding area was catastrophic. Power outages covered a fifty-mile radius around the Tri Bio laboratory. All of the buildings in the area had significant wind and water damage; very few windows were intact and at least four inches of water flooded the basements of buildings in the region impacted by the hurricane. The devastation resulted in the governor of Texas placing the area under a state of emergency. Tri Bio executed and initiated its EDP to the fullest extent.

In brief, the communication plan included provisions for all key staff members to communicate using a Tri Bio-issued cell phone. The appropriate connections were made and relevant individuals were situated at each animal facility. As part of the EDP, Tri Bio secured a contract with Unlimited Rent All, a company that guaranteed them first access to two high-voltage generators. Gina, the Tri Bio animal facility manager and EDP coordinator, stopped to pick up the generators on her way to the vivarium. The Unlimited Rent All manager explained that, due to the state of emergency, the generators were not available because the state had secured them for use at temporary shelters. Gina immediately phoned her emergency responder team and informed them that the generators were not available and that they should arrange to initiate phase 2 of the EDP, which involved relocating the animals.

Upon arrival at the vivarium, Gina found that four members of her five-member team were in the basement of the tower building checking on the status of the animals. She tried to contact her team without success and later learned that their cell phones had no reception in the basement or tower. Consequently, she was unable to provide instructions to her team members that were physically on the scene. To make matters worse, her team leader, Alan, informed her that the animals could not be relocated to the temporary housing facility because the state had converted it into an emergency shelter for displaced families. At this point Gina realized that Tri Bio's most valuable asset—the transgenic mice—could be lost, which could result in a financial catastrophe for the company. Ultimately, phase 3 of the EDP, which included euthanizing the animals, had to be initiated. This loss cost Tri Bio millions of dollars: Granted funds had to be returned to the NIH, revenue was lost due to contracts with biomedical companies being cancelled, and the transgenic strains were lost.

Fortunately, Tri Bio had taken a final precaution that ultimately saved it from bankruptcy. The company had stored frozen embryos of almost all of the strains of its precious transgenic mice at a distant location. Tri Bio recovered the transgenic strains and, although it did sustain some loss of income, the company was able to survive. However, Tri Bio did learn the importance of trying to consider all of the ramifications that could have an effect on a carefully developed EDP.

Emergency situations can have a tremendous impact on an organization's animal care and use program (ACUP). Ensuring the health and safety of the staff, as well as that of the animals, must be the primary goal of each EDP. The plan should consider complications associated with natural disasters such as tornados, hurricanes, blizzards, floods, fires, and electrical storms. In addition, most organizations should plan to address other concerns such as acts of violence, vandalism, and arson. Plans should be as inclusive as possible and cover all feasible situations that an organization could possibly imagine or experience, including what to do if the first disaster plan does not work. Some institutions prepare 1 or 2 back-up actions in case the first action is unsuccessful (often the last back-up action is to euthanize the animals).

Regulatory Requirements and Guidance Documents

- *Guide for the Care and Use of Laboratory Animals* (**henceforth referred to as the *Guide*) (p. 46)** states that organizations should integrate an emergency disaster plan into their overall safety program. The plan should include primary components that protect the safety of both personnel and animals in the event of an unforeseen disaster. Risks to animals and animal users can be associated with disasters resulting from criminal acts or from natural occurrences. In most cases, IACUC administrators play an active role in developing and maintaining an organization's EDP.
- **Office of Laboratory Animal Welfare (OLAW) frequently asked questions (FAQ #3, Section G, Institutional Responsibilities):** OLAW has established a FAQ that summarizes the primary objectives and components of an organization's EDP, summarizes the issues that each organization should consider when developing its EDP, and provides regulatory citations.

DEVELOPING AN EMERGENCY DISASTER PLAN

During past IACUC administrators' best practices (BP) meetings, attendees indicated that before an organization establishes an EDP, it should establish a team of first responders that will help to develop the nuts and bolts of the program. Those in attendance agreed that staff from veterinary services, animal facility management, the animal diagnostic lab, the IACUC, the Institutional Biosafety Committee (IBC), Environmental Health and Safety (EHS), Occupational Medicine, Police Services, Public Relations, Physical Plant, and Regulatory Compliance—at a minimum—should comprise the team committee of experts developing the EDP. Since research compliance often coordinates the activities relating to the ACUP, the director of that area or sometimes the IACUC administrator generally chairs this committee of first responders.

The IACUC administrators at past BP meetings agreed that representatives from each area bring their own unique skill set to the group, which ultimately helps to ensure that the EDP can address a wide array of disasters. For example, representatives from EHS, veterinary staff, and Occupational Medicine staff can help to develop a plan that would address outbreaks of zoonotic and enzootic disease. Police Services and Public Relations staff can develop methods to address criminal activities such as vandalism. In addition, physical plant staff could assist in developing plans to address power outages, heating and air conditioning failures, and other issues relating to building integrity.

The emergency disaster communication plan should also include notifying OLAW, the Association for Assessment and Accreditation of Laboratory Animal Care (AAALAC) International and the US Department of Agriculture if a disaster occurs. These organizations are often able to provide assistance and advice.

Identifying Potential Disasters

Since there are a variety of things that could potentially affect an animal care and use program, institutional representatives should develop guidance criteria to assist them with the development of the disaster plan. For example, BP meeting attendees suggested using the following definition when considering the components of EDP: *Emergency disaster plans should cover any adverse event that could have a negative impact on the health and/or safety of animals and personnel.*

BP meeting attendees indicated that once the appropriate group of institutional representatives has been established, the first task should be to conduct a risk assessment. The process for risk assessment is to first conduct a brainstorming session to identify the potential adverse events (e.g., natural disasters, power outages, equipment malfunctions, or events as a result of a criminal act) that could negatively impact the ACUP. Many BP meeting attendees indicated that it is helpful to ask the following questions when conducting the brainstorming exercise. Probably the simplest, but the most important question to ask when trying to identify risks is "What if?" For example, consider the following questions:

What if the power in our animal facility was out for 1 day, 5 days, or 10 days?

What if our colony of macaques acquired an active B-virus infection?

What if the military assumed control of our emergency animal housing facilities as part of a hurricane relief effort?

The IACUC administrators discussing the issues agreed that one of the biggest deficiencies in an EDP is that organizations forget to consider concerns that are unlikely to happen. For example, questions such as the following should be considered:

What if an organization's research records are lost?

What if standard methods of communication are not functioning?

What if funding is unavailable for a prolonged period of time?

Another point that organizations should consider is the level of security of the animal facilities. For example, points to consider should (or could) be the ease in which a group of individuals could break into and vandalize a building. The unit could determine if adequate alarms are in place to notify authorities if unapproved access to secure buildings should occur.

Organization representatives should also consider whether there is the potential for first responders to acquire a zoonotic disease; if this is possible, could measures of control be developed? The organization should also develop methodologies to prevent zoonotic disease infection of other animal resources. In all cases, the emergency preparedness team should consider all options to identify as many potential negative incidents as an ACUP could experience.

THE EDP PROGRAM STRUCTURE

BP meeting IACUC administrators identified certain key components of an effective EDP: For a program to be functional, an effective team of response members must be strategically identified and trained; once these members are identified, the second most important component of the program is an effective communication plan.

BP meeting attendees also decided that once the key individuals have been identified and the communication plan has been discussed and established by the group, the next step is to develop a written plan. This plan should include a process to test the effectiveness of the disaster plan.

Communication Plan

BP meeting attendees agreed that identifying an initial point of contact is a critical component of the communication plan. Since the EDP is associated with animal care and use, those in attendance agreed that the director of animal resources (e.g., attending veterinarian [AV]) is often the first to become aware of an incident impacting the ACUP. Consequently, the AV or animal resources director contacts the chair of the first-responder team, who initiates the relevant components of the EDP.

The first-responder team chair determines the need for a meeting or may specifically identify which team members should be contacted and deployed. For example, if the incident involves a potential zoonotic infection of a staff member, the chair may contact the occupational medicine physician, EHS staff members, and staff veterinarians. The chairperson may also identify individuals to disseminate information to pertinent staff members and decide when to contact Public Relations and ask that office to prepare a press release. The chairperson should inform local emergency responders (e.g., the local fire company or hazardous material team) if there is a threat to the community relating to a biohazardous material.

The Written Plan

Once the first responders have been identified, developed a communication plan, and identified the potential issues that could have a negative effect on each component of the ACUP, a written plan should be developed.

BP meeting attendees agreed that the written communication plan should not only identify methodologies for dealing with disasters but also be organized in a manner that specifically describes the area of interest. For example, organizations should consider including a brief description of the facility, a list of key personnel and their contact information, a description of the facility resources, a description of the security system, and the potential threats, with measures to address any identified risks. With regard to the animals, in the event that provisions cannot be made to minimize the welfare threats to the animals, the written plan should also include provisions for relocating and/or euthanizing the animals. The written plan for a transgenic rodent facility may be organized as follows:

Facility: Transgenic mouse facility, tower building

Facility description: The tower building is the primary housing location for the organization's transgenic mouse colonies. The animals are maintained in a barrier facility and entry into that facility should be according to established procedures as posted on entry doors and listed in the SOP on the institution's server. The mice are a critical component of over 250 research projects and are a vital asset of the company.

Key personnel:
L. Johnson   Phone: 123-456-7890
A. Anderson   Phone: 098-765-4321

Facility resources: The following equipment is vital to the operation of the tower building. The cage washer and related equipment, autoclaves, ethylene oxide sterilizer, ventilated housing racks, and HVAC systems are all pieces of critical equipment required to ensure that the animals can receive proper care.

Facility security system: The tower facility is locked down between 5:00 p.m. and 7:00 a.m. Although the building is constantly monitored using a sophisticated video security system, during this time period silent alarms are employed. This system automatically notifies police services when an unauthorized entry has occurred. In addition, the animal rooms are secured using a key card access system that docu-   \ ments who entered the facility and at what time.

Potential disasters (weather related):

- **Loss of power (e.g., storm related):** The loss of power can have a detrimental effect on the operation of the facility. The primary concern when there is a loss of power is the operation's heating and ventilation system. Emergency generators will automatically engage 2 minutes after power has been lost. The generators operate on diesel fuel and an adequate supply of fuel for over 2 weeks of continued operation is held in reserve.

- **Staff members are unable to travel to the vivarium (e.g., hazardous road conditions from a blizzard or flood):** The tower facility must be visited daily to check on and feed the animals. All of the animals in the facility can be checked on and cared for by two individuals in approximately an 8-hour period. The facility is equipped with overnight accommodations for at least two individuals. In the event that weather conditions suggest the possibility that caretakers may not be able to get to the building in the morning, at least two assigned staff will spend the night at the facility.

- **Building damage due to wind-related problems (e.g., hurricanes and tornados):** The tower facility is constructed with concrete blocks and bricks. Consequently, it is unlikely that wind will damage the facility. However, in the event that high winds damage the building (e.g., broken windows on upper floors), physical plant personnel will be notified by computerized monitoring equipment and dispatched immediately to access the damage.

- **Deliveries of feed or other needed supplies are stopped due to hazardous road conditions (e.g., blizzards, floods):** The feed is stored in room 2 on the bottom floor of the tower. Food supplies are delivered weekly and the inventory of food maintained in this facility would sustain the animals for 2 months. In the event that food supplies are short, facility managers would discuss decreasing the number of animals being maintained or use alternate sources of nonstandard feeds.

In addition to weather-related hazards, some organization attendees indicated that organizations should also consider covering potential criminal acts, animal rights protests, equipment failures, enzootic and zoonotic outbreaks, and problems with conventional communication methods, research records, and accessing financial resources.

Animal Disposition

Relocation

An example of how the relocation procedure may be worded is as follows:

The tower II building is approximately twenty-five miles north of tower I and is a conventional vivarium that can be used to house any animals in the event they need to be moved as a result of an emergency. If animals need to be relocated, the plastic mouse cages will be moved in the environmentally controlled transport truck. Prior to moving the animals into the truck, its environment will be adjusted and the temperatures stabilized to approximately 70°F–75°F. Once the animals arrive at the facility, they are moved into the animal rooms and cared for according to standard protocols.

Euthanasia

In the event that all methods for ensuring the health and welfare of the animals have been exhausted, the animals will be euthanized. The mice will be euthanized by either veterinarians or animal care technicians that have been appropriately trained to conduct the procedure. The method of euthanasia used will be CO_2 narcosis followed by an appropriate secondary method (e.g., bilateral thoracotomy, exsanguination, etc.).

Evaluating and Testing Emergency Disaster Plans

BP meeting attendees agreed that the team of EDP developers should run mock exercises to evaluate the effectiveness of the program and meet at least annually to assess the program. BP meeting attendees reported that the IACUC administrator may serve as a central coordinator for the first responders. The administrator provides relevant information to the team of responders such as information on activist activities, advisories relative to potential storms, and concerns relating to potential pandemics (e.g., H1N1 and bird flu), and recent lessons learned from disaster events at other institutions.

At some organizations, the IACUC administrator also arranges for a mock exercise; this helps to evaluate the program effectiveness. For example, the IACUC administrator initiates the scenario by announcing a lightening strike has interrupted power to the tower building. In accordance with the communications plan, the Disaster Committee chair is notified by the Office of Physical Plant facilities director. The Disaster Committee chair immediately contacts the Animal Resource Program director, who confirms that the emergency generator has been started and the housing facilities have power. The Disaster Committee chair then decides to contact the physical plant representative, who reports that the generator is functioning appropriately and that there is sufficient fuel in reserve to run the facility for at least a week. In addition to physical plant personnel and the director of animal resources, the Disaster Committee chair decides to contact Police Services and inquire whether the power outage had any negative effects on the building's security system. Once Police Services indicates that the facility is secured, the first-responder chairman asks the IACUC administrator to communicate the relevant information to the entire group of first responders that participated in the development of the EDP.

In addition to mock exercises, the team of first responders should meet at least annually to evaluate the program. The IACUC administrator typically schedules a meeting and asks all interested parties to read through the EDP prior to the meeting. Team members are asked to report by e-mail any concerns with the exercise to the IACUC administrator, who forwards a collated version of the issues to all team members. Once the committee meets, its members evaluate the successes and failures of any mock exercises, assess the comments made by the team of responders, and make any necessary changes to the program. In addition, the team also evaluates any concerns that were made by the IACUC during semiannual program reviews.

Continuing Program Business during a Disaster

Some disasters do not disrupt buildings or supply lines but may be just as traumatic. For example, if there were a flu outbreak and 40 percent of the campus population were at home unable or unwilling to come to work, the ability of the IACUC to perform its required oversight tasks could be severely tested.

BP meeting attendees suggested that such a scenario involving mass illness or personnel restrictions was much more likely than a natural disaster and could have as significant an adverse outcome. Many organizations had arranged for telecommittee meetings by establishing an account with a secure web hosting service so that, if the need arose, the IACUC could have a meeting using the Internet or telephone access. Some attendees noted that their organizations had established an account with a secure data delivery service so that protocols, minutes, or other documents could be shared with IACUC members or other program leaders as appropriate. These services are often necessary since most institutional e-mail accounts limit attachment sizes to 5 or 10 MB and frequently an IACUC package or several consolidated documents will exceed 40 or 50 MB in size.

To assure effective utilization during a disaster, many organizations rely upon these services one or more times a year to develop familiarity with and confidence in their ability to use alternate methodologies for conducting business during epidemics or other disasters that impede the ability of individuals to come to campus.

Additionally, several organizations have established policies that authorize a single member of the animal program (e.g., veterinary technician) to perform an inspection on behalf of the IACUC during a disaster event. Organizations should also have backup veterinary support in the event that their own veterinarian may not be available during an emergency. Organizations having a single veterinarian, or maybe two veterinarians, should consider identifying a nearby veterinarian who would be willing to serve in an emergency capacity were a disaster to occur and the primary veterinarian(s) be unavailable; this may not be a concern for facilities having several veterinarians on staff. Once identified, the backup veterinarian should receive the occupational health and safety program training and be familiarized with the campus program in the same manner as the primary veterinary professionals. The backup veterinarian may be expected to participate as an IACUC member, perform one facility audit each year (for continued familiarization), or meet with the AV on an infrequent basis to review program changes and enhancements. Such policies and practices regarding unusual methods of assessment or emergency methods of animal care are acceptable but should be determined beforehand and approved by the IACUC for use in disaster conditions.

CHAPTER 19

The Role of a Primary Grantee

SCENARIO

GEU's animal care and use program is accredited by the Association for Assessment and Accreditation of Laboratory Animal Care (AAALAC) international. In addition, GEU maintains an animal welfare assurance that was negotiated with and approved by the Office of Laboratory Animal Welfare (OLAW). Dr. Dale Heaton, a GEU molecular biologist, was funded by the National Institutes of Health (NIH) to develop a novel protein to be used in diabetes therapy. Although his research involves in vivo experiments, Dr. Heaton will conduct only the in vitro research at GEU. Dr. Heaton's protein will be evaluated in rabbits by his colleagues at Meyer's Biologics.

The rabbits are a unique strain used in diabetes research, and they were developed and are owned by GEU. Meyer's Biologics is a US Department of Agriculture (USDA)-registered research facility but is not accredited by AAALAC International. It does not currently have Public Health Service (PHS) funding and consequently is not PHS assured.

What actions must take place before this collaboration between GEU and Meyer's Biologics can occur? This particular scenario identifies noteworthy complications that must be addressed before collaboration can occur. Since only GEU's program is AAALAC accredited, the rules of accreditation must be considered. The complications grow since a PHS-assured organization will be partnering with a nonassured organization. Let us consider the regulatory and accreditation components of this scenario and determine how the two organizations can collaborate on this important project.

AAALAC ACCREDITATION

Since GEU retains ownership of the animals, under the rules of accreditation, it is responsible for the care of the rabbits regardless of housing location. There are only a few specific solutions for a scenario in which animals are owned by an accredited organization and housed at a nonaccredited organization.

AAALAC Option 1

GEU could extend its AAALAC accreditation coverage to those components of the Meyer's Biologics program where the animals will be housed and animal activities will be conducted. In other words, GEU would describe Meyer's Biologics facilities in its written program description. Such an arrangement may include consultation with legal counsel and the development of a clear and defined memorandum of understanding (MOU). Meyer's Biologics may require IACUC approval for the work being performed, but as the accredited entity, GEU's IACUC will oversee the project and require IACUC approval of the activities.

AAALAC Option 2

Meyer's Biologics can have its program for animal care and use accredited by AAALAC International. In this scenario, since GEU has the expertise and skills, it could effectively guide Meyer's through the accreditation process. This is not generally a rapid process, so both organizations should allocate time to complete the accreditation process. Meyer's Biologics will need to prepare a written program description using the AAALAC template, submit it to AAALAC for review, and schedule time for AAALAC representatives to evaluate the program. Once the program description is submitted to AAALAC, the evaluation visit is typically scheduled 1 to 4 months later. Consequently, depending on the time Meyer's Biologics needs to prepare a written program description, the company should be prepared for a 6- to 9-month delay before the collaboration can be initiated.

PHS ASSURANCE RAMIFICATIONS

Funding originating from PHS agencies (e.g., NIH) cannot be used for animal care or animal use at Meyer's Biologics until it is covered by an OLAW-approved animal welfare assurance.

Since GEU would like to collaborate with Meyer's Biologics, a nonassured organization, GEU should contact its OLAW assurance officer and explain the circumstances. In this scenario, OLAW will help Meyer's Biologics establish an assurance or include the company as a covered component under GEU's assurance. As a covered component, GEU's assurance would be extended to cover Meyer's Biologics.

OLAW and GEU will revise the assurance to include Meyer's Biologics as a performance site. This process involves GEU formally submitting a request to OLAW to add Meyer's Biologics as a performance site on the GEU assurance. Once Meyer's Biologics becomes an approved GEU performance site, as the prime grantee GEU can use PHS funds to support animal activities at Meyer's Biologics. However, GEU is responsible for ensuring that the PHS policy and the terms and conditions of the grant for PHS-funded animal activities conducted at Meyer's Biologics are satisfied. For example, GEU must ensure that PHS-supported animal activities receive IACUC approval and oversight. In addition, GEU is required by

PHS policy to notify OLAW if any reportable incidents (e.g., significant noncompliance) occur.

Since Meyer's Biologics will be conducting PHS-supported activities as a performance site under GEU's assurance, OLAW would expect GEU's IACUC to oversee the covered activities occurring at Meyer's Biologics. In some cases, GEU may choose to develop an MOU that defers the oversight of animal activities to the Meyer's Biologics IACUC. Although the MOU establishes provisions for IACUC oversight, the prime grantee (i.e., GEU) is still ultimately responsible for ensuring that PHS funds are utilized according to the provisions of the PHS policy, the terms and conditions of the grant, and any other federal policies and regulations.

Animal Welfare Assurance

The PHS financially supports research conducted at organizations by awarding grants. Scientists interested in competing for these awards apply directly to PHS agencies (e.g., NIH, the Food and Drug Administration [FDA], and the Centers for Disease Control and Prevention [CDC]). The recipient of an award is known as the prime grantee.

Once an award has been granted, the prime grantee must agree to satisfy the terms and conditions associated with receiving the award. In addition, before the funds are distributed, the prime grantee must meet the requirements of the PHS policy when the research includes vertebrate animal activities. The PHS policy requires prime grantees conducting animal activities to be covered by an animal welfare assurance.

The assurance is a written document that fully describes the details of an organization's program of animal care and use. It describes, for example, how the prime grantee's IACUC will approve and oversee animal activities. It also describes the organization's training and occupational health and safety programs.

Only organizations receiving PHS agency awards are eligible to establish and maintain an assurance. Scientists from nonassured organizations that are prime grantees are notified by their grant manager, the Office of Extramural Research, that they must establish an assurance to receive the award. Organizations secure an assurance by negotiating their own, establishing an interinstitutional assurance with an assured organization where their PHS-funded animal activity will occur, or by being covered under another organization's assurance.

SECURING ANIMAL WELFARE ASSURANCE COVERAGE

Negotiating an Assurance

Once a nonassured prime grantee is notified by the Office of Extramural Research that it must establish an assurance, the organization starts the process of negotiating its assurance with OLAW.

To initiate the process, the organization prepares a detailed written description of its animal care and use program. Once prepared, the written program description is the organization's assurance. OLAW has made available through its website a sample assurance template for organizations' convenience. The assurance may be collectively prepared, for example, by the IACUC administrator, IACUC chair, and attending veterinarian (AV). While the PHS policy requires the assurance to cover animal activities only funded by PHS agencies, some organizations broaden the coverage of the assurance to include all animal activities conducted at the organization.

Before deciding to cover all animal activities under the organization's PHS assurance, the institutional official (IO) should consider the advantages and the disadvantages. Best practices (BP) meeting attendees agreed that organizations are ethically responsible for treating all research animals the same, irrespective of the funding source. In other words, what is good for animals on PHS-funded projects should also be good for animals on non-PHS-funded projects. This also avoids any appearance that one group of animals (NIH funded) is treated better than another group of animals (not NIH funded). In addition, some organizations chose to cover all animal activities under their assurance, irrespective of the funding source, because it is easier to apply a single set of standards to all animal activities. Many organizations believed that this is a more efficient arrangement and better facilitates committee obligations and activities. This philosophy is still strongly rooted in the foundations of many animal care and use programs, and consequently many organizations still cover all of their animal use activities under their PHS assurance.

IACUC administrators at past BP meetings noted that some organizations develop an assurance that covers only PHS funded activities—especially when the organization's program is accredited by AAALAC International. In addition, BP meeting attendees also agreed that whether or not their organizations' assurance covered only PHS funded projects, there was no effect on how business was conducted. In other words, animal welfare incidents or noncompliant activities that occurred under an organization's program were given the same level of scrutiny and oversight regardless of the funding source. The primary difference is that OLAW-defined reportable events and noncompliance associated with specific non federally funded activities are not required to be reported to OLAW unless they are functional or programmatic in nature or have an impact on the PHS-funded animals.

This concept is important since reports made to the federal government (e.g., OLAW) are attainable by anyone under the federal Freedom of Information Act (FIOA).

To further expand on the attainability of a document provided to the federal government, consider the following scenario: If significant noncompliance or another reportable event occurs, OLAW requires prompt reporting. This report can be by phone or e-mail. "The IACUC, through the IO, shall promptly provide OLAW with a full explanation of the circumstances and actions taken with respect to noncompliance, serious deviation and suspensions." Consequently, a final written report summarizing the concerns is prepared and submitted to OLAW through the IO.

The report should include the organization's animal welfare assurance number, the associated grant or contract number, a description of the potential effect the incident may have on a PHS-supported activity or program, a full explanation of the incident (i.e., where and when it happened, the species involved, the number of animals involved, the immediate corrective actions taken by the organization, changes to the program to prevent recurrence, and the long-term corrective plan). The detailed procedures for preparing the report may be reviewed on the OLAW website in the document "OLAW Guidance on Prompt Reporting, NIH Guide Notice NOT-OD-05-034." Once the report is prepared, it is filed with OLAW and becomes a document maintained by a federal agency, which can be accessed through FOIA.

BP meeting attendees noted that once the document is on file with OLAW, individuals can request it through FOIA. To acquire copies of reports, an individual must prepare and submit a formal request to a government agency. FOIA access to documentation associated with federal funding leads organizations to cover only PHS-funded activities under their assurance. Following this practice ensures that only reports associated with federally supported activities are filed with government agencies.

Once the organization decides whether the assurance will cover only the PHS-supported activities or all activities, the IACUC administrator or organization designee (e.g., IACUC chair, AV, IO) completes the assurance template (provided on the OLAW website for the convenience of the organization) and submits it to OLAW. The assurance template provided by OLAW is used to gather details associated with the organization's animal care and use program. For example, section III D of the sample document asks the organization to describe the activities of the IACUC. In many cases, organizations have written policies and standard operating procedures (SOPs) that describe their practices. They may have a SOP on conducting the semiannual program review. It may provide the methodology of identifying subcommittees, subcommittees' meetings to discuss their assigned section, subcommittees' reports to the committee during a full IACUC meeting, and approval of subcommittees' assigned sections of the program. Consequently, the IACUC administrator or designee may find it helpful to copy information directly from the organization's SOPs and best practices documents directly into the assurance.

Once the IACUC administrator adds all of the relevant information to the assurance, he or she will list the organization as being either a category 1 or 1 organization. Category 1 organizations are those with their entire program accredited by AAALAC and category 2 organizations are either not or only partially accredited. Organizations with programs that are not fully accredited are required to provide a copy of their most recent semiannual report and program review as an appendix to their assurance; accredited organizations do not have this additional requirement. Once the document is complete, the organization submits the assurance to OLAW for review.

To ensure that the organization's program adheres to the requirements of the PHS policy, the assurance (i.e., written program description) is assigned to a senior OLAW assurance officer for review and approval. The review and approval process, identified as "negotiating the assurance," is a collaborative effort between the organization and the assurance officer. Negotiating may involve verbal or written communications between the organization and the assurance officer to clarify ambiguous points described in the assurance. For example, an assurance officer may request a more thorough explanation of the organization's designated member review process. Once the organization and the assurance officer agree that the assurance accurately describes the organization's program and complies with the PHS policy, it is approved by OLAW. The organization receives official notification from OLAW that the assurance has been approved for a 4-year period. In some cases it may be short as 4 years is the maximum period for an active assurance.

After 4 years, OLAW requires organizations to renegotiate their assurance. This process requires organizations to review and update their assurance and submit it to OLAW for the renewal assurance to be negotiated.

Interinstitutional Assurance

A scientist from a research organization without an animal care and use program can receive a PHS award to support animal activities. Since the funds can only be awarded to an assured organization, assurance coverage must be arranged. Consequently, when the grants manager from the Office of Extramural Research notifies the scientist of the award, he or she must explain the circumstances. For example, a prime grantee may explain that his or her animal activities will be conducted at a colleague's organization that maintains an animal welfare assurance.

This scenario would require the prime grantee's organization to establish an interinstitutional assurance with the colleague's assured organization. The interinstitutional assurance is project specific (i.e., an interinstitutional assurance must be in place for each grant or project on the grant by number). The process for acquiring an interinstitutional assurance begins with the prime grantee. In this scenario, the prime grantee may choose to establish an interinstitutional assurance with multiple organizations. For example, the prime grantee could establish an interinstitutional assurance to conduct mouse studies at GEU and another to conduct swine studies at Great Western University (GWU).

The grants manager from the Office of Extramural Research notifies the scientist from the nonassured organization that he or she has been awarded, for example, an NIH grant. The principal investigator indicates the intent to conduct the animal activities at the colleague's assured organization. Grants Management contacts OLAW and requests that OLAW negotiate assurances for the prime grantee and all performance sites on the grant. OLAW contacts the prime grantee and provides the link to the sample interinstitutional assurance form for completion. The prime grantee completes his or her portions of the form and copies it to the organization's letterhead. Once completed, it is forwarded to the collaborator at the assured organization for completion.

A representative (e.g., the IACUC administrator) from the assured organization completes the form by providing the required information (e.g., organization name, contact information, assurance number, etc.). The form must then be signed by representatives from the assured organization. It must be signed by the IO and the IACUC chair or IACUC administrator with authority to confirm that the protocol(s) covering the granted activities have been reviewed and approved by the IACUC. Once the form is complete, it is returned to the prime grantee, who returns it to OLAW for review and approval. Once OLAW approves the interinstitutional assurance, the funds can be dispersed to the prime grantee organization.

Once the assurance is negotiated and approved, the prime grantee's organization must comply with the collaborating organization's assurance. In addition, the assured organization agrees to oversee the animal activities supported by the specific PHS funds (each interinstitutional assurance is grant specific) received by the organization covered under the interinstitutional assurance. This oversight includes, ensuring that animal activities are reviewed and approved by the IACUC, that the animals under study receive veterinary care, and that any required reports relative to the grant occur. In addition, any noncompliance associated with the project must be addressed, reported, and overseen by the organization maintaining the assurance.

Performance Site

In certain circumstances, scientists from assured organizations may wish to collaborate with scientists at nonassured organizations. In these situations, the nonassured organizations may become performance sites on the assured organizations' assurances.

Consider the following scenario: A scientist at assured GEU wants to conduct animal activities at his colleague's nonassured organization Great Center University (GCU). In this scenario, the PHS funds have been awarded to GEU (i.e., assured organization). As part of the collaboration, PHS funds will be transferred to GCU to finance PHS-supported animal activities. Since the funds cannot be transferred to a nonassured organization, an authorized individual (e.g., the IACUC administrator) from GEU should contact an OLAW assurance officer to make arrangements for GCU to be listed as a GEU performance site. The addition of a performance site on a grant requires approval from the program official on the grant. GEU can contact Grants Management for the grant that this animal activity will be conducted under and request approval. Once approval is obtained, OLAW can add the covered component to the GEU assurance.

Upon contacting OLAW, the GEU IACUC administrator, for example, identifies the prospective performance site and other relevant information (e.g., the activities to be conducted, the grant number, and any other information requested by the assurance officer). To become a performance site on an organization's assurance, the nonassured agency must agree to function in accordance with the assured agency's animal welfare assurance. Once this process is completed and the performance site addition is approved by OLAW, the nonassured organization with GCU can be used by the assured organization to conduct PHS-supported animal activities.

COLLABORATIONS AND CONTRACTS

Collaborations When Research Is Funded through a PHS Agency

Since the PHS policy (V, (B), p. 19) states that an awardee organization and any collaborating organizations (where PHS-funded animal research activities will occur) must have an approved PHS assurance on file with OLAW, certain precautions should be taken when establishing collaborations. Prior to receiving PHS support for animal activities, organizations must provide verification that the animal activities have been approved by the IACUC. Consequently, the *Guide for the Care and Use of Laboratory Animals* (henceforth referred to as the *Guide*) indicates that collaborating organizations should establish an MOU prior to participating in collaborative research to eliminate ambiguities. The MOU provides a conduit for the prime grantee to assure that colleagues will ensure that the terms and conditions of the grant and PHS policy are satisfied. For example, the MOU will allow the grantee to verify that the collaborator has an OLAW-approved animal welfare assurance. It will document which IACUC will approve, oversee, and report (e.g., significant noncompliance) when necessary to OLAW.

OLAW does not require animal activities to be approved by more than one IACUC. Consequently, organizations establishing MOUs must decide which organization's IACUC will oversee the animal activities. The majority of past BP meeting attendees indicated that the IACUC at the organization conducting the animal activities oversaw them. Attendees suggested that this best practice is most practical and allows more efficient and thorough oversight, although not all BP attendees agreed.

Although MOUs should be in place anytime organizations collaborate on animal research activities, they are particularly important when the collaboration involves PHS funding. The prime grantee is ultimately responsible for ensuring that PHS funding is utilized in accordance with the PHS policy and the terms and conditions section of the grant. Consequently, the grantee should always initiate an MOU when collaborating with other organizations.

The MOU can be very simple. At a minimum, it should identify the collaborating organizations, the specific research it covers, recognition of the approving and overseeing IACUC, and reporting requirements, and provide signatures of individuals authorized to commit on behalf of the organization. Since the prime grantee is ultimately responsible for the animal activities, he or she may also want to ensure that methods are established to provide to the prime grantee copies of relevant information (e.g., postapproval monitoring reports, inspection findings, and IO reports) as they relate to the project.

In addition to establishing the MOU, the prime grantee must ensure that an IACUC has reviewed and approved all of the animal procedures listed in the grant. Once the MOU is in place, this task can be satisfied in one of two ways:

- The first option is for the prime grantee to conduct his or her own IACUC review on the project even if the activity will not occur at his or her site. This process allows the grantee to do a protocol grant congruence review and approve all of the animal activities in the grant.
- Many BP meeting attendees use a second option. At their organizations, they require the collaborating organization to submit a copy of the approved IACUC protocol as an MOU attachment. This process allows the prime grantee to conduct a congruence comparison between the grant and the collaborator's IACUC-approved protocol (i.e., approved animal activities) to confirm that the animal activities described in the grant have been reviewed and approved by the collaborator's IACUC.

AAALAC International Accreditation and Collaborations

When an AAALAC-accredited organization collaborates with other organizations, the accredited organization has two points to consider. As suggested in the *Guide*, an MOU should be established to define the responsibilities of both organizations.

The initial point can be addressed in the MOU. As part of the accreditation process, AAALAC International ensures that all animals owned by the accredited organization are maintained and utilized according to the PHS policy, the Animal Welfare Act regulations, and other AAALAC-accepted standards (e.g., the *Guide for the Care and Use of Agricultural Animals in Research and Teaching*). For the purpose of accreditation visits, AAALAC must know as part of the assessment process, which organization owns the animals and where they are being held. Therefore, when an accredited organization collaborates with another, the MOU should clearly note who owns the animals, who reports adverse events, and who provides oversight of the animal activities. The AAALAC encourages the collaborating organizations to enter into a concise and binding agreement that clearly defines which organization provides IACUC oversight.

The second situation for an accredited organization to consider is the ramifications of collaborating with an organization that is not AAALAC accredited. On occasion, accreditation may be lost or the organization may be placed in a probationary accreditation status. Therefore, one way to determine an organization's accreditation status is to document it as part of the MOU. If the accreditation status has changed, the MOU may require notification of the collaborator.

When an AAALAC-accredited organization retains ownership of animals used for research at a nonaccredited organization, the accredited organization must cover the nonaccredited organization's animal housing facility as part of its own accreditation. For example, if an accredited organization plans to have antibodies produced at a nonaccredited facility in a special breed of rabbits the organization owns, the accredited site must describe the antibody producer's rabbit housing facility in its animal care and use program description. Consequently, the accredited organization's IACUC must inspect the antibody producer's rabbit housing facility at least semiannually. In addition, it must provide oversight and require deficiency corrections as though the facility were its own. AAALAC will consider the rabbit housing area a satellite facility and will visit the off-site rabbit holding area during an accreditation review. Satellite facilities need not be reviewed by AAALAC if the accredited organization decides that satellite facilities will not be accredited. Consequently, AAALAC-accredited organizations may consider adding a question to the MOU that confirms that the collaborating organization is AAALAC accredited.

USDA and Collaborations

Although the AWAR do not include requirements for collaborators to establish MOUs, the USDA stresses the importance of identifying which organization's IACUC will oversee the project before the research is initiated. In addition, the

USDA has identified the requirement that only one organization should report the animals used on its annual USDA report. In general, the organization housing the animals and providing IACUC oversight should be the reporting agency.

International Collaborations

The USDA does not oversee international research. However, if a national organization collaborates with an international organization, the same rules of AAALAC accreditation apply. In other words, as long as both organizations are accredited, they satisfy the rules of accreditation. In the event that the international organization is not accredited, the national organization must cover the nonaccredited organization in its animal care and use program if the accredited organization retains animal ownership.

CHAPTER 20

FOIA, Sunshine Laws, and Confidential Information

SCENARIO

The GEU IACUC chair received a letter from a special interest group requesting copies of the organization's IACUC meeting minutes for the past 3 years. In addition to the meeting minutes, the requester indicated that he also wanted a copy of each animal research proposal cited in the minutes. The notice was sent directly to the organization's institutional official (IO) by registered mail, and it indicated that the request was being made under the Freedom of Information Act (FOIA). The letter stated that since GEU uses federal funds to conduct research, they are required to comply with the request. The correspondent also indicated that the organization had 10 business days to respond.

The IO sent the letter to the IACUC Administration for processing. The IACUC administrator informed the IO that he would respond to the request. He explained to the IO that the first request of this type sent to GEU was forwarded to the university's legal counsel for guidance. GEU's attorneys advised that FOIA requests only applied to federal agencies and that, since GEU was not a federal government agency, complying with the request was not required. GEU attorneys prepared the following memo and informed the IACUC administrator it could be slightly modified (i.e., dates) and used to respond to all requests made under the FOIA:

In our capacity as general counsel for the Great Eastern University, we are writing in response to your letter to the **<insert position/person>** dated **<insert date>** in which you requested certain documents under the provisions of the Federal Freedom of Information Act (FOIA).

Kindly be advised that Great Eastern University is not subject to FOIA. Since the information requested routinely includes proprietary information that is not made available to the general public, we cannot release the requested information.

Please direct any further inquiries or comments you may have to me. Your cooperation in this matter is appreciated.

Great Eastern University General Counsel, Perry Mason

FREEDOM OF INFORMATION ACT

FOIA is a federal law enforceable in federal court that applies only to federal agencies. Nongovernment organizations are not required to comply with FOIA. As it relates to animal research, the most relevant government agencies that must comply with FOIA include the Office of Laboratory Animal Welfare (OLAW), the US Department of Agriculture (USDA), the Department of Defense (DOD), and the Department of Veterans Affairs (VA). Since the Association for Assessment and Accreditation of Laboratory Animal Care (AAALAC) International is a nongovernment agency and is not subject to the FOIA.

"The Freedom of Information Act (FOIA) 5 U.S.C. 552. provides individuals with a right to access to records in possession of the federal government." If an organization or individual requests specific records in possession of the federal government, the agency is required to release those records. FOIA requests must specifically identify the documents requested. The request must be made in writing (e-mail) to the relevant government agency. The requester can be required to cover the expenses associated with the preparation of the requested materials including labor and copies. The agency has 20 business days to respond to the request. An agency can extend the 20-day deadline an additional 10 business days, but the notice to extend the deadline must be provided in writing to the requester.

Documents submitted by an animal program to the OLAW or the USDA can be requested under the FOIA—for example, the animal welfare assurance, noncompliance reports, and annual reports. Unless one of the FOIA exemptions or exclusions applies, the government agency must provide the records in their entirety.

The USDA utilizes an electronic FOIA method (E-FOIA). It posts requested documents on the USDA website so that the documents can be accessed by interested parties.

In certain circumstances, an FOIA request can be denied. Generally, upon receiving a FOIA request, the federal agency will contact the institution which provided the FOIAed documents and ask if the institution can identify cause to redact any or all of the materials on file. BP meeting attendees agreed that FOIA clarification requests from federal agencies should be forwarded to the organization's legal counsel for evaluation. If the information requested falls into one of nine specific exemptions, the information can be withheld. The exemptions include

1. Information about national security
2. Information exempt under other laws
3. Confidential business information
4. Personal privacy
5. Law enforcement records
6. Financial institutions
7. Geological information
8. Internal personnel rules
9. Inter- or intra-agency communication that is subject to deliberation or litigation

As a general statement, most animal protocols and animal care documentation do not fall under any of the exempted categories. However, specific information in selected documents may be exempt from FOIA and could be redacted or deleted from the provided documents. Protocol information regarding individuals may be exempted as protected personal information, and specific techniques or procedures that are unique to the researcher or the laboratory may be exempted as confidential business information. However, rarely will protocols contain other narrative information that may be covered under the exemption categories. BP attendees concluded that institutions should carefully consider what documents are provided to federal agencies. All legally required documents must be provided, but additional documents or expanded narrative in the federal files may not be in the best interest of the institution.

RESPONDING TO FOIA REQUESTS

In some cases, providing the information that is requested under FOIA may be in the best interest of the organization. For example, an individual may request under FOIA that an organization provide copies of all its protocols involving nonhuman primates. The response may be, "At GEU we have no animal care and use activities that involve nonhuman primates." A lack of response may imply that the organization is not being transparent and is trying to hide something. If an organization decides to respond and demonstrate compliance, the response may be initiated and the following statement may be used in reply to a FOIA request: "Our organization is not a government agency and consequently not subject to FOIA; however, as a courtesy, the requested information is being provided."

STATE SUNSHINE LAWS

State-operated organizations are subject to state laws; some states have public access laws known as "sunshine laws." If information is requested under a state regulation, consult with the organization's attorneys. Meeting minutes can be requested under sunshine laws, so it is not advisable to include names, protocol titles, addresses, or other revealing or identifying information in the minutes. Private organizations operating in a particular state are generally not subject to the state sunshine laws; consult the organization's legal counsel for specific details on state and location regulations.

INSTITUTIONAL ACCESS TO INFORMATION

IACUC administrators and office staff members have access to significant quantities of important and confidential information regarding research models, projects, and activities. BP meeting attendees strongly encourage organizations to develop a policy of information management. Items that should be considered for inclusion in such a policy include the facts that (1) the IACUC office staff can only provide protocols or protocol information to the principal investigator or a member of the protocol participants, the IO, the department chair of the PI, and regulatory and accreditation officials; (2) the PI of the protocol may share information with whomever he or she chooses; and (3) media or animal rights organization requests for information must be vetted by general counsel, the Public Affairs office, and the IO.

In many cases, there are several parties on campus who will require specific protocol information as part of their daily duties (e.g., grants and contracts staff, veterinarians, veterinary technical staff, etc.). Before they are provided either information or access to protected information, these individuals should receive a security briefing and sign a confidentiality agreement to protect any information they obtain from the IACUC administrative office.

CHAPTER 21

DOD and VA Regulations
Know the Differences

SCENARIO

Dr. Smith from GEU received an additional appointment at the Department of Veterans Affairs (VA) Medical Center. She wanted to conduct her animal work at the VA for several reasons. Housing is plentiful there, per diems are less, and the US Department of Agriculture (USDA) does not inspect the VA site. Dr. Smith had been awarded a National Institutes of Health (NIH) grant to support the research she would conduct at the VA. As part of the research, she proposed to conduct survival surgery on guinea pigs. Dr. Smith prepared and sent the VA's IACUC a protocol for review and approval. GEU and the VA have an established memorandum of understanding (MOU) that indicates that when Public Health Service (PHS) funds (e.g., NIH) are used to support animal activities conducted at the VA, GEU will be the primary recipient (i.e., grantee) and the funding will be rerouted to the VA to support activities performed at the VA.

Dr. Smith prepared and submitted a protocol to the VA IACUC, but was informed that the GEU IACUC must review and approve the protocol before it can be considered by the VA committee. Since Dr. Smith was not familiar with the VA systems, she was frustrated with the process. However, she believed that conducting her research at the VA would afford her benefits not available at GEU. Consequently, she sought advice from the VA IACUC administrator, who helped her understand and guided her through the VA IACUC approval process.

Regulatory References and Guidance Documents

- **Veterans Administration medical centers** are subject to the same federal regulations as any other organization (i.e., Animal Welfare Act regulations [AWAR], PHS policy, and *Guide for the Care and Use of Laboratory Animals* [henceforth referred to as the *Guide*]). In addition, the VA animal care and use community has adopted and implemented the *VA Handbook* 1200.7 ("Use of Animals in Research"), which prescribes animal care and use management and oversight. Overall, the *VA Handbook* is similar to other regulatory guidance, but it implements additional requirements that must be applied when animal activities are conducted at VA facilities. An interesting twist to the VA system is a mandate that the VA animal care and use program must maintain Association for Assessment and Accreditation of Laboratory Animal Care (AAALAC) accreditation.

- **Department of Defense (DOD):** Both internal and external DOD animal care and use programs are bound by the USDA regulations and the PHS policy (if assured).
  - Similarly to the *VA Handbook* 1200.7, the DOD community has its own regulatory documents. The core document is DOD Instruction 3216.01 ("Use of Laboratory Animals in DOD Programs"). This is a "super-service document," which means that it originates in the Pentagon and applies to research conducted by the Army, Navy, Air Force, or Marines. The DOD Instruction 3216.01 sets the overarching parameters for animal use, defining IACUC membership, training, and applicability.
  - A secondary animal care and use program document is titled "The Care and Use of Laboratory Animals in DOD Programs." This document is also a service-level document, holding different designations depending upon which unit of the service is the current user—for example, AR 40-33 (Army), AFMAN 40-401 (Air Force), SECNAVINST 3900.38C (Navy), DARPAINST 18 (DARPA), and USUHSINST 3203 (Uniformed Services University).
  - Referred to as the "implementation document," the DOD Instruction 3216.01 provides additional detail on the structure and operations of DOD animal care and use activities. Each organization or facility may issue additional operational language, as long as those organizations do not deviate from the overarching specifications found in the DOD Instruction 3216.01 or "The Care and Use of Laboratory Animals in DOD Programs."

FEDERAL FUNDING AND PHS ASSURANCE

While either VA or DOD researchers may compete for PHS funds, there is a federal prohibition of transferring funds among federal agencies. Consequently, funds cannot be transferred from the NIH directly to a VA or DOD facility. Therefore, a receipt of government funds requires an affiliation with a civilian organization that will serve as the primary recipient of those funds.

Within the VA community, *VA Handbook* 1200.7 stipulates that all VA research must be performed as described in the organization's PHS assurance. As a result, a VA must have or be associated with a PHS animal welfare assurance. A VA either may have its own PHS assurance or may be identified as a performance site on a civilian organization's assurance. If a VA decides to secure its own PHS assurance, the center's director serves as the institutional official (IO). If the VA appears as a performance site on a civilian organization's assurance, then the affiliate's IO serves as the IO for both programs. The requirements and stipulations for DOD to secure a PHS assurance are consistent with the requirements of civilian organizations.

USDA OVERSIGHT

On occasion, a VA or DOD activity will involve USDA-covered species. The USDA overseeing activities at a VA or DOD facility primarily reflects the organization's own mandates (e.g., *VA Handbook* 1200.7 or DOD agency decision). Since a government position exists that prohibits one federal agency from inspecting another, the USDA does not inspect either VA or DOD facilities under federal law. However, the VA and DOD requirements and agency decisions require all animal activities conducted to be fully compliant with the AWA and the AWAR. Although federal law does not require either agency to file an annual report with the USDA, both choose to do so as a result of agency policy. In the case of the VA, the USDA inspectors (veterinary medical officers) are authorized by *VA Handbook* 1200.7 to inspect its facilities. The DOD regulatory documents include no provisions for USDA representatives to inspect its facilities. Unlike civilian organizations, VA and DOD facilities are required to provide information that will become part of the congressional annual work summary report. This animal use report is required of all federal operations and details the numbers of animal used, the category of pain or discomfort in which the animals participated, the alternative search strategy, and the presumptive value of animal use for each activity.

DEFINITION OF TERMS

Each agency has several specialized personnel terms and responsibilities. The VA uses some very unique terms, while many of the DOD's terms are identical to those used by civilian organizations. Those that may be of particular interest to IACUC administrators include:

- **Chief executive officer (CEO).** This is defined as the highest ranking administrator at the organization.
- **Associate chief of staff (ACOS).** In the VA, this is the individual who is the day-to-day functional senior manager and represents the IO. The ACOS is not authorized to dedicate resources as necessary to resolve any issue within the animal care and use program and consequently cannot serve as the IO.
- **Veterinary medical unit (VMU).** Within the VA, the VMU includes animal housing and husbandry personnel.
- **Veterinary medical officer (VMO).** Within the VA, the VMO is the veterinarian with a VA appointment who is a specialist in lab animal medicine. *Note*: VMO is the same term used for the USDA inspector.
- **Veterinary medical consultant (VMC).** Within the VA, the VMC is a veterinarian with a contractual VA appointment who is a specialist in lab animal medicine.
- **Chief veterinary medical officer (CVMO).** Within the VA, the CVMO is the senior veterinary medical officer within the system.
- **Colonel.** Within the DOD, the colonel is a member of senior management. He or she would generally equate to a department chair or professor at a civilian organization. Regulatory officials are generally either a colonel or a senior lieutenant

colonel. Individuals serving as a colonel generally have 15–20 years or more of experience in animal care and use.

- **Lieutenant colonel.** Within the DOD, the lieutenant colonel is also a member of senior management who would generally equate to a department vice chair or associate professor at a civilian institution. Most animal care and use program managers are lieutenant colonels and have an average of 15 years of service.
- **Major.** Within the DOD, the major is a member of middle management who would generally equate to a Division Head and be similar to an Assistant Professor at a civilian organization.
- **Captain.** Within the DOD, the captain is also a member of middle management who would generally equate to a clinic manager or section manager and be similar to an instructor at a civilian organization. Most captains have 5–8 years of military experience and may not have any focused research animal experience.
- **First lieutenant/second lieutenant.** Within the DOD, first and second lieutenants are junior members within the DOD structure who would generally equate to a section manager having a similar status to an instructor at a civilian organization.

UNIT STRUCTURE

The DOD has service-specific oversight offices, generally referred to as the Clinical Investigation Regulatory Office (CIRO). These units are headed by senior DOD members who provide oversight for internal use of animals (generally medical centers).

The Animal Care and Use Regulatory Office (ACURO) will conduct audits and provide higher headquarter approval for the use of primates and dogs, which are protected species within the DOD regulatory framework. The ACURO also reviews potentially painful activities in all species and is the proponent for regulatory changes for DOD activities. In addition, the ACURO audits organizations using DOD funds for animal activities.

The DOD uses a consolidated activity for developing consistent processes and methodologies. This group, the Joint Technical Working Group (JTWG), includes all services and DOD agency representatives. The JWTG meets quarterly to recommend policy and harmonize guidance.

While most organizations refer to the animal oversight body as the IACUC, the DOD occasionally uses the term LACUC (Laboratory Animal Care and Use Committee). While some DOD operations continue to refer to the oversight body as the LACUC, others have shifted to the more common term IACUC.

The Department of the Navy Bureau of Medicine and Surgery (BUMED) provides audits, higher order taxonomy approval, and process development for Navy activities.

The Air Force Office of the Surgeon General (AFOSG) provides audits, higher order taxonomy approval, and process development for Air Force activities.

DOD AND VA TRAINING

While all organizations using animals provide a variety of different types of required animal user training, the VA and the DOD are far more prescriptive than most civilian organizations. For example, at VA organizations, the same mandatory training applies to all VA sites. Through IACUC oversight, each VA medical center must ensure that all personnel involved with animal research receive training to competently and humanely perform their duties related to animal research. This includes IACUC members, veterinarians, veterinary technicians, husbandry staff, research technicians, investigators, and all others that perform procedures or manipulations on laboratory animals. For example, investigators and research staff who utilize laboratory animals must complete the modules and successfully pass the exam for the module "Working with the VA IACUC." Additionally, VA researchers must complete a species-specific web course for each species that they plan to use. While not a requirement, the VA system recognizes the value of American Association for Laboratory Animal Science (AALAS) certification and specifies the desirability of AALAS certification in *VA Handbook* 1200.7.

IACUC members also have required training. All members must complete the training and successfully complete the examination covering "Essentials for IACUC Members."

The VA requires annual training of IACUC members and research staff members. A web-based training with an examination is used to demonstrate compliance to federal and unit training requirements. Education goals for web-based training will be considered satisfied when personnel successfully pass the exam. The VA requires that any examination be of sufficient difficulty to provide some assurance that important concepts have been mastered.

In the DOD system, training is generally consistent with civilian organizations. The DOD requires IACUC members to receive 4 hours of protocol review training that covers regulatory requirements and animal use techniques. In addition, each member receives at least 4 hours of additional training on humane care and ethical treatment of research animals. Nonaffiliated members receive an additional 8 hours of training to assist them with developing knowledge based on important issues and review activities within the DOD system. The training requirements are system-wide and generally through the AALAS Learning Library.

SEMIANNUAL REVIEW PROCESSES

The semiannual review processes are very similar to those of civilian organizations, but with some unique aspects.

VA

The VA has developed and utilizes a number of unique forms to guide it through and to document the semiannual review process. The VA utilizes the VA IACUC program and facility self-assessment form (VA form 1) when conducting program reviews and facilities inspections. This document guides the IACUC through each process and documents the fact that a complete and thorough review occurred. The VA also allows the use of the OLAW semiannual program and facility review checklist or a similar form, providing that the form incorporates all of the elements found in VA form 1. The VA also uses form 2 ("Table of Program and Facility Deficiencies") when evaluating program deficiencies. The form ensures that the VA IACUC is consistent when applying corrective action plans and timetables for each deficiency. In addition, form 2 serves to document deficiency discussions and the actions for resolution taken by the committee. Upon completion of the program review and facility inspections, the VA committee prepares the IO report. The VA uses VA form 3 ("Post Review Documentation") to document a meeting with the IO to discuss the status of the program. During this meeting, any issues requiring attention are defined.

DOD

Similarly to the VA, the DOD uses a form specifically developed to document facility inspections and program reviews: facility inspection and program review (FIPR). This DOD-specific checklist is heavily based upon the OLAW checklist, but includes DOD-specific items and issues. The management of the FIPR and the outcomes of the review are identical to those in civilian organizations.

IACUC MINUTES

VA

The VA structure identifies the IACUC as a subcommittee of the organization's Research and Development Committee, but the IACUC retains the same overarching authority found in civilian IACUCs. The VA system has a specific outline and has required inclusions in the IACUC minutes. The *VA Handbook* 1200.7 stipulates that the minutes must be written and published within 3 weeks of the meeting date, the use of abbreviations is not acceptable, and subsequent pages must be numbered. The IACUC administrator uses the following format when preparing the meeting minutes.

The information on the cover page must appear in separate lines. Boldface, large typeface must be used. The name of the facility and facility number, the official address, the official committee name, and the date of the meeting are included on the first page. The names of all voting members present and absent are listed. A separate list of nonvoting members is also added. The appointed role of each individual is part of each member list, which establishes that the IACUC is properly constituted to conduct business. The use of the term "ex officio" is only appropriate when the member's office or legal role (e.g., organization veterinarian) dictates that he or she will be on the committee. The minutes must indicate and document that a quorum is present, which is defined as a majority (more than 50 percent) of voting members.

The minutes are arranged into three sections: review of previous minutes, old business, and new business. In addition, at each meeting, the IACUC conducts and documents into the minutes a review of the semiannual review "schedule for deficiency corrections." The minutes are used to monitor progress toward completing corrections of deficiencies previously identified during facility inspections and reviews.

Business items are retained under old business in subsequent minutes until the final approval is given by the IACUC, the project is disapproved by the IACUC, or the project is withdrawn from consideration by the investigator. The final disposition of each project is clearly stated in the minutes.

For each project under consideration, the first and last name of the principal investigator and the complete name of the project are listed in the minutes. For each new project, the motion voted on by the committee (i.e., approved, approved pending clarification, deferred, disapproved) and the exact vote (i.e., the number voting for the motion, the number voting against, and the number abstaining) is recorded into the minutes.

Once IACUC minutes are approved at the following meeting, the IACUC chairperson signs and dates them. The *VA Handbook* 1200.7 stipulates that no local official may alter the IACUC minutes once they are signed by the IACUC chairperson. The *VA Handbook* 1200.7 also stipulates that the Research and Development Committee must review (not approve) a copy of the signed minutes as an item of business at its next meeting.

DOD

For the DOD, the minutes are generated using a local format that is frequently similar to that of most civilian academic organizations. The DOD meeting minutes require no specific citations or information and the content is consistent with that found in civilian organization meeting minutes.

PROTOCOLS

Both the DOD and VA have their own specialized protocol formats.

VA

The VA protocol form is called the animal component of research protocol (ACORP). The ACORP must be used when the project involves direct VA-funded animal research. However, the VA IACUC may review the civilian organization's protocol for other types of funding (e.g., PHS funding). The VA may require a supplemental appendix for protocols on non-VA forms, to assure that all pertinent VA issues have been sufficiently addressed. Depending on the circumstances, the VA may or may not require its IACUC to review and approve a protocol that has already been approved by a civilian organization. Conversely, a civilian organization serving as the primary grantee for animal activities that will be conducted at the VA may or may not accept the VA IACUC's review and approval. In either case, to avoid approval delays, the best practice may be to have a protocol first approved by the primary grantee's organization and then refer it to the VA for acceptance or an additional review and approval by the VA IACUC. It is often helpful to have an MOU negotiated between the two organizations to assure that both recognize the rights and obligations of the other while protecting the risk and investments of each.

DOD

A common protocol template is used for all DOD organizations. Each organization may develop a supplemental page for local issues or guidance procedures, but the core protocol must remain consistent with the DOD template. The DOD template is generally consistent with most civilian organization protocol forms, but it does include specific questions, such as, "Please explain the military relevance." In accordance with DOD policy, each DOD IACUC must establish a policy for prolonged restraint, food and fluid restriction, multiple major survival procedures, enrichment of primates, and the exercise of dogs.

DOD protocols are generally concise and include only the minimal information needed for the IACUC to conduct a thorough review since DOD protocols are subject to the Freedom of Information Act. DOD protocols consider the four Rs of research, rather than the more commonly recognized three Rs of replacement, reduction, and refinement (the fourth R is responsibility). DOD protocols must be written in language that can be easily understood by individuals with a high school education. The nontechnical synopsis and the objective of the DOD protocol are included in the DOD annual report to Congress (also referred to as the "work unit summary"). The DOD literature search requires a minimum of three databases. One of the three must be a database that covers federally funded projects such as federal research in progress (FEDRIP).

UNIQUE ASPECTS OF VA AND DOD PROGRAMS

VA

The VA Handbook 1200.7 requires the VA animal care and use program to oversee all animals housed on VA property, regardless of ownership, regardless of whether the animals were purchased with VA funds, and regardless of the housing location. In addition, the *VA Handbook* 1200.7 requires the VA IACUC to oversee VA-owned animals independent of the housing location. Many civilian organizations may be affected by these requirements since VA-owned animals are often used at local non-VA organizations. In a situation such as this, the VA IACUC must review the semiannual program and facilities inspection report of the organization housing the VA-owned animals. In addition, the VA IACUC may wish to perform its own assessment to validate that the processes used at an organization housing VA-owned animals are consistent with VA standards. *Note*: Animals purchased with VA or VA Foundation funds may not be housed in facilities not accredited by the AAALAC without a waiver from the chief research and development officer (CRADO).

The VA considers an animal as being "used" on an approved protocol once it is delivered to the organization. It allows for the adoption of excess animals as pets, in specific circumstances, and requires a minimum of two reviewers for each proposed use of animals.

When the VA IACUC is a stand-alone committee, it must include a minimum of five members. The committee must include a chair, attending veterinarian, scientist, nonaffiliate, and a lay member (often facility manager). If a veteran volunteers or is receiving care at the VA, he or she cannot be considered a nonaffiliate or lay member. The lay and nonaffiliate members should not be the same person.

A local veterinarian without laboratory animal training or experience may supplement but not replace the Veterans Administration VMO/VMC. The local veterinarian must function within the VMO/VMC established plan for a program of adequate veterinary care and must be approved by the CRADO. A VA veterinarian may be hired or contracted, as a VA foundation employee, through an affiliated civilian organization or as part-time VA employee.

The VA central office requires VA units to use a secondary veterinary medical review program (SVMR). The SVMR process is used to review meeting minutes, facility inspection and program review reports, and proposals submitted for VA funding.

DOD

The DOD defines an animal as any living or dead vertebrate animal, including birds, cold-blooded animals, rats of the genus *Rattus*, and mice of the genus *Mus*. Offspring of egg-laying vertebrate species are not counted as vertebrate animals until they hatch from the egg, with the exception of fish and amphibians, whose larval offspring are considered vertebrate animals. A dead animal is defined as any animal euthanized for the purpose of conducting training or for experimental reasons. However, this does not include dead animals or parts of dead animals purchased at grocery stores or slaughterhouses. The definition differs only slightly from the USDA and PHS definitions, but it is unique and applicable to all activities when DOD funding is used to support the use of vertebrate animals in research, testing, or teaching activities.

The DOD excludes from the definition of vertebrate animals those animals used strictly for ceremonial and/or recreational purposes, as well as working animals, such as military working dogs. In addition, farm animals used for food and fiber studies, animals used in disease surveillance (unless the disease screening procedure harms the animal), and animals involved in field studies are excluded. *Note*: The DOD uses the USDA definitions of field study (no manipulation of animals or change in their behavior) and field research (an activity that causes a change in an animal or its behavior).

The DOD requires animal suppliers to be licensed by the USDA unless specifically exempted by the ACURO from the licensing requirements. *Note*: The actual exemption is from the USDA but it is coordinated through the ACURO or other service agency.

All DOD-sponsored proposals (worldwide) must be reviewed by an IACUC consisting of at least five members with a nonscientific and nonaffiliated member present (a primary and alternate nonaffiliated member is required). The DOD also stipulates that IACUC members must be government employees. Since the DOD IACUCs perform a government function, the nonaffiliated and the nonscientific member shall either be a federal employee with demonstrated commitment to the community or a consultant consistent with the requirements established by reference.

The DOD requires a headquarters-level (ACURO) review for the use of nonhuman primates, dogs, cats, and marine mammals. This secondary review, subsequent to the IACUC review, emphasizes the importance of meeting the regulatory requirements and assures that all of the provisions for special species have been approved and included in the protocol. The purpose of the process is to create a consistent systemwide review process.

Extramural (non-DOD) animal use requires a headquarters-level review through the ACURO or appropriate oversight office. A site visit to extramural sites conducting activities with nonhuman primates, dogs, cats, or marine mammals is required worldwide. During the review, the extramural organization's most recent USDA inspection reports are reviewed. In addition, the USDA inspection reports are reviewed annually for the duration of the DOD award to the organization.

The DOD also stipulates that IACUC members must be government employees. Since the DOD IACUCs perform a government function in an approval process and do not serve merely as advisory bodies, the nonaffiliated and the nonscientific members of DOD IACUCs shall either be a federal employee, with demonstrated commitment to the community, or a consultant consistent with the requirements established by reference.

VA MEMORANDUM OF UNDERSTANDING

VA facilities and their affiliated organizations should have an MOU to assure that each organization understands the obligations of the other. There are four principal areas that require agreement: (1) which organization's IACUC will conduct the program review and inspect the housing facilities, (2) which organization's IACUC will conduct the protocol reviews, (3) how animal transfers between organizations will be monitored, and (4) how animal use will be documented.

VA MODELS AND RELATIONSHIP TO SISTER
CIVILIAN ANIMAL CARE AND USE PROGRAMS

Four models have been used to affiliate with a VA organization. Each of the four methods has its own unique advantages and disadvantages.

Model 1

The VA and affiliate share the IACUC and facilities. In this model the VA and the affiliated organization agree to share resources and costs. This arrangement is the least complicated of the four models and is commonly used at smaller VA sites. The arrangement provides access to all of the resources available to both organizations. A critical component of this relationship is that the joint IACUC must agree to follow the *VA Handbook* 1200.7 (Sections 6.n. (3), (5), and 6.n. (11)).

Model 2

The VA and affiliate share the animal facility but have separate IACUCs. This model has the advantage of maintaining a single animal facility, which tends to be the most costly component of any organization's program. This model could be beneficial because the same veterinary and husbandry staff is used. However, it is critical that approval reciprocity is fully transparent and that the activities comply with the *VA Handbook* 1200.7 (Section 6.n. (10)(a)(2)).

Model 3

The VA and affiliate share the IACUC but have separate animal facilities. This model is rarely used as it requires the greatest degree of agreement. The model requires that the veterinary and the animal care staff cooperate and use uniform IACUC-approved policies.

Model 4

The VA and affiliate have separate IACUCs and separate animal facilities. This model can be both the easiest and the most complicated, but it is also the most common arrangement used. This model also requires clearly written and detailed MOUs to assure that both organizations know each other's obligations and expectations.

While any model or hybrid of multiple models may work effectively, the key for any model is communication, clarity, and understanding.

CHAPTER 22

Data Management and Electronic Systems

SCENARIO

A scientist at GEU acquired PHS funding to investigate why large numbers of Tasmanian devils were dying. The research proposal noted that an oncogenic virus appeared to be infecting the animals and causing large tumors preferentially in the mouths and other parts of their heads. GEU wildlife expert Dr. Green and GEU's attending veterinarian (AV) decided to collaborate on the study, and they submitted a proposal to the IACUC for review and approval.

The IACUC discussed the proposal and approved the animal procedures, which included identifying infected animals, capturing them, and studying individual animals through the duration of the infection. The animals would be held in field housing stations and identified using ear tags. At specific time points, biopsies of tumors as well as blood samples would be collected and analyzed for the presence of a viral suspect. A challenge for the IACUC during the approval process was deciding how to inspect the remote housing location that was miles away in Tasmania. The committee members discussed various methods for satisfying the facility inspection requirement.

After networking with colleagues at the IACUC administrators' best practices (BP) meetings, the GEU IACUC administrator discovered that other organizations had experienced similar challenges. Several best practices were shared, which included the use of video technology, digital photography, and secure e-mail systems to complete the inspection process. The IACUC administrator shared these ideas with his committee members, who considered several options for conducting the inspections and decided to conduct the inspection using video technology. The IACUC asked the principal investigator (PI) to record a video tour of the proposed housing facility, which could then be reviewed by the committee. Dr. Green agreed to video the housing area and volunteered that he had the technology to broadcast a live video of the area. He indicated that he could give the IACUC a virtual tour of the housing facilities, which would allow them to ask questions during the virtual inspection. During the inspection, committee members were able to direct Dr. Green to specific locations within the housing unit. In addition, Dr. Green also provided still photos of feeding and watering systems to be included as part of his protocol file. After conducting the live video inspection and reviewing the photographs of the feeding/watering devices, the IACUC requested copies of animal care standard operating procedures, which were provided via e-mail. The IACUC was satisfied that the housing requirements met minimum standards and agreed that the animals could be housed without creating adverse welfare concerns. The committee also agreed that future semiannual inspections would be conducted in the same manner.

Regulatory Guidance on the Use of Electronic Communications

In March of 2006, the Office of Laboratory Animal Welfare (OLAW) released a guidance document on the use of electronic communications titled "Guidance on Use of Telecommunications for IACUC meetings under the PHS Policy on Humane Care and Use of Laboratory Animals" (notice number: NOT-OD-06-052). OLAW and the US Department of Agriculture (USDA) concur that in certain circumstances electronic communication methods can be used to conduct IACUC business. For example, video conferencing equipment can be used to allow committee members to join meetings from remote locations in real time. In addition, various forms of technology (e.g., digital still photography and videotaping) can be used to enable facility inspections. A key aspect about using telecommunication techniques is to be able to effectively use the equipment without compromising the quality and the effectiveness of IACUC actions or deliberations.

The following identified criteria, as determined by OLAW, must be satisfied when using methods of telecommunications:

1. All members are given notice of the meeting.
2. All documents normally provided to members during a physically convened meeting are provided to all members in advance of the meeting.
3. All members have access to the documents and the technology necessary to fully participate.
4. A quorum of voting members is convened when required by PHS policy.
5. The forum allows for real-time verbal interaction equivalent to that occurring in a physically convened meeting (i.e., members can actively and equally participate and there is simultaneous communication).
6. If a vote is called for, the vote occurs during the meeting and is taken in a manner that ensures an accurate count of the vote. (A mail ballot or individual telephone polling cannot substitute for a convened meeting.)
7. Opinions of absent members that are transmitted by mail, telephone, fax, or e-mail may be considered by the convened IACUC members but may not be counted as votes or considered as part of the quorum.
8. Written minutes of the meeting are maintained in accord with the PHS policy IV.E.1.b.

METHODS OF TELECOMMUNICATIONS

Telecommunication is a resource that has developed in recent years to be both economical and practical. BP meeting attendees provided several examples of appropriate and inappropriate uses of telecommunication systems (e.g., Skype, telephone, e-mail, and live transmission of video).

E-mail as a Method of Communication

Appropriate and Effective Use of E-mail

A convenient way for IACUC administrators to communicate with committee members is through e-mail. For example, IACUC administrators frequently use e-mail to facilitate designated member review (DMR) assignments. The IACUC administrator may use e-mail to recommend DMR appointments to the IACUC chair, who then approves the recommendation by responding to the notice. The members are then contacted via e-mail to alert them to their DMR appointment.

In addition, e-mail is often used to distribute submission information to committee members and to provide each member the opportunity to request a submission be reviewed by the full committee. In this particular situation, the IACUC administrator regularly e-mails a list of submissions to be reviewed to all of the IACUC members. At a minimum, the e-mail list includes the name of the PI, the project title, the protocol number, and a brief summary of the project; each committee member is able to access additional information when needed. The e-mail may include a note that committee members have 5 days (or some other preapproved time) to call for full committee review and that the lack of response would be assumed to be affirmative for DMR.

IACUC administrators may also use e-mail to distribute information that will serve to provide continuing education. For example, the IACUC administrator may prepare a PowerPoint presentation on conducting semiannual inspections or on reviewing protocols and distribute it to committee members using e-mail.

Inappropriate Use of E-mail

BP meeting attendees warned that e-mail cannot be used as a means for voting on official IACUC actions. For example, the IACUC cannot use e-mail to vote to suspend a protocol based on a perceived incident of noncompliance. Consider the following scenarios:

Scenario 1. While walking through a housing facility, an animal care technician noticed four mice with excessive tumor burden. As a result of his discovery, he reviewed the approved IACUC proposal and discovered that the PI had allowed the animals to progress significantly beyond the humane end points specified in his IACUC-approved protocol. The animal care technician contacted the IACUC administrator, who prepared an e-mail notice that included a detailed description of the technician's findings and sent it to the committee members. The IACUC administrator's e-mail prompted immediate responses from almost all of the committee members. Seven of the ten members, including the AV, responded to the IACUC administrator's e-mail by indicating that the PI should be immediately notified that his research activities must be halted until the IACUC had adequate time to meet and discuss the issue. Two of the remaining three committee members did not respond to the notice, and the final one indicated that an immediate meeting should be scheduled to discuss the concerns.

Based on the e-mail responses, the IACUC administrator decided that a quorum of the IACUC agreed that the incident was noncompliance and that the project should be suspended. The PI was informed that, since he was not following his approved protocol, his project has been suspended pending further IACUC investigations. This process was an inappropriate use of e-mail. There were several problems with the use of e-mail in this manner:

- Polling (as a decision-making tool) is specifically prohibited by the regulatory agencies.
- The IACUC members did not actively discuss the concerns.
- An official meeting including access to relevant materials was not scheduled and announced.
- Official actions such as maintaining meeting minutes did not occur.

A preferred approach could have been that the IACUC administrator made the following announcement: "An urgent matter has evolved that requires immediate attention by the IACUC. As a result, a telephone conference meeting has been scheduled for May 21 to discuss the issues. During the meeting, the committee will review and discuss an alleged incident of noncompliance that has animal welfare implications."

A best practice could have been for the IACUC administrator to initiate the conference call with a summary of the alleged incident. The deliberation could have then been turned over to the IACUC chair, and he would have opened the floor for discussion. Upon completion of the discussion, the IACUC chair could have called for a motion and asked for a vote. Following the vote the IACUC members may have discussed the necessary corrective actions, which too may have resulted in a motion and a vote. The results of the votes would then have been recorded in the meeting minutes, and a draft of those minutes would have been reviewed and approved at a future meeting. The IACUC administrator could then have officially notified the PI that his project had been suspended due to a protocol deviation until specified corrective measures had been engaged.

Scenario 2. The IACUC administrator at GEU regularly e-mailed committee members a list of protocols. The list included each protocol number, the PI's name, and a brief project summary. Committee members were given 2 weeks either to request that the protocol be reviewed before the full committee or to recommend approval. In the past 3 years, only four protocols have been reviewed by the full committee. All other protocols have been approved based on the number of recommended approvals received through e-mail replies. The IACUC administrator records the number of recommended approvals, disapprovals, and abstentions and applies the standard rules for majority voting. The votes are documented in the protocol file, and the approval letter is distributed by the IACUC administrator.

There are several problems with using e-mail in this manner:

- Polling is being used as a decision-making tool, which is specifically prohibited by the regulatory agencies.
- The IACUC can approve protocols during full committee meetings or by using the designated member review process; e-mail cannot be used.
- The IACUC members did not actively discuss the concerns.
- An official meeting including access to relevant materials was not scheduled and announced.

A preferred approach would have been for the IACUC administrator to have e-mailed the protocol list to the committee members and given them 5 days to identify which projects they wished to be reviewed at the full committee level. For those projects not identified for a full committee review, the IACUC chair could have appointed specific qualified committee members as DMRs to approve the protocols. The DMR could then have communicated with the IACUC administrator via e-mail and indicated which proposals had been reviewed and approved. The IACUC administrator would then have notified the PI that his or her project had been reviewed and approved by the committee and reported the DMR outcomes to the IACUC at the next scheduled meeting.

Use of Video Conferencing Equipment

Organizations can effectively use telecommunication systems (e.g., Skype, FaceTime) to facilitate official IACUC business. BP meeting attendees noted specific ways in which this technology can improve the efficiency of the IACUC oversight and review processes. Consider the following scenarios.

Scenario 1. GEU has multiple experimental stations about eighty-five miles north of its main campus. Approximately 35 percent of GEU's animal care and use protocols are conducted at the remote research stations. Since these facilities are very active, the IACUC chair asked a station director to serve on the IACUC. The director expressed a keen interest but declined because he did not want to travel 170 miles (round trip) to attend monthly meetings. The IACUC chair responded by encouraging IACUC membership and participation in the monthly meetings using the "Skype" program on his computer. After a few months of using Skype, the IACUC chair and the research director reconsidered the process. Both agreed that Skype was allowing for video and voice participation in the IACUC meetings, but the quality of the video was not always the best and the audio dropped on occasion. The IACUC chair presented the situation to the institutional official (IO), who was supportive of the research director's continued participation. Consequently, the IO provided the funding to purchase enhanced video conferencing equipment, which facilitated more effective participation by remote committee members.

The resulting equipment upgrade created additional opportunities for GEU's IACUC. The research station director could simplify the semiannual inspections of the research stations and save IACUC members the 170 miles of travel. The research director could now walk through the facilities with a video camera and transmit a live video stream for all committee members to view. During the walk-through, committee members were able to ask questions and direct the videographer to particular areas within the facility.

Scenario 2. GEU has a state-of-the-art BSL3 animal facility. Scientists in the facility work with infectious agents that can be lethal to anyone who is infected by accidental exposure. The facility security room is equipped with multiple surveillance monitors that receive video feeds from key locations throughout the facility. A surveillance camera is mounted in each animal room. The camera permits the observation of animals, procedures, cage activity, watering systems, and environmental controls. Cameras are also mounted in the feed storage area, surgical suites, cage wash facilities, and research labs; they also transmit video feeds to the security room.

Since physically entering the facility can pose unnecessary risk to IACUC members, the committee members effectively perform the BSL3 facility inspection using the security room monitors. During the inspection, the BSL3 facility director is equipped with a mobile telephone to communicate with the IACUC members conducting the inspection. The committee members are provided a map of the lab, which allows them to ensure that the facility is thoroughly inspected. The director visits each animal housing room and answers committee members' questions. Even though all of the animals can be thoroughly reviewed using the surveillance monitors, the facility director can also hold cages of animals in front of the camera, which offers a close-up view of the animals. To facilitate the inspection, the program director also holds copies of the daily observation logs in front of the camera, which provides an opportunity to review the records using the security monitors. The same process is used for the review of surgery, cage wash, and sanitation records.

RECORDKEEPING

A primary responsibility of IACUC administrators is the development and maintenance of effective record management systems. There are generally two types of record keeping systems: hard-copy and electronic records.

Hard-copy records tend to be used in smaller programs while electronic record-keeping tends to be used at larger organizations with more complex programs. Either system can be effective or either system can be a nightmare. Many organizations that have historically used hard-copy records are gradually transitioning to various electronic processes, in part to increase efficient reporting and monitoring practices.

Hard Copies

The use of paper records requires the development of spreadsheets, checklists, or data systems to track dates and processes. Hard copies are generally kept in a file cabinet, sorted by department or PI name. Specific concerns about hard copies include that records checked out may be misplaced, records on one person's desk may not be accessible to other staff members requiring the document, misfiled records may be difficult to find, and hard-copy records may have a backup copy at a second location. Although these concerns exist, many organizations very effectively use a hard-copy document management system.

Electronic Systems

There are several options for electronic records—some simpler than others and each having its own advantages and disadvantages. Consider the following examples:

PDF-Based Systems

Many organizations use Adobe systems to develop dynamic interactive PDF IACUC submission templates and other animal care and use program forms. PDF templates may be stored on web servers for easy access, which allows for standardization across multiple locations. The PDF documents can be completed and submitted by e-mail central servers, which can later be accessed and processed by the IACUC administrator. One advantage of using PDF forms is that they can be electronically signed. However, many organizations agree that e-mails sent through a secure e-mail system are equivalent to an electronic signature. If an e-mail is sent from Dr. Smith's password-secured e-mail account, organizations have agreed that the e-mail is equivalent to an electronic signature.

Since PDF documents can be electronically stored by organizations at local and distant sites, the use of PDF forms can be an integral part of the disaster plan. In addition, PDF documents can be e-mailed to IACUC members, permitting easy access to IACUC submissions. A common best practice for many organizations is to use a secure document processing server to share large files. Organizations engaging this form of electronic data management can share entire documents that are several hundred megabytes and eliminate (or significantly reduce) the need for paper documents.

Microsoft Software

Many organizations commonly use Microsoft products. As a result, BP meeting attendees indicated that they have developed Microsoft-based data management systems. IACUC administrators frequently use programs such as Microsoft Excel to maintain and monitor program and protocol data at minimal costs. IACUC administrators frequently develop spreadsheets and enter data to track protocol-related activities such as animal use and expiration dates. While enterprise consultants who can build reporting templates for Excel or Access are available, a novice computer user can easily develop spreadsheets to track expiring protocols and run relevant reports.

Table 22.1 has formulas built into the column headed "Expiration Status" that displays the "Check" notice once the project is 60 days away from expiring. When a project expires, "Expired" is displayed directly adjacent to the PI's name, and an "Expired Protocols" notice is displayed across the top of the spreadsheet.

Table 22.1　Sample Data Management Spreadsheet

				There Are Expired Protocols					
Expiration Status	Principal Investigator	Protocol Number	Expiration Date	Sponsor	USDA Pain Category	Protocol Type	Species	Species Type (Ag/Lab/Wildlife)	Surgery (Sur/NonSur)
	1 Alex	33544	3/9/2011	NSF	D	Research	Birds	Wildlife	Sur
	2 Alice	29145	8/8/2011	PSU	C	Teaching	Deer	Wildlife	No
Expired	3 Alison	29605	10/5/2010	PSU	D	Teaching	Mice	Lab	NonSur
	4 Curt	33684	4/6/2011	NIH	D	Research	Mice	Lab	Sur
	5 Darcy	31738	8/1/2011	PSU	D	Research	Hamster and gerbils	Lab	No
	6 Darcy	30743	3/15/2011	PSU	D	Research	Mice	Lab	No
	7 Eric	30201	1/18/2011	USDA	D	Research	Chickens	Ag	NonSur
	8 Erica	30661	3/15/2011	Pharma	D	Research	Swine	Ag	NonSur
Check	9 Frank	32327	10/19/2011	PSU	C	Breeding	Cattle—dairy	Ag	No
	10 Greg	31415	6/21/2011	Pharma	E	Research	Swine	Ag	No
Expired	11 John	26800	9/29/2010	PSU	C	Research	Mice	Lab	No
	12 Jon	28178	4/19/2011	NSF	D	Research	Peromyscus leucopus	Lab	No
	13 Les	31061	4/25/2011	NIH	D	Research	Mice	Lab	NonSur
Check	14 Mary	29692	11/16/2010	PSU	D	Research	Mice	Lab	Sur
	15 Melanie	30198	1/12/2011	PSU	D	Teaching	Rats	Lab	Sur
	16 Melissa	34174	6/1/2011	PSU	D	Research	Bull calves	Ag	Sur
Expired	17 Peter	29566	9/29/2010	USDA	D	Research	Cattle—dairy	Ag	Sur
	18 Robert	30866	4/11/2011	PSU	D	Diagnostic	Rabbits, peromyscus, and Rats	Lab	No
Expired	19 Sally	29287	10/11/2010	PSU	C	Breeding		Lab	No
	20 Tony	31297	6/16/200	NIH	E	Research	Peromyscus leucopus	Lab	No

Note: The "Expired" notice appears the day after the project expires. The "Check" notice appears 60 days before the protocol expires.

Commercial Off-the-Shelf Systems

There are many commercial systems available that can meet the data storage and reporting needs of an organization. Compliance software organizations such as Click Commerce, Edstrom, eSirus, and IRBnet have elaborate commercial systems that can be customized to serve the needs of almost any organization. Most commercial products have built-in application wizards, which allow direct proposal submission to the IACUC office. Once the proposals are in the database, critical information such as protocol expiration dates, the location of surgery suites, and yearly animal use can be managed by the IACUC administrator.

Although desired by some organizations, BP meeting attendees indicated that their organizations' budgets have been unable to accommodate the cost of purchasing and implementing commercial compliance data systems. In addition to the cost of the software system, charges associated with implementation, staff training, and software licensing and maintenance fees must be considered.

It is worth noting that the implementation phase of a new electronic system may take months to years, depending on the complexities of the program.

RECORDS MANAGEMENT

Veterinary Care Records

Veterinary care records are maintained by the veterinary unit and are not generally held or managed by the IACUC administrator. These records include medical records, daily observation records, purchase records, and euthanasia records, for example—generally any record describing direct patient care or direct subject use. Veterinary care records should be reviewed as part of the IACUC's semiannual program and facilities inspection. The purpose of the review of these records is to show IACUC oversight of all activities involving animals at the campus and to confirm adequate care of the organization's animals.

Program Records

Records that document program activities (e.g., compliance deliberations, the IO report, accrediting documentation, and the USDA/PHS/AAALAC annual reports) must also be maintained. These records document required program activities and are maintained for regulatory reasons and to facilitate ongoing program consistency. These records are generally kept by the IACUC administrator and provided to the IACUC, IO, or regulatory agencies as requested.

Records are usually collected on templates. Most BP meeting attendees indicated that the templates are often created using Microsoft Word and include IACUC member appointment letters, meeting minutes, and noncompliance notices.

In addition to Word template documents, IACUC administrators also noted that they frequently store and secure documents that include signatures such as the IO's signature on the OLAW report. In all cases, such documents should be maintained on a secure server with regular backup to prevent loss of critical working files or templates and data files.

Meeting Minutes: What Must They Contain?

BP meeting attendees discussed the contents of meeting minutes and the ramifications of detailed versus meager minutes. The controversy exists since an organization's meeting minutes can be accessible to the public through open records laws. BP meeting attendees agreed that IACUC administrators must understand their states' open records laws so that they are able to satisfy these legal requirements. Having a working knowledge of that information will best position the IACUC administrator to comply with open records laws while minimizing risk to the organization.

In some cases, IACUC administrators use recordings to assure accuracy of the minutes. Audio recordings are subject to certain open records requirements; therefore, as a matter of general practice, the recording should be destroyed after the minutes are approved.

Before preparing meeting minutes, IACUC administrators should be aware of what is required to be in the meeting minutes. The specific names of meeting attendees need not be listed in meeting minutes. Regulatory agencies permit organizations to use coding systems to identify attendees. The code key can be provided to federal regulators upon request, and the two documents are generally maintained separately. While there may be operational reasons to maintain the details of designated member reviews and approvals, it is not required.

The details of the IACUC's discussion of each protocol must be recorded in the meeting minutes. Organizations make every effort to keep the minutes compliant with the regulatory expectations. While not necessarily the preferred option of the regulatory agencies, it is generally in the best interest of the organization to maintain meager minutes that are accurate and complete. A documented discussion may include a method to identity the protocol (e.g., the protocol number), a superficial summary of the IACUC discussion, and the final disposition of the protocol.

To supplement the minutes, many organizations maintain working documents, which are not IACUC approved but rather are administrative documents. These records frequently convey the intent of the IACUC on specific matters. They may include reviewer sheets or specific information provided to the IACUC to support or supplement a discussion. Generally, working documents that have not been approved by the IACUC are not subject (or less subject) to states' open records requirements. However, each organization must be familiar with its state's open record laws to verify which records are accessible under the law.

Animal Care and Use Protocol Records

Records that document animal care and use activities must also be maintained by organizations. Protocol records include, for example, the animal care and use protocol, annual progress reports, postapproval monitoring records, and approval letters. These records are regularly monitored to ensure that a protocol does not expire before the research has been completed. IACUC administrators also monitor animal use records to ensure that a PI does not exceed his or her approved animal numbers. These records are also used to monitor projects that have approved *Guide* deviations as well as projects that involve more than momentary pain or distress, for example.

Duration of Record Maintenance

Veterinary medical records for USDA-covered species must be maintained for 2 years past the death or disposition of the animal. The PHS policy requires medical records to be maintained for 3 years, so if a PHS-funded project involves USDA-covered species, the medical records for those animals must be maintained for 3 years. All other veterinary records (e.g., daily observation sheets, purchase records, etc.) should be maintained for 3 years after the conclusion of the project or the death of the animals, whichever is later. Program records such as meeting minutes, IO reports, and annual reports must be maintained by the organization for 3 years after the project has been completed. For example, if a project continues for 12 years (four subsequent protocol renewals at 3 year intervals), then all materials must be maintained for 15 years—the 12 years of the project plus 3 years as required by regulation.

USING ELECTRONIC SYSTEMS
FOR THE PROTOCOL REVIEW PROCESS

Protocol review efficiency may be improved by the use of computer protocol management software. BP meeting attendees discussed using commercial software systems and homegrown electronic processes for conducting and documenting the protocol review and approval process.

Commercial Software

Organizations that purchase commercial compliance software usually have the selectivity to ask only those questions relative to the scientist's research. For example, a researcher may be asked to indicate whether his or her project includes survival surgery. A "yes" response would prompt additional questions relative to survival surgery (e.g., questions on anesthesia, analgesia, or postoperative procedures). For example, when application wizards are appropriately designed and employed, a research application that involves raising mice as a source of tissue for in vitro work may be five pages long, whereas one involving multiple test reagents, survival surgery, and dietary deviations may be twenty-five pages.

Commercial software frequently permits IACUC members to log in to a secured network and review recent submissions to the committee. They are able to enter comments about the proposed activity, request full committee reviews, or access grant applications associated with the project.

Organization-Specific Electronic Processes

Some organizations have developed ways to use technology that is already in place, instead of purchasing commercial protocol management software.

In almost all cases, IACUC administrators use networks, e-mail, and other computer technologies at their disposal to facilitate the protocol review and approval process. Organizations have developed very effective methods for using these resources. Considering the regulatory requirements associated with protocol review, one method discussed by BP meeting attendees included the use of e-mail and a shared network specifically secured for IACUC business. The IACUC administrator accepts submissions from PIs in the form of an e-mailed Word document. The administrator saves the protocol to a secure server where committee members can log in to access the submissions. The IACUC administrator accumulates the submissions for 10 days and then e-mails a list of submissions ready for review to the committee members—for example:

To: <List of IACUC Members>
Sent: Tuesday, December 31, 2012, 9:05 a.m.
Subject: IACUC Protocols—Scheduled for January 10, 2013, REVIEW
IACUC Members,
    The following protocols will be reviewed by the IACUC 10 days after this e-mail correspondence (i.e., January 10, 2013). If you would like any of the listed submissions reviewed by the full committee, identify them by responding to this e-mail prior to the scheduled review date. Those protocols not identified for full committee review will be assigned to a designated reviewer.

1. IACUC #12345 (3-year renewal)—Dr. Peters
2. IACUC #67890 (annual renew)—Dr. Jones
3. IACUC #54321 (modification)—Dr. Allen
4. IACUC #09876 (new submission)—Dr. Williams

Note: Written descriptions of the proposed activities are accessible through the share drive and can be found in the IACUC folder (file name: January 10, 2013 review).
Best regards,
<IACUC administrator>
Great Eastern University

The e-mail template satisfies many regulatory requirements associated with the protocol review process, including that

- The entire IACUC was provided a list of projects for review
- Committee members were given the opportunity to ask that the submission be reviewed by the full committee
- Committee members were given access not only to the title of the proposed activities but also to the complete submission through the shared drive

The shared drive is established in conjunction with the IACUC office's information technology services. IT staff members established a shared LISTSERV that is available through the Internet. The system includes security levels accessible through passwords. Many organizations provide community members with "guest accounts" to ensure that they can access the shared drive or mail/e-mail review packets to community members every 10 days. Once committee members have access to the shared drive, they are able to log in to the system and review the projects in a timely and efficient manner.

CHAPTER 23

Policies, Guidelines, and Standard Operating Procedures

SCENARIO

GEU's animal care and use program (ACUP) has grown exponentially over the past 3 years. In early 2010, the GEU IACUC oversaw only nine protocols for two scientists conducting animal research. In 2014, the IACUC oversaw over two hundred protocols for more than one hundred practicing scientists.

As a result, the IACUC administrator encouraged GEU to formalize its program by preparing a written description of the GEU ACUP. He suggested developing a set of policies to serve as the foundation for the program and standard operating procedures (SOPs) to identify standard practices within the program. The IACUC chair appointed a subcommittee and asked the IACUC administrator to serve as the subcommittee chair. The task of the subcommittee was to establish a written program of animal care and use for GEU.

After approximately 6 months, the subcommittee brought its first round of documents to the committee for review and approval. The first set of documents included policies were titled:

"The Use of Vertebrate Animals in Research, Teaching, and Testing"
"The Occupational Health and Safety Program (OHSP) for Animal Users"
"Required Safety Equipment When Working with Laboratory Animals"

As the subcommittee prepared to discuss the proposed policies, members explained to the IACUC that a networking system of IACUC administrators who have well-established ACUPs had been designated by the IACUC Administrators Association to help with the process. The subcommittee indicated that benchmarking with colleagues allowed it to customize policies that had already been tried and tested as part of fully functional academic animal care and use programs for GEU. The IACUC administrator noted that the subcommittee had also evaluated the proposed policies using the regulations (e.g., PHS policy, Animal Welfare Act regulations [AWAR], and the *Guide for the Care and Use of Laboratory Animals* [henceforth referred to as the *Guide*]) as the standards.

As the subcommittee began to develop the policies, its members began to identify significant program deficiencies. For instance, they knew that animal users were not formally trained to recognize and protect themselves from animal-related risks, and that they were not required to wear personal protective equipment when working with animals. To compensate for the deficiencies, a GEU policy was developed that requires all animal users to complete a comprehensive training program on animal-related risks and to use defined personal protective equipment (e.g., gloves, lab coat, and safety glasses) when working with animals. In addition, GEU's OHSP did not include a personal risk assessment for animal users that entailed health care professionals (e.g., physician, clinician, or nurse) assessing an individual's health status to determine if his or her exposure to animals would place the individual at an elevated level of risk. The GEU process was to simply notify animal users that GEU employs a physician who is available to answer any questions relating to medical issues associated with animals. Through the policy development process, the subcommittee resolved the regulatory concern by requiring all staff and students conducting animal care or use activities to complete a medical surveillance questionnaire, which would be reviewed by a health care professional. The health care professional might be part of the GEU's medical staff or the animal handler's personal physician. In either case, a physician or nurse must provide clearance for each animal handler to work with animals before beginning any animal activities.

In addition, GEU's IACUC established informal practices for overseeing all animal research activities, but no process was in place for overseeing animals used in biology or agriculture animal production courses. Because the PHS policy requires IACUCs to oversee the use of vertebrate animals in research, teaching, or testing, the subcommittee proposed a policy requiring all activities involving vertebrate animals to be reviewed and approved by the IACUC before they were initiated.

Upon completion of the formal policy preparation process, the IACUC and the subcommittee realized the importance of establishing written documents to ensure that oversight is applied consistently and to document that all of the regulatory standards are being satisfied. Based on the subcommittee's explanations and justifications, the IACUC unanimously approved the three submitted policies. The policies were provided to the institutional official (IO), who immediately endorsed them, and the policies were distributed campuswide.

The "ink was still wet" on the policies when the IACUC administrator received a phone call from the Occupational Medicine (OM) physician, who asked who had come up with a policy that required him to see over one thousand faculty, staff, and students before they could begin their work with animals. He indicated that, based on what he had learned from the policy, he would be reviewing approximately fifty personal risk assessments a week and scheduling appointments with animal workers accordingly. Prior to ending his discussion with the IACUC administrator, the doctor indicated that he would need additional staff to satisfy the newly created policy requirements. As a result, the physician immediately contacted GEU's IO and explained how the new policy requirements would affect his department and that he would need funds to hire another clinician. The IO responded that no funds were available to accommodate a new position, but that a solution must be found.

Soon after his conversation with the OM physician, the IACUC administrator received a call from the Environmental Health and Safety (EHS) director. She asked who had drafted a policy that required animal users to wear safety glasses when handling animals. The IACUC administrator explained that, as a result of a subcommittee review, the IACUC deemed the use of safety glasses necessary. The EHS director objected, noting that it is the practice of the EHS office to conduct environmental risk assessments in all areas at least annually. She indicated that there were no eye injury hazards associated with the species of animals used at GEU that warranted the use of safety glasses. She indicated that the policy created by the IACUC directly conflicted with a GEU health and safety policy and that the IACUC policy could not be enforced.

Where did the best intentions go wrong? Was there a better way to address the very real issues of policy development and effective oversight? The moral of this story is to communicate, communicate, and communicate to all appropriate organization representatives that can be affected by the development of ACUP policies. Although offices such as OM and EHS may affect only a portion of the program, decisions by the IACUC may have serious and consequential impact upon other departments across campus. Affected activities outside the ACUPs must be considered during programmatic discussions.

In this scenario, the IO, the OM physician, and the EHS director were negatively affected by a decision made by the IACUC. The confusion may have been avoided if the IACUC subcommittee had invited organization representatives to participate in the development of the policies. If the EHS director had been involved in the discussion regarding safety glasses, the IACUC may have concluded that safety glasses were not needed when working with research animals, or the EHS director might have seen value in safety glasses that were previously not required. If the OM physician had been aware of the new workload requirements, he could have initiated measures, whether internally or externally, to support the animal program and fulfill the expectations of qualified medical surveillance oversight.

Whenever animal program leaders consider the development of a new policy or requirement, it is generally best to vet the draft policy through other agencies on campus that may be impacted or have special knowledge on the issue.

POLICIES, SOPs, AND GUIDELINES—WHAT IS THE DIFFERENCE?

Best practices (BP) meeting attendees have reported confusion as a result of organizations using the terms *policy*, *guideline*, and *standard operating procedure* interchangeably. In some cases, researchers have told IACUC administrators that they do not have to follow SOPs, because those are not strict policy documents. Another BP meeting attendee reported that the IACUC's guidelines were not necessarily something for which the principal investigator (PI) could be held accountable. BP meeting attendees believed that it was generally a good practice to explain the purpose, authority, and scope of the various policy documents used at an organization. Meeting attendees formulated sample definitions.

Policy

A policy is a plan or course of action intended to influence and determine decisions, actions, and other matters. A policy is the formally voted and approved position of the IACUC. Even though there is no requirement for the IO to endorse IACUC-approved policies, the committee may consider requesting an IO endorsement for those policies that may meet with a degree of resistance. A policy usually does not describe a process or define required steps. For example, a policy may be written that states that "all animal use conducted under the auspices of GEU must be reviewed and approved by the IACUC," rather than the steps for submitting a project for IACUC review. A policy of the organization is enforceable by federal oversight bodies (and the Association for Assessment and Accreditation of Laboratory Animal Care), in the sense that the federal and accrediting agencies expect organizations to adhere to their own policies. Local policies may not be less stringent than federal regulations and policies, but they may restate or be more stringent than the federal expectation.

Policies are exceedingly helpful in assuring consistency over time and across situations. Policies provide foundation and fences so that IACUC administrators will know how to consistently assist the researcher. The following are situations that are typically governed through the use of policies:

- The type of training that personnel must undergo before working with animals
- The requirement for the IACUC to review and approve animal care and use activities before they are initiated
- The requirement for a safety review when using biohazards in conjunction with animal experiments
- The expectations of IACUC members in fulfilling their duties and roles
- The criteria for a retired research animal to be eligible for adoption

Standard Operating Procedure

An SOP generally consists of detailed written instructions to achieve uniformity of performance of a specific function. SOPs are developed to achieve consistent results. SOPs may or may not be voted on by the IACUC, but should at a minimum have an authorizing signature from senior personnel (e.g., AV, Facility Manager). For example, an SOP may be developed to ensure that a policy is regularly satisfied. An SOP that outlines the steps for submitting a proposed animal use activity for review and approval will help to ensure that the policy requirements are satisfied and that no animals may be used at GEU before IACUC approval is granted. For example, an SOP for starting animal use may be the following:

1. Complete the application and submit it to the IACUC for review.
2. Receive the approval letter from the IACUC.
3. Provide the approval letter to the animal resource program staff member.
4. Order animals for the project.

Organizations using SOPs generally require close adherence to the SOP to ensure program consistency, quality outcomes, and efficiency. Examples of processes that are typically outlined using SOPs include the protocol review and approval process, the process for conducting semiannual inspections, investigating and reporting non-compliance, and accounting for the number of animals used in a specific experiment.

BP meeting attendees noted that some organizations permit PIs to reference an SOP within the IACUC proposal (instead of requiring all of the details to be included in the protocol application). Some organizations develop SOPs for conducting routine animal use activities and encourage their use by PIs. Let us say, for example, that ten PIs at one organization are genotyping mice using a tail snip. The IACUC decides to develop and approve a step-by-step procedure for genotyping mice. To further facilitate efficiency, the IACUC agrees that when a PI is going to genotype, the PI only needs to list in the protocol that: "Genotyping will be conducted according to the IACUC-approved 'Genotyping Mice' SOP." Although this is feasible, most BP meeting IACUC administrators agreed that the information should be put directly into the protocol; otherwise, the process becomes too automated and PIs begin to rely too much on SOPs and do not think about what they are doing or how it should be done.

IACUC administrators agreed that if IACUCs allow referencing SOPs, then IACUCs must establish a way to ensure that the SOPs are periodically reviewed by the IACUC, much in the same way that protocols are annually or triennially reviewed, and that when a search for alternatives is part of the SOP, the search is regularly updated. In addition, an organization following this process must establish a method to ensure that scientists receive and are working according to the most recent edition of the SOP.

Guideline

A guideline is a suggested approach to an activity or issue. Generally considered a local "best practice" based upon experience or past successful outcomes, guidelines rarely have a formal IACUC vote or approval. The goal of a guideline is to standardize a particular process according to a best practice. Organizations often develop guidelines to ensure that specific animal procedures are of the highest quality. In situations where guidelines are routinely used, the IACUC may endorse the guidelines as part of its programmatic review process. For example, the organization may have developed a guideline, "Performing Blood Collection in Mice." The guideline may describe one or more acceptable methods of collecting blood and the specific steps in each method.

Guidelines are routinely developed for scientists by veterinarians and the IACUC members. Guidelines typically encourage and facilitate consistency between research groups and provide new researchers an opportunity to learn what is generally considered an acceptable practice on campus, or perhaps in the industry. Examples of activities that are typically governed using guidelines include blood collection volumes for research animals, survival surgeries per procedural activities, aquatic animal management and care practices, and tissue collection methods.

POLICY FORMATS

Organizations use diverse document structures. BP meeting attendees indicated that document structure is often driven by the intended audience.

A policy used by the IACUC may be one or two sentences and define a concept to validate decision consistency. For example, an IACUC policy may simply state, "Pharmaceutical reagents must be used in research when they are available. Nonpharmaceutics can only be used when scientific justification is provided."

A policy written for the research community may be a page or two and include substantially more detail while remaining sufficiently global for application across the campus research environment. This style of policy may include a purpose, definition, policy statement, and reference section. For example, a policy titled, "The Use of Vertebrate Animals in Research, Teaching, or Testing," in the purpose section may discuss an organization's commitment to the humane care and use of animals as well as the federal regulations. This policy's definition section may define vertebrate animals and what is considered research, teaching, and testing. The actual policy section of the document typically defines the requirement—for example, "All animal use activities must be approved by the IACUC."—and the reference section may list additional institutional documents, web links, federal mandates, and policies.

SAMPLE POLICIES

Any list of polices would be incomplete as each program must be established to complement its research portfolio. However, the following sample policies were provided by BP meeting attendees.

Adoption Policy

IACUC administrators agreed that the premise of allowing retired research animals to be adopted was an ethical position and fully supported by the three Rs (replacement, reduction, refinement). However, any such policy should include considerations for the retired research animal's temperament or health status, as well as potential risk to the adopter and to the organization (e.g., whether the organization could be liable if a retired animal harmed its new owner). An additional concern expressed by many BP meeting attendees was whether laboratory animals such as lab mice and rats should be eligible for adoption since many of them (e.g., mice with genetic abnormalities) were bred specifically for research and could have potential impact upon wild mice if they escaped or might have specialized health concerns that would require specialized skills to manage. The BP meeting attendees suggested two policy types.

The simplest included only the policy statement and read, "Research and the research environment can have undetermined effects on animals. As a result, no animal used for research or maintained in a research setting is eligible for adoption."

A more complex version of the adoption policy is less conservative and was written for the scientific audience and general public. Consider the following policy format:

Title: The Adoption of Retired Research Animals at GEU, policy 1234.

Purpose: To establish criteria that will be used to determine whether a retired research animal is eligible for adoption. The policy will also identify the appropriate parties that can facilitate the adoption process.

Definitions:
- **Retired research animal:** A retired research animal is any dog, cat, rabbit, guinea pig, hamster, or bird that has not been physically affected by research as determined by the facility veterinarian. Animals that do not qualify for adoption include laboratory mice and rats, as well as animals that are transgenic, disfigured, or disabled or those having pre-existing conditions.
- **Eligible candidate:** An eligible candidate is anyone wanting to adopt a retired research animal who is willing to certify that he or she will give the animal appropriate care and provide to the IACUC the name of a clinical veterinarian that will care for the animal after adoption.

Policy: It is the policy of GEU to allow eligible candidates to adopt a retired research animal providing that it is approved for adoption by a staff veterinarian. Eligible candidates interested in adopting retired research animals should contact the attending veterinarian directly at <contact information>.

Use of Hazardous Materials in Conjunction with Animal Activities Policy

At past BP meetings, IACUC administrators discussed policy needs as they related to the use of biohazardous materials in conjunction with animal research. Since the *Guide* discusses the organization's responsibility for a formal process that evaluates potential risks and ensures the safety of animal users, it is therefore prudent for organizations to have formal programs that look at risks such as infectious agents, genetically modified organisms, and chemical hazards. In this situation IACUC administrators agreed that a policy on the use of biohazardous materials is typically a document that must be employed throughout the institution. The IACUC administrators' collaborative effort established a sample policy:

Title: The Use of Hazardous Materials in Animal Research

Purpose: To define the criteria that must be satisfied in order to use hazardous agents in conjunction with animal research

Definitions:
- **Animal activity:** Animal activity is any research, teaching, or testing event that involves an animal.
- **Hazardous material:** A hazardous agent is any biological, chemical, or radioactive material that increases an individual's personal risks.
- **Biological hazard:** This hazard is a bacteria, virus, or parasite known to cause disease in humans or animals. In addition, a transgenic animal may be considered a biological hazard under certain circumstances.

- **Chemical hazard:** This hazard consists of any carcinogens, tumor promoters, toxins, or toxicants.
- **Radioactive:** Any ionizing agent is radioactive.
- **Biosafety Committee:** This group comprises safety professions that review and approve the use of hazardous agents.
- **Radiation Committee:** This group of radiation professions review and approve the use of ionizing radioactive materials.

Policy: It is the policy of Great Eastern University that animal research involving hazardous materials be reviewed and approved by either the Biosafety or the Radiation Committee before the IACUC approves any project involving animal activities.

IACUC administrators are encouraged to contact their colleagues at similar organizations or visit the IACUC Administrators Association website for examples of more policies.

Appendix 1

Position Announcement—Postapproval Monitor

Great Eastern University is interested in hiring a postapproval monitor to join its research compliance office team.

DUTIES

The postapproval monitor will be coordinating and conducting reviews of approved animal research activities. This activity may include reviewing clinical records, postsurgical records, and other documents for all laboratories in the university. During postapproval monitoring visits, the monitor will review protocols and records for performance, accuracy, and completeness and report to the Director of Compliance Office on any improvements to be made. The monitor may also work with principal investigators to ensure that regulatory standards are satisfied. The candidate may serve as a member of the IACUC and be involved in the review of the animal care and use program and participate in facilities inspections. He or she may also be asked to review research projects involving vertebrate animals.

QUALIFICATIONS

The applicant should have a bachelor's degree in a relevant discipline (e.g., the biological or veterinary sciences). He or she should have 3–5 years' experience relating to Institutional Animal Care and Use Committees (IACUCs) and animal care programs. This may include experience with animal experimentation or supporting an organization's IACUC. Detailed knowledge of applicable federal regulations governing animal use activities is preferred. The applicant should be able to work independently and have relevant certifications (e.g., certified professional IACUC administrator, American Association for Laboratory Animal Science certification). The individual must have excellent written and oral communication, customer service, and advanced computer software skills.

Appendix 2

Position Announcement—IACUC Administrator

POSITION SUMMARY

Reporting to the Research Compliance Office director, the Institutional Animal Care and Use Committee (IACUC)/Institutional Biosafety Committee (IBC) administrator will provide research compliance administration for the animal subjects and biosafety research programs. The selected candidate will provide the direct support needed to administer the daily activities of the IACUC and IBC programs.

DUTIES

- Manage protocol-related activities, such as conducting administrative pre-reviews; serve as the principal investigator's liaison to the committees, tracking animal use and expiration dating on protocols; and ensure that protocols are receiving effective compliant reviews
- Develop relevant educational and outreach programs and deliver them to faculty, staff, and students
- Develop and execute quality assurance programs to ensure that programmatic activities are compliant and effective
- Assist with the daily management of the IBC program
- Participate in the development, and implementation of policies and standard operating practices in the research compliance areas

QUALIFICATIONS

The candidate should have a bachelor's degree in a relevant field (e.g., biological or veterinary sciences). He or she should have a minimum of 3 years' experience working in research compliance. The applicant should have a working knowledge of animal care and biosafety compliance administration and have relevant certifications, such as the certified professional IACUC administrator and/or American Association for Laboratory Animal Science certifications. Knowledge of federal rules and regulations regarding animal care and biosafety, particularly at a higher educational setting, is highly desirable.

Appendix 3

Position Announcement—
Research Compliance Coordinator

The qualified individual will ensure that Great Eastern University's animal care and use and biosafety programs comply with federal, state, and local regulations and policies governing research. The selected candidate will interact with researchers by prescreening and assisting with research proposal development, serving as their advocate at compliance committee meetings, and guiding them to maintaining compliance.

DUTIES

The selected candidate will serve as compliance committee liaison to the university (i.e., researchers, administrative personnel, and other relevant staff). The individual will be responsible for executing and implementing actions identified by the compliance committee(s) as necessary to maintain research program compliance. In addition, the individual will perform the daily activities required to maintain compliant research programs. These activities include (but are not limited to) maintaining written and electronic records, reviewing and approving research proposals, maintaining a compliant occupational health and safety program as it relates to animal use, conducting facility inspections, preparing and distributing correspondences to university research personnel, monitoring approved research proposals, preparing reports, providing appropriate guidance to office personnel, and completing other duties as assigned.

QUALIFICATIONS

Interested candidates should have a bachelor's degree or equivalent in a related field, plus 3 years of work-related experience. The candidate should be proficient with Microsoft Office and must demonstrate effective administrative and technical skills. Candidates must be able to exercise a wide range of independent judgment and discretion, be able to maintain confidentiality, have excellent interpersonal and communication skills, and have demonstrated the ability to work independently.

Knowledge of federal, state, and local regulations and policies governing research is preferred. The relevant documents include the National Institutes of Health guidelines for research involving recombinant DNA molecules, the Animal Welfare Act regulations, and the Public Health Service policy on the humane care and use of laboratory animals.

Appendix 4

Sample Organization Chart

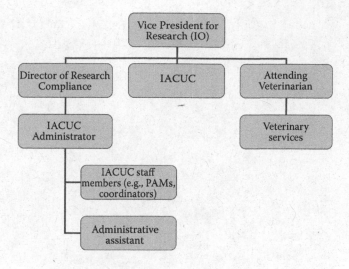

Appendix 5

Sample Organization Chart

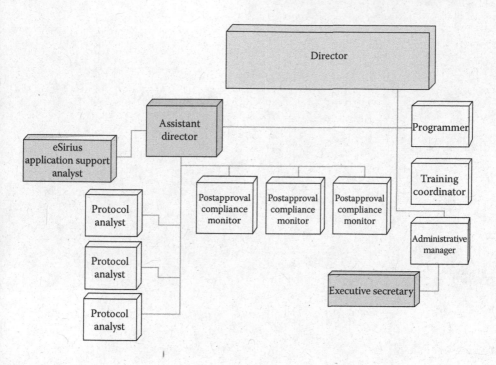

IACUC Member Agreement (Issued at Initial Appointment and Reappointment)

ATTENDANCE

Regular meeting attendance by IACUC members is necessary in order for the IACUC to achieve and maintain the quorum required to conduct official business. Full IACUC members are expected to attend **75 percent of the scheduled full committee meetings**. If unable to attend a meeting, an IACUC member should notify the IACUC administrator at the Research Integrity Office as soon as possible.

Participation in semiannual facility inspections for a minimum of **8 hours** per calendar year is expected. The full day expectation may be met by participating in partial days, as needed.

OTHER EXPECTATIONS

Availability to conduct reviews as a primary, full, or designated member is expected. This expectation extends throughout the term of each member. Agreeing to serve implies agreement to be regularly available, except for extenuating circumstances.

Prompt response to inquiries regarding attendance to upcoming meetings, reviewing all materials provided, and being prepared to discuss all information at full committee meetings is expected.

Regulatory standards (*Guide for Care and Use of Laboratory Animals*, *US Government Animal Care Policy 15* [institutional official and IACUC membership], the Public Health Service policy, and the Animal Welfare Act regulations [AWARs]) require institutions to appropriately train and qualify IACUC members to perform their duties. It is expected that all IACUC members (i.e., full and alternate voting members) maintain current IACUC training, renewing/refreshing as necessary and availing themselves of various training/educational opportunities offered by the Office for Resource Protections, professional groups, etc. It is also expected that IACUC members will participate in subcommittee activities as assigned by the IACUC chair throughout their terms of service.

CONFIDENTIALITY REQUIREMENTS

IACUC members are required by federal law (7 U.S.C. § 2157) to maintain confidentiality. Specifically:

1. Release of confidential information prohibited

     It shall be unlawful for any member of an Institutional Animal Committee to release any confidential information of the research facility including any information that concerns or relates to

   a. the trade secrets, processes, operations, style of work, or apparatus; or
   b. the identity, confidential statistical data, amount or source of any income, profits, losses, or expenditures, of the research facility.

2. Wrongful use of confidential information prohibited

     It shall be unlawful for any member of such Committee

   a. to use or attempt to use to his advantage; or
   b. to reveal to any other person, any information which is entitled to protection as confidential information under subsection of this section.

CONFLICTS OF INTEREST/SIGNIFICANT FINANCIAL INTERESTS DISCLOSURE

Specific to IACUC membership (and in addition to required annual disclosures):

IACUC members are required to disclose the existence of any conflict of interest they may have related to the research study under review. No IACUC member may participate in the IACUC's review of any research study in which the member has a conflict of interest, except to provide information requested by the IACUC. IACUC members are automatically considered to have a conflict of interest if they, their spouse, or any dependent children

- Are involved as research personnel in the study under review or
- Have significant financial interest in the research

Appendix 7

IACUC Voting Member Appointment Letter

(Great Eastern University's IACUC letterhead)
<DATE>
<Member name>
<Member address>

Dear **<Member name>**:
I am pleased that you have accepted my invitation to serve as a **<scientific, non-scientific, community, veterinary>** voting member on **<institution's name>** Institutional Animal Care and Use Committee (IACUC). This committee must make decisions regarding many controversial issues surrounding the use of animals in research and teaching. Your appointment represents a critical role on the IACUC and your involvement will significantly enhance committee meeting discussions.

Your appointment to the IACUC will be for the period of **<start>** through **<end>**.

I want to express my appreciation, personally and on behalf of the organization's administration, for your service on this committee. The protection of animals in research is a serious obligation, and I am grateful for your participation in the review process.

Sincerely,

<Chief executive officer>
(OR)
<Institutional official> (if different and the authority has been delegated by the CEO)

Appendix 8

Chief Executive Officer Delegation
of Institutional Official Responsibilities

(Great Eastern University's IACUC letterhead)
<DATE>
To: **<IO name>**
        <IO address>
From: **<CEO name>**
        <CEO address>
SUBJECT: Institutional Official Appointment, Animal Care and Use Program

On behalf of **<organization's name>**, I hereby designate you as the Institutional Official for purposes of ensuring the implementation and maintenance of the Animal Care and Use Program, assign to you the responsibilities associated therewith, and delegate to you the authority to carry out such responsibilities, which includes appointing Institutional Animal Care and Use Committee (IACUC) members.

Appendix 9

IACUC Vice Chair Appointment Letter

(Great Eastern University's IACUC letterhead)
<DATE>
<Member name>
<Member address>

Dear **<Member name>**:
I am pleased that you have accepted my invitation to serve as the vice chair of **<organization's name>** Institutional Animal Care and Use Committee (IACUC). This committee must make many controversial decisions regarding many issues surrounding the use of animals in research and teaching. Your experience in animal research adds to the value of your participation on this committee. In the event that the IACUC chair is unavailable to serve in his or her official capacity, you will assume the responsibilities of chair, including assigning designated member reviewers and facilitating IACUC meetings.

Your appointment as chair of this committee will be for the period of **<start>** through **<end>**.

I want to express my appreciation, personally and on behalf of the university's administration, for your continued service on this committee. The protection of animals in research is a serious obligation, and I am grateful for your continued participation in the review process.

Sincerely,

<Chief executive officer>
(OR)
<Institutional official> (if different and the authority has been delegated by the CEO)

Appendix 10

IACUC Alternate Voting Member Appointment Letter

(Great Eastern University's IACUC letterhead)
\<DATE\>
\<Member name\>
\<Member address\>

Dear **\<Member name\>**:
I am pleased that you have accepted my invitation to serve as a **\<scientific, nonscientific, community, veterinary\>** alternate voting member on **\<institution's name\>** Institutional Animal Care and Use Committee (IACUC). This committee must make decisions regarding many controversial issues surrounding the use of animals in research and teaching. You appointment represents a critical role on the IACUC and your involvement will significantly enhance committee meeting discussions.

Your appointment to the IACUC will be for the period of **\<start\>** through **\<end\>**.

I want to express my appreciation, personally and on behalf of the organization's administration, for your service on this committee. The protection of animals in research is a serious obligation, and I am grateful for your participation in the review process.

Sincerely,

\<Chief executive officer\>
(OR)
\<Institutional official\> (if different and the authority has been delegated by the CEO)

Designated Member Review Subsequent Full Committee Review Template

(Great Eastern University's IACUC letterhead)

DESIGNATED MEMBER REVIEW (DMR) SUBSEQUENT FULL COMMITTEE REVIEW (FCR) (COMMITTEE MEMBERS' AGREEMENT)

We as committee members agree that the quorum of members present at a convened IACUC meeting may decide by unanimous vote to use DMR subsequent to FCR when modification is needed to secure IACUC approval. However, any member of the IACUC may, at any time, request to see the revised protocol and/or request FCR of the protocol.

	Voting Members	**Signature**	**Date**
1	IACUC chair <e-mail>		
2	Attending veterinarian <e-mail>		
3	IACUC member 3 <e-mail>		
4	IACUC member 4 <e-mail>		
5	IACUC member 5 <e-mail>		
6	IACUC member 6 <e-mail>		
7	IACUC member 7 <e-mail>		
8	IACUC member 8 <e-mail>		
9	IACUC member 9 <e-mail>		
10	IACUC member 10 <e-mail>		

Index

Printed in the United States
by Baker & Taylor Publisher Services

Graduate Texts in Physics

Graduate Texts in Physics publishes core learning/teaching material for graduate-and advanced-level undergraduate courses on topics of current and emerging fields within physics, both pure and applied. These textbooks serve students at the MS- or PhD-level and their instructors as comprehensive sources of principles, definitions, derivations, experiments and applications (as relevant) for their mastery and teaching, respectively. International in scope and relevance, the textbooks correspond to course syllabi sufficiently to serve as required reading. Their didactic style, comprehensiveness and coverage of fundamental material also make them suitable as introductions or references for scientists entering, or requiring timely knowledge of, a research field.

More information about this series at http://www.springer.com/series/8431

Walid Younes • Walter D. Loveland

An Introduction to Nuclear Fission

 Springer

Walid Younes
Livermore, CA, USA

Walter D. Loveland
Department of Chemistry
Oregon State University
Corvallis, OR, USA

ISSN 1868-4513   ISSN 1868-4521   (electronic)
Graduate Texts in Physics
ISBN 978-3-030-84594-0   ISBN 978-3-030-84592-6   (eBook)
https://doi.org/10.1007/978-3-030-84592-6

This Springer imprint is published by the registered company Springer Nature Switzerland AG
The registered company address is: Gewerbestrasse 11, 6330 Cham, Switzerland

Preface

It has been over 80 years since the discovery of fission by Hahn and Strassman. During the intervening years, there have been virtually countless studies of the fission process and related phenomena. The complexity of the fission process and its practical importance have attracted some of our best minds. The field has enjoyed the guidance offered by numerous review articles and monographs, especially the iconic books by Vandenbosch and Huizenga, Wagemans, and Krappe and Pomorski. That poses the question of why another book on fission is needed. The first answer to that question is to point out that Vandenbosch and Huizenga was published in 1973 and the Wagemans book was published in 1991, while the more recent book by Krappe and Pomorski in 2012 presents a more advanced treatment of the theory of fission. There has been significant progress in our understanding of fission in recent years, leading to a renaissance in the field. Presentation and discussion of this recent research seems important. We have written this book in a manner that we hope will be attractive to students. The first chapter contains a discussion of fission and the work of some early pioneers in the field and some elementary models of fission. The second chapter gives an overview of the general characteristics of fission and the important quantities involved. The third chapter reviews various models of fission and illustrates their application. Chapter 4 contains a discussion of the products of fission. Chapter 5 contains a discussion of the neutron and gamma rays emitted in fission. Chapters 6 and 7 contain a more sophisticated treatment of the structure and dynamics of fission. It is our hope that beginning students and advanced professionals will find something of interest in this book. The book contains a number of solved problems that illustrate important points and lead the reader to investigate fission in the time-tested manner of solving real problems. Some of these problems are difficult but all promise a suitable pedagogic reward.

Livermore, CA, USA   Walid Younes

Corvallis, OR, USA   Walter D. Loveland

Acknowledgments

The authors gratefully acknowledge the support of their fission research programs by the U.S. Department of Energy, Office of Science, Office of Nuclear Physics, the National Science Foundation, the National Nuclear Security Agency, and our home institutions, the Lawrence Livermore National Laboratory (WY, prior to his retirement) and Oregon State University (WL). The contributions to this manuscript by WY were written by the author acting on his own independent capacity, and not on behalf of the Lawrence Livermore National Laboratory. We thank Dr. Mark Stoyer for his careful review of the manuscript and for many useful suggestions. We thank our numerous colleagues for their comments and patient tutoring.

Contents

List of Figures

List of Tables

Chapter 1
History

Abstract Fission studies involve a wide range of phenomena that are important for nuclear structure as well as nuclear astrophysics. This chapter is a discussion of the history of studies of the fission process, from the very first models of the nucleus leading up to a remarkable paper published by Bohr and Wheeler shortly after the discovery of fission, which gave a first comprehensive model of the phenomenon. In this paper, they outline the central questions to be addressed in understanding the fission process. The chapter explains how they estimated the magnitude of the energy released in fission, detailed what happens to the nucleus during fission, estimated the cross section for the fission reaction, predicted the mass and charge distributions of products, explained the origin of neutrons produced by fission, and gave insightful descriptions of many other aspects of this phenomenon. The chapter also contains a unique discussion of the pivotal role played by three scientists, Ida Noddack, Irène Joliot-Curie, and Lise Meitner, in understanding fission.

1.1 Fission as Basic Science

Nuclear fission, the process whereby an atomic nucleus divides into fragments, is an amazingly rich phenomenon that continues to challenge theorists and experimentalists alike. This book is an exploration of the physics of fission intended to convey the depth and breadth of the subject and an appreciation for the central role it plays in so many aspects of nuclear science. For the physicist interested in nuclear structure, fission can provide a wide array of nuclei far from stability to study. To the many-body theorist, fission represents an extreme example of large-amplitude collective motion of the nucleus and a proving ground for state-of-the-art computational and mathematical tools. For the nuclear astrophysicist, fission is a process that limits the heaviest elements that can exist in nature and that may seed the formation of the elements heavier than iron.

This chapter gives an overview of many of the basic concepts in fission, disguised as a history of its discovery. To guide the reader toward a better appreciation of the concepts, the discovery of fission is presented from three different perspectives: (1)

with a chronological account from 1934 to 1939, (2) focusing on three scientists who made it possible, and (3) through the first theoretical model that explained it.

1.2 The Discovery of Fission

It can be argued that modern nuclear physics was born in 1932, when James Chadwick discovered the neutron by bombarding beryllium with alpha particles [12]. Soon after, Enrico Fermi realized the potential of using neutrons to form elements beyond uranium, the heaviest known element at the time, by using them to bombard elements with increasingly higher atomic number (Z). Indeed, for elements up to thorium $(Z = 90)$, the capture of the incident neutron would often be followed by a beta decay, changing the target's atomic number from Z to $Z + 1$. From 1934 to 1935, Fermi and his group in Rome tried this approach with a uranium target $(Z = 92)$, and although they succeeded in producing beta activity following the neutron irradiation, they were unable to account for the unexpectedly large number of separate activities they had observed [24, 33]. The search for transuranic elements (i.e., elements with $Z > 92$) through neutron bombardment of uranium was then picked up by two other groups: one in Paris, with Irène Joliot-Curie and Pavel Savić, and another in Berlin with Lise Meitner, Otto Hahn, and Friedrich "Fritz" Strassmann. From 1935 to 1938, the Paris and Berlin groups worked diligently to try to make sense of the puzzling beta activities that followed the uranium irradiations [42]. Their frustration grew as increasingly complicated explanations were needed to account for their observations. Eventually they reached a crisis state as each neighboring element to uranium was systematically ruled out as a possible product of the reaction [33]. The uranium puzzle was finally solved in December 1938, when Hahn and Strassmann conclusively showed that one of the elements produced in their experiment, and which they had mistakenly believed to be radium $(Z = 88)$, was in fact barium $(Z = 56)$. Their paper confirming the discovery of fission following the neutron bombardment of uranium was published shortly after, in January 1939 [22].

It may seem surprising that it took so long (from 1934 to 1938) to recognize fission, when all the experimental evidence unequivocally pointed to it, and it is instructive to examine the causes for this delay with the benefit of hindsight. As it turns out, the success of experimental techniques and physics models commonly used at the time became a pernicious obstacle to the discovery of fission. From the point of view of physics as it was understood in the 1930s, the α particle was the heaviest object that could be expected to be emitted by a nucleus. There were several reasons for this belief. The first reason was simply that nothing heavier than an α-particle had ever been observed being emitted by a nucleus. Second, the prevailing view of reaction mechanisms until 1936 was one of a direct process where the incident particle (e.g., a neutron) interacts with a particle cluster inside the nucleus, effectively knocking it out [33, 44]. In this picture, it is easy to visualize light particles being ejected, but it is harder to imagine anything much heavier than an α particle being knocked out of the target nucleus. Third, George Gamow had developed a very successful theory of α decay in 1928 [20], which explained the process as quantum tunneling through a barrier. That same model applied to uranium shows that the probability of division into two equal fragments is impossibly small [33]. For these reasons, the concept of fission would have been unthinkable before 1938. To make matters worse, neutron sources available at the time had very low intensities. As a result, radiochemical techniques had to be employed to isolate and identify the products of the reaction [43]. The chemical separation of radioisotopes is facilitated by the addition of a carrier element belonging to the same chemical group (same column in the periodic table). Choosing the right carrier element was therefore critical to the successful identification of the reaction products [43]. In 1938 Meitner, Strassmann, and Hahn were investigating the production of Ra from the neutron bombardment of Th via the (erroneously assumed) reaction: $^{232}\text{Th} + n \rightarrow ^{229}\text{Ra} + \alpha$. This turned out to be a fortuitous mistake because barium could be used as a carrier element for radium and would play a crucial role in the events to follow. After the forced emigration of Lise Meitner, Hahn and Strassmann turned their attention to the neutron bombardment of uranium, which they expected to proceed via the reaction $^{238}\text{U} + n \rightarrow ^{235}\text{Th} + \alpha$ followed by the decay chain $^{235}\text{Th} \xrightarrow{\alpha} ^{231}\text{Ra} \xrightarrow{\beta^-} ^{231}\text{Ac}$. They chose to look directly for the Ra isotopes by chemical means, rather than the emitted α particles. Using barium as a carrier, they were unable to separate any Ra from the Ba in the last step, known as fractional crystallization, of the chemical process. After repeated checks and tests, they were led to the remarkable and inevitable conclusion that the element they had isolated was in fact the same as the carrier, barium, and not radium [24, 43].

1.3 Focus: Three Important Scientists

Three scientists in particular figure prominently in the crucial period from 1934 to 1938 that led to the discovery of fission: Ida Noddack, Irène Joliot-Curie, and Lise Meitner. We give a brief account of their pivotal contributions to solving the fission puzzle.

1.3.1 Ida Tacke Noddack (1896–1978)

Ida Noddack was a German chemist and physicist. In 1925, working with Walter Noddack and Otto Berg, Ida Tacke discovered element 75, rhenium. They also claimed the discovery of element 43, which they named masurium, but which we know today as technetium. Unfortunately later work could not confirm this result, and the credit for the discovery of element 43 eventually went to Emilio Segré and Carlo Perrier in 1937. Ida Tacke married Walter Noddack in 1926, and in 1934 the two of them turned their attention to the search for transuranic elements. In a paper she wrote that same year, Ida Noddack offered a prescient critique of Fermi's results and correctly formulated the remarkable hypothesis that the bombardment of heavy nuclei, such as uranium, could cause them to break into smaller nuclei, instead of producing a neighboring element to the target [31]. Unfortunately, her paper did not provide the experimental evidence to support the fission hypothesis. As a result her conclusions were largely disregarded, and an early opportunity to discover fission was missed.

1.3.2   Irène Joliot-Curie (1897–1956)

The daughter of Marie and Pierre Curie, she received the 1935 Nobel Prize in Chemistry with her husband, Frederic Joliot-Curie, for their discovery of artificial radioactivity. In 1937 she joined the search for transuranic elements. Working with physicist Pavel Savić, she came close to identifying lanthanum as one of the elements produced following the neutron bombardment of uranium. Joliot-Curie and Savić had found an element with a 3.5-h β-decay half-life that could be precipitated using lanthanum as a carrier. From a modern perspective, we can speculate that the 3.5-h activity they observed could have resulted from a mixture of the $^{141}La \xrightarrow{\beta^-} {}^{141}Ce$ and $^{92}Y \xrightarrow{\beta^-} {}^{92}Zr$ decays that occur with half-lives of 3.92 and 3.54 h, respectively [25]. Unfortunately, they, like many of their scientific peers, were unprepared to accept the concept of such a large change in Z, from U ($Z = 92$) to La ($Z = 57$), and surmised instead that the reaction had produced actinium ($Z = 89$), the chemical homolog[1] of lanthanum. When the 3.5-h half-life could not be reconciled with the known properties of Ac, the mystery only deepened.

1.3.3   Lise Meitner (1878–1968)

Long before her contributions to the discovery of nuclear fission, Lise Meitner started a 30-year collaboration with Otto Hahn. In 1917, they discovered element 91, protactinium. In 1918, she became the head of the physics department of the Kaiser Wilhelm Institute (KWI) in Berlin. Otto Hahn would become the director of the KWI in 1928. In 1934 she and Hahn embarked on the search for transuranic elements, but initially they drew the same erroneous conclusions that Fermi and many others had, and failed to recognize the signatures of fission. With Adolf Hitler's accession to power in 1933, Lise Meitner witnessed many of her Jewish colleagues, including her own nephew Otto Frisch, being forced to resign from

---

[1] Homologous elements belong to the same group (column) in the periodic table and have similar chemical properties.

their positions. Her Austrian citizenship protected her from the same fate until the annexation of Austria by Nazi Germany in March 1938. With her situation becoming increasingly difficult, she fled Germany in July 1938 and eventually took a position at the Nobel Institute in Stockholm, Sweden. Meanwhile, Hahn and his student, Friedrich Strassmann, were left to continue the transuranic research in Berlin without her. Dubious about the results of Joliot-Curie and Savić, and spurred on by Meitner in a meeting with Hahn in Copenhagen in November 1938, they repeated their experiments and finally identified barium and solved the 3.5-h activity puzzle by recognizing lanthanum among the daughter products of barium. Meitner and Frisch gave the first physics interpretation of Hahn and Strassmann's chemical results in February 1939, introducing the word "fission" to describe the process which behaved similarly to a drop of liquid dividing into smaller drops when its vibrations are sufficiently violent [28]. In the paper, Meitner and Frisch also estimated the energy released in fission to be about 200 MeV. As the paper was being completed, Otto Frisch also performed a simple but crucial ionization chamber experiment in Copenhagen that gave the final confirmation of the emission of fission fragments. Ironically, similar experiments had been conducted by others, including Fermi, but had failed to detect the fission fragments because the actinide targets were routinely covered with an absorber in order to suppress their natural α-activity backgrounds. This seemingly innocent practice had the unintended effect of also stopping most fission fragments inside the absorber [24].

Otto Hahn received the chemistry Nobel Prize in 1944 for the discovery of fission. The decision by the Nobel committee to exclude Meitner, Strassmann, and Frisch from the prize remains controversial. Some of the reasons behind that decision are examined in the article by Crawford, Sime, and Walker [14]. One of those reasons is that fission was seen as a discovery in chemistry, as was the topic of radioactivity, while Meitner and Frisch's contribution was viewed as belonging to the domain of physics. Strassmann was probably dismissed from consideration for being a more junior member of the collaboration. For her part, Meitner never expressed any bitterness for being excluded from the prize and went on to win many prestigious awards for her work. In 1997, element 109, meitnerium, was named after her.

1.3.4 Closing Remarks

It should also be mentioned that although Marie Sklodowska Curie was not directly involved in the work that started in 1934 (she died in July of that year) and which culminated in the official discovery of fission in 1938, her work on radioactivity— which earned her the 1903 Nobel Prize in Physics along with her husband, Pierre Curie, and Henri Becquerel—laid the groundwork for that discovery. For a more extensive discussion of the contributions of women to the field of fission and the discovery of new elements, see, for example, the article by J. L. Spradley and references therein [42].

1.4   A First Model of Fission

1.4.1   The View of the Nucleus Before 1932

The periodic table of the elements is such a familiar sight in classrooms and textbooks that it is easy to forget that, before 1913, the atomic numbers used to sort chemical elements into its rows and columns were little more than a convenient label without a precise physical meaning. In the original version of the periodic table devised by Dmitri Mendeleev, the elements were arranged according to their atomic mass. Then in 1913, Henry Moseley discovered a mathematical relation between the wavelengths of x-rays produced by various elements and their atomic number. Thus "Moseley's law" equated the atomic number of a chemical element with its nuclear charge [29]. This critical discovery set the stage for Ernest Rutherford's famous gold foil experiment in 1919, which led him to the identification of the proton as a fundamental constituent of the nucleus. In light of these discoveries, the atomic number Z of the element could now be identified with the number of (positively charged) protons in its nucleus. However, it was soon realized that adding up the charged mass of the nucleus (number of protons times the hydrogen mass) only accounted for roughly half of its total mass. Another constituent particle was needed to make up the difference. We recognize now the additional particle as the neutron, discovered in 1932 by James Chadwick, but before then the fundamental building blocks known to theorists were the proton and the electron. Thus in 1919 Wolff proposed that the extra particles were protons neutralized by "nuclear electrons" [48]. Wolff also addressed the puzzle of what held the Z protons in the nucleus together by speculating that the Coulomb force behaved differently at the short distance scales of the nucleus. This assertion did not hold for long as alpha particle scattering experiments showed no deviations from the one-over-distance-squared form of the Coulomb force, down to nuclear length scales.

Such was the nascent picture of the nucleus at the beginning of the 1920s. The flawed concept of nuclear electrons in particular would endure for the better part of that decade and ultimately throw the entire model of the nucleus into crisis, as we shall see.

1.4.2   The Liquid-Drop Model

By the time Hahn and Strassmann published their seminal paper in 1939 announcing the discovery of fission, one of the essential pieces of the theoretical framework needed to describe it, the liquid-drop model, was already in place thanks, in large part, to the works of George Gamow and of Niels Bohr.

We begin the story of how the liquid-drop model became (and still is) one of the most useful descriptions of fission with Ernest Rutherford's theory of alpha decay. After his discovery of the proton in 1919, Rutherford had come to view the nucleus as being composed of hydrogen, mass-3 nuclei, α particles, and the requisite number of nuclear electrons to give the nucleus its expected positive charge [36]. In 1927 he published a paper detailing his "satellite model" of α decay [38, 44]. In this model, Rutherford envisioned a radioactive nucleus as a central core orbited by neutral α particles. Alpha decay could then be explained as the result of one of the α satellites losing its two electrons, which would plummet into the core, and the remaining positively charged α particle would then be expelled from the nucleus by the Coulomb repulsion from the positively charged core. This, of course, was not the correct picture of the nucleus or of the α-decay process, as George Gamow would soon show.

In the summer of 1928, while visiting Max Born's institute in Göttingen, Gamow happened upon Rutherford's paper on α decay and instantly recognized the process as an example of quantum-mechanical tunneling. By the time he left Göttingen and published a paper on the subject [20], his theory of α decay had supplanted Rutherford's [44]. In September 1928, Gamow traveled from Göttingen to Copenhagen to meet with Niels Bohr. By December of that year, he had formulated his next great contribution to nuclear physics: the liquid-drop model of the nucleus.[2] Gamow spent the period from January to February 1929 visiting Rutherford's Cavendish laboratory in Cambridge. Rutherford was so impressed by Gamow's work that he invited him to a discussion on the structure of atomic nuclei at the Royal Society in London in February 1929, where Gamow gave the first recorded description of his liquid-drop model of the nucleus [21].

The starting point for this model was Rutherford's own picture of the nucleus as a collection of α particles [44]. Although the idea of α particles as fundamental building blocks of the nucleus would turn out to be wrong, the validity of the liquid-drop model itself did not depend on this assumption. Gamow reasoned that the assembly of α particles could be described in an analogous manner to a drop of water held together by surface tension [37]. By January 1930, Gamow had quantified these ideas and presented them in a paper on the "Mass Defect Curve and Nuclear Constitution" [21]. As the title of the paper suggests, Gamow had used the liquid-drop model to express the mass defect (the difference between the mass of the nucleus, in atomic mass units, and the number of nucleons) as a function of the number of α particles for nuclei with atomic weights that are a multiple of 4 [40, 44]. Gamow found that the calculated mass defects could be made to more or less agree with experimental values through a judicious choice of the number of α particles and of nuclear electrons. The first signs of real trouble came when Gamow tried to apply Dirac's newly developed relativistic equation to the problem.

The mid- to late 1920s witnessed the maturation of quantum mechanics. In 1926, Erwin Schrödinger published his famous wave equation [39], and in 1928 Paul Dirac derived a relativistic version of the wave equation [16]. Gamow realized

---

[2] As noted in [40], Gamow referred to his model as the "water drop" rather than "liquid-drop" model, which might be more closely associated with Bohr in our modern perspective. We will not make such a fine distinction in our discussion.

that the nuclear electrons would have to move at relativistic speeds and that their energy levels should therefore be calculated using Dirac's wave equation. When he attempted to do this however, he quickly became mired in the "Klein paradox." In 1928, Oskar Klein had solved the Dirac equation for the textbook problem of the scattering of a relativistic electron by a step function potential barrier [26]. The counterintuitive result that Klein obtained is that for a sufficiently high potential barrier step, and an incident particle with enough energy but still below the barrier, there is a non-vanishing probability of finding a transmitted particle anywhere below the barrier and with negative kinetic energy! The physical interpretation of this phenomenon is that, at the potential wall, a particle-antiparticle pair is created. (see, e.g., [11] and Section 1.2 in [1] for a more detailed explanation). For all the models that relied on nuclear electrons, the Klein paradox spelled doom because it implied that electrons could not be confined indefinitely inside the nucleus by any potential. The nuclear electrons would (quickly) give rise to positrons and escape the confines of the nucleus [40, 44]. After the discovery of the neutron in 1932, the nuclear electron hypothesis was discarded, and Gamow's drop model prediction of the nuclear masses underwent a series of reformulations and improvements by Werner Heisenberg in 1934 [23], then in 1935 [46] by his student, Carl Friedrich Von Weizsäcker, and finally by Hans Bethe in 1937 [4].

While Gamow and his successors were developing a liquid-drop model of the nucleus to account for its static properties (e.g., its mass defect), a separate and apparently independent effort [44] was being pursued by Niels Bohr, Fritz Kalckar, and John Wheeler to construct a liquid-drop model of the dynamical properties of the nucleus and of its excitation mechanisms in particular. This parallel line of development began with Bohr's model of compound-nucleus formation, which he delineated in a 1936 paper [8]. In this model, Bohr proposed that reactions induced on nuclei by various types of incident particles proceed in two distinct stages: the formation of a compound system where the energy of the captured incident particle is shared among the nuclear constituents, followed by a subsequent emission when enough energy happens to be concentrated on a particle near the surface of the nucleus (this is commonly referred to as the Bohr hypothesis). This contrasted with the direct-reaction view mentioned earlier. Within the context of the compound-nucleus picture, Bohr also examined the distribution of energy levels in the nucleus which could be inferred from neutron capture experiments [5].[3] Starting from the compound-nucleus picture and the distribution of energy levels presented in [5, 8], Bohr and Kalckar reasoned that the pattern of energy levels drawing ever closer with increasing excitation energy was not consistent with an independent particle picture but resembled more what would be expected from the collective motion of the nucleons, in analogy with the behavior observed in solids and liquids [6]. They

---

[3] A word of caution is warranted regarding the terminology used here. In a more modern context, the term "neutron capture reaction" designates a process wherein the incident neutron is absorbed and the compound nucleus subsequently decays by gamma-ray emission. In this chapter, we use the term in the broader sense of neutron absorption by the target, regardless of the subsequent mode of decay.

used this analogy to calculate the collective vibrational excitations of the nucleus. All these developments had set the stage for the first quantitative model of fission that would be advanced by Bohr and Wheeler in 1939.

1.4.3 The Bohr-Wheeler Fission Model

Hahn and Strassmann's paper heralding the discovery of fission was published on January 6, 1939 [22]. Meitner and Frisch's brief paper, likening the process to the division of a liquid drop and making the first estimate of the associated energy release, was published on February 11, 1939 [28]. On September 1st of that year, Niels Bohr and John Wheeler published an extensive 25-page paper outlining in detail a first quantitative model of fission [7].[4] After giving credit to Meitner and Frisch for emphasizing the picture of fission as that of a liquid drop stretching to its breaking point [7], Bohr and Wheeler embark on a veritable tour de force (given the limited amount of available experimental data at the time) to address an impressive number of central questions (and related sub-questions):

- How much energy is released by fission?
- What happens to the nucleus during fission?

  - How much energy must be supplied in order to trigger the process?
  - Can the nucleus fission spontaneously, without this energy input?

- How can we estimate the cross section for neutron-induced fission?
- What is the predicted cross section for the neutron-induced fission of uranium?

  - Which isotope in natural uranium is primarily responsible for the observed fission cross section?

- What is the origin of neutrons observed several seconds *after* fission?
- What is the expected distribution in size of the fission fragments?
- What are the predicted fission cross sections for other types of incident particles (deuterons, protons, photons)?

It is easy to underestimate the difficulty in answering these questions within the context of the nascent field of nuclear physics at the time. The main theory tools at their disposal were:

- A model of the nucleus as a charged liquid drop, which allowed them to view the fission process as a competition between the Coulomb repulsion of the positively charged protons and the stabilizing effect of the surface tension of the drop.
- The principle from statistical mechanics that the probability—or the *width*, to use the more relevant term—of a nuclear process (fission, neutron emission, etc.) is

---

[4] For a modern perspective on Niels Bohr's contributions to nuclear physics in general, and to fission in particular, see, for example, [3].

proportional to the number of quantum states available to the system undergoing this process.

- The theory of resonant neutron capture developed by Breit and Wigner in 1936 [10], which gives a formula for the cross section in terms of widths and the incident energy.

On the empirical side, they were aware of several crucial pieces of information:

- Atomic weights of the nuclei from the paper by Dempster [15] giving the packing fraction (= mass defect per nucleon) as a function of atomic number from light elements up to uranium
- Numerical estimates of the nuclear radius and surface tension parameter obtained by Feenberg [19] from fits of the liquid-drop model to mass defects
- The excitation energy ($\sim 6\,\mathrm{MeV}$) required to induce fission in a $^{239}\mathrm{U}$ compound system (formed by a neutron incident on a $^{238}\mathrm{U}$ target)
- The typical spacing d between levels in uranium nuclei: at low energies ($d \sim 100\,\mathrm{keV}$), around 6 MeV in excitation ($d \sim 20\,\mathrm{eV}$), and for capture of 2.5-MeV neutrons ($d \sim 0.2\,\mathrm{eV}$), from some of Bohr's earlier work
- The cross section for a resonant neutron capture at 25 eV on uranium ($\sigma \sim 2.4 \times 10^{-21}\,\mathrm{cm}^2$ at the resonance energy)
- The abundances of the various uranium isotopes found in natural uranium

Using these theoretical and empirical pieces of information, they addressed each one of the questions above in turn. We summarize their conclusions below.

How Much Energy Is Released by Fission?

Thanks to Einstein's famous mass-energy equivalence formula [17], $E = mc^2$, this question could be easily answered once the masses of the parent nucleus and its daughter products were known. Bohr and Wheeler (hereafter referred to as B&W) considered the fission of $^{239}\mathrm{U}$ into nearly identical fragments $^{119}\mathrm{Pd}$ and $^{120}\mathrm{Pd}$ but readily admitted to having no experimental measurements of the masses of these unusually neutron-rich nuclei ($^{110}\mathrm{Pd}$ is the heaviest stable isotope of palladium) [7]. However, using the liquid-drop model and the packing fraction curve of Dempster [15], they were able to estimate the difference in mass between parent and daughters and deduced an equivalent energy of 200 MeV, in agreement with Meitner and Frisch [28]. Using modern tabulations of nuclear masses, we find 191.9 MeV for the $^{239}\mathrm{U} \rightarrow {}^{119}\mathrm{Pd} + {}^{120}\mathrm{Pd}$ division, which is remarkably close to B&W's original estimate. Using their mass formula, they were also able to calculate that an additional 31 MeV would be released by subsequent β decays of the Pd fragments.

What Happens to the Nucleus During Fission?

Based on Bohr's earlier work [5, 8], the excitations of the parent nucleus could be understood by analogy to the vibrations of a drop of liquid, with the surface tension providing the restoring force and the Coulomb repulsion driving the system toward fission. They imagined the nucleus driven by these vibrations and deforming through a series of increasingly elongated shapes.

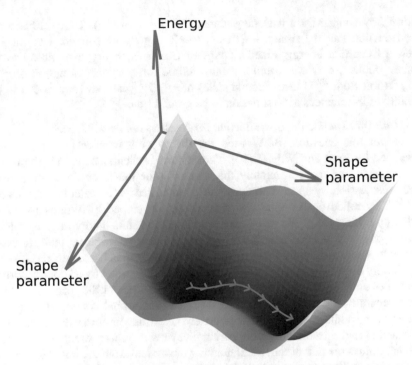

Fig. 1.1 Three-dimensional surface consisting of two minima (wells) separated by a saddle shape. This represents an energy surface plotted as a function of two parameters that specify the shape of the nucleus. The arrows show the reaction direction (for example, the direction of increasing elongation of the nucleus) going from one minimum, over the saddle, and to the other minimum

Starting from this liquid-drop picture, B&W calculated the energy of the vibrating nucleus as the sum of contributions from the surface tension (proportional to surface area) and Coulomb repulsion (volume integral of the one-over-distance potential). By parameterizing the shape of the nucleus, they could therefore plot its energy as a function of its shape. For two shape parameters (one of them giving the elongation of the nucleus), they found an energy surface with a saddle point feature separating two wells or "hollows," similar to the shape displayed in Fig. 1.1.

They could then explain the fission process in a classical picture as a ball in one of the wells, stimulated by the incident particle into rolling around the bottom of the well and eventually rolling over the saddle and into the adjacent well (provided the initial "kick" from the projectile was sufficiently energetic) [7]. This saddle point feature provided a barrier—whose height E_f they called the critical fission energy—that had to be overcome for the nucleus to fission (or for the ball to roll from one well into the other). Using the shape-dependent energy formula they had derived, B&W also found that there had to be a critical value of the squared-charge-to-mass ratio (Z^2/A) beyond which the nucleus is no longer stable against deformation. Coupled with the knowledge that $E_f \sim 6$ MeV for ^{239}U, they were able to estimate the critical Z^2/A value for a fissioning nucleus ($Z^2/A \sim 47.8$). They further made the deduction that the fission fragments would usually be formed with residual internal excitation energy. Finally, adapting Gamow's theory of α decay to the fission problem, they calculated a mean lifetime for spontaneous fission of ^{239}U (i.e., fission from ^{239}U in its ground state) of $\sim 10^{22}$ years, thereby revealing the stability of this nucleus against fission in its ground state.

How Can We Estimate the Cross Section for Neutron-Induced Fission?
To answer this question, B&W used the theory of resonance capture recently developed by Breit and Wigner [10]. The main ingredients needed for the theory are the rates—or more accurately the widths[5]—for the three competing decay processes: fission, neutron emission, and gamma decay. To calculate the fission width, they relied on the transition state theory developed by Wigner [47] (and earlier by Eyring [18]). We will explore the transition state theory in greater detail in later chapters, but for now it will be sufficient to describe its main features. This theory was originally developed to calculate chemical reaction rates, but it is generally applicable to the evolution of a system described by an energy surface like the one in Fig. 1.1. For this type of system, a barrier in the form of a saddle shape separates two wells representing an initial and a final configuration of that system. The saddle-like shape of the barrier (i.e., a maximum in the direction of the reaction and a minimum in the perpendicular direction) presents a bottleneck that determines the rate of the reaction taking the system from the initial to the final configuration. Whatever trajectory the system follows from the initial well to the final one, the system must cross the saddle. Assuming each system only crosses the saddle once, the reaction rate is proportional to the number of such crossings by an ensemble of systems. In a quantum-mechanical picture, passage over the saddle is mediated by a set of *transition states*. Thus the fission width could be taken as proportional to the number of transition states over the saddle, at a given excitation energy. Using a similar line of reasoning, the neutron emission width can be shown to be proportional to the number of states reached in the residual nucleus from that same initial excitation energy, which B&W could estimate from their experimental knowledge of the level spacing d (see above). Making the argument that, for sufficiently energetic incident neutrons, the gamma decay width was negligible, B&W could then express the fission probability using the formalism of Breit and Wigner [10].

What Is the Predicted Cross Section for the Neutron-Induced Fission of Uranium?
Having established the formulas needed to calculate fission cross sections, B&W applied them to the uranium problem. Given that the existing fission data at the time had been obtained using natural uranium, which consists primarily of the isotopes ^{235}U and ^{238}U with a trace amount of ^{234}U, the first task was to disentangle their respective contributions. Based on the known abundances of these isotopes, and

---

[5] The width for a process is defined as $\hbar$ divided by its lifetime; in other words, it is proportional to its average rate.

the measured radiative neutron capture cross section at 25 eV (neutron absorption followed by gamma emission), B&W first deduced that the radiative neutron capture occurs primarily with the ^{238}U isotope, producing a ^{239}U compound. From this, they extracted an estimate of $d \sim 20$ eV for the level spacing at the capture excitation energy of ~ 6 MeV. They deduced the expected level spacings in ^{235}U and ^{236}U. On the other hand, they could use the measured cross section for fission induced by thermal neutrons (0.025 eV) and the cross-section formulas they had obtained earlier to calculate the corresponding level spacing in the fissioning isotope, whichever it may be. Thus, by comparing level spacings, they correctly deduced that the major contribution to that measured cross section had to come from the isotope ^{235}U, which forms the compound system ^{236}U before fissioning.

What Is the Origin of Neutrons Observed Several Seconds after Fission?
A few months earlier, Roberts et al. [34] had published their observation of neutron emission up to 1.5 min after neutron bombardment of uranium. In their paper, Roberts et al. speculated that the phenomenon might be somehow caused by photodisintegrations. They did however note that the emission half-life of the observed gamma rays and neutrons closely matched that of a 10-s β decay reported earlier by Meitner et al. [27]. B&W correctly concluded that these delayed neutrons were indeed the result of β decays of the fission fragments [7]. Using Fermi's theory of β decay, they calculated a probability distribution of excitation energy left in fission fragments after β decay and argued that, for a daughter product with sufficiently small neutron binding energy, enough excitation energy could be left in some nuclei to allow neutron emission following the β decay of its precursor. They also considered the process where neutrons are emitted as the parent nucleus tears itself apart, anticipating the possibility of so-called scission neutrons, which continue to be a topic of ongoing research but for which the experimental evidence remains weak (see also Sect. 2.8).

What Is the Expected Distribution in Size of the Fission Fragments?
Based on the shape of the energy surface (see, e.g., Fig. 1.1), B&W argued that the system could fan out into a wide range of fragment sizes once it passed the saddle. They also recognized that the details of the fragment distribution would be determined by the dynamics of the fission process beyond the saddle, as the potential energy of deformation is converted in kinetic and excitation energy of the fragments.

What Are the Predicted Fission Cross Sections for Other Types of Incident Particles (Deuterons, Protons, Photons)?
Finally, B&W considered reactions leading to fission other than those initiated by a neutron. For incident deuterons, in addition to the usual capture process, they also allowed for the possibility that the deuteron could break up in the Coulomb field of the target, thereby leaving the neutron to be absorbed and the proton to be deflected away. In the case of compound-nucleus formation for incident deuterons and protons, they modified their fission cross-section formula by a multiplying factor to account for the electrostatic repulsion between projectile and target. Using the modified formula, they estimated cross sections for 6-MeV deuterons and

protons incident on ^{238}U of $\sim 10^{-29}$ cm^2 and $\sim 10^{-28}$ cm^2, respectively, which are consistent with more recent measurements. Using the Breit-Wigner formalism, they also predicted a photofission cross section of $\sim 10^{-27}$ cm^2 for ^{238}U and $\sim 10^{-28}$ cm^2 for ^{233}Th. The photofission cross sections depend sensitively on the energy of the incident gamma ray because of the role played by giant resonances in the capture process, and it is therefore difficult to compare the values obtained by B&W to modern-day measurements.

1.5 Fission in the Cosmos

Although we will be mainly concerned with study of fission here on earth, it may play an important role far beyond this planet, in the nucleosynthesis of elements found in nature. The rapid neutron capture process, or r-process, is responsible for the production of about half the elements heavier than iron. As the r-process builds increasingly heavier nuclei through successive neutron capture reactions, fission eventually becomes an important competing process. Fission impacts the abundance of elements found in nature both by placing a limit of heaviest nuclei that can be formed and by providing lighter elements (the fragments) that can serve as seeds for further capture reactions. This "fission cycling" process [13, 32, 35] may take place in neutron star merger events [45].

1.6 Structure of the Book

This textbook covers both experimental and theoretical aspects of nuclear fission. It is assumed that the student is familiar with calculus, quantum mechanics, and some nuclear physics. In order to keep the material accessible, the more advanced mathematical techniques have been deferred to the final chapters (Chaps. 6 and 7), relegated to advanced exercises at the end of the chapter, or omitted altogether (with references given for further reading). Chapter 2 gives an overview of the general properties of fission to familiarize the reader with concepts that will be revisited in later chapters. Chapter 3 is a first introduction to models used to describe fission and in particular to the liquid-drop model. Chapter 4 covers properties of fission fragments and products. Chapter 5 examines the particles most often emitted by the fragments, neutrons, and photons. Chapter 6 goes beyond the liquid-drop model to single-particle models and more advanced treatments of the static quantities relevant to fission. Finally, Chap. 7 tackles the theory of fission reactions and fission dynamics.

1.7   Additional Resources

Programming will also be a valuable asset for some of exercises (and beyond the classroom as well!). Although there are many freely available resources on line to learn almost any programming language, along with good programming practices, we also recommend a few published titles

- L. R. Nyhoff and S. C. Leestma, "Fortran 90 for Engineers and Scientists", Pearson (1996).
- H. Ruskeepa, "Mathematica Navigator", Academic Press (2009).
- E. J. Billo, "Excel for Chemists", Wiley (2011).
- H. P. Langtangen, "A Primer on Scientific Programming with Python", Springer (2016).
- L. R. Nyhoff, "Programming in C++ for Engineering and Science", CRC Press (2012).

For Python, we also recommend the very educational video by David Beazley, "Learn Python through public data hacking," which can be found on YouTube. To help with the style and organization for future researchers in the field, we recommend "On preparing a manuscript for publication," R. E. Adelberger, J. Undergrad. Phys. 18, 32 (1989).[6] In addition to the included problem sets, we recommend the book by A. A. Kamal "1000 Solved Problems in Modern Physics," Springer (2010), especially Chapters 7 and 8, which deal with nuclear physics, as a source of shorter exercises and problems. For the treatment and analysis of data, P. R. Bevington and D. K. Robinson's "Data Reduction and Error Analysis," 3rd ed., McGraw-Hill (2002), and the paper by D. L. Smith and N. Otuka, "Experimental Nuclear Reaction Data Uncertainties: Basic Concepts and Documentation," Nucl. Data Sheets 113, 3006 (2012), are useful resources.

On the topic of fission, there are three very comprehensive books that might also be of interest:

- R. Vandenbosch and J. R. Huizenga, "Nuclear Fission," Academic Press (1974)
- C. Wagemans, "The Nuclear Fission Process," CRC Press (1991)
- H. J. Krappe and K. Pomorski "Theory of Nuclear Fission," Springer (2011)

Finally, we also recommend some websites: the lectures on various aspects of fission given at the 2014 and 2017 FIESTA summer schools held in Santa Fe and organized by the Los Alamos National Laboratory at http://t2.lanl.gov/fiesta2014/school.shtml and http://t2.lanl.gov/fiesta2017/school.shtml, and the American Institute of Physics (AIP) site at http://history.aip.org/exhibits/mod/fission/fission1/01.html where the reader can find further background information and some of the historical scientific papers connected with the discovery of fission.

---

[6] Also available as R. E. Adelberger, International Amateur-Professional Photoelectric Photometry Communications, No. 40, p. 32 (1990).

1.8 Questions

1. Why is the study of nuclear fission interesting from a basic science perspective?

2. Why did Enrico Fermi study the neutron bombardment of actinide targets?

3. Besides Fermi and his collaborators in Rome, what other groups were working on the same problem?

4. When was the concept of fission first formulated, and by whom?

5. What are some of the reasons for the delay in the discovery of fission despite the evidence for it in Enrico Fermi's experiments?

6. Which element, produced by neutron-induced fission of uranium, did Irène Joliot-Curie and Pavel Savić come close to identifying?

7. What are some things that the 1939 paper by Meitner and Frisch contributed to the fission problem?

8. What is the concept of nuclear electrons, and why was it abandoned in the late 1920s?

9. What are some important questions addressed by the 1939 Bohr and Wheeler paper on fission?

10. How did Bohr and Wheeler compensate for the fact that few nuclear masses where known when they wrote their 1939 paper on fission?

1.9 Exercises

Masses needed in exercises can be retrieved from the Reference Input Parameter Library https://www-nds.iaea.org/RIPL-3 for example.

Problem 1 Using modern mass values, calculate the energy released by the $^{239}U \rightarrow \, ^{119}Pd + \, ^{120}Pd$ division to confirm the value in the text. How does this compare to the energy released by α decay of ^{239}U?

Problem 2 Approximating the Pd fragments at scission in the $^{239}U \rightarrow \, ^{119}Pd + \, ^{120}Pd$ division as two touching spheres, calculate the Coulomb energy at scission and compare to the result in problem 1? (use $1.3A^{1/3}$ for the radius of a nucleus in fermis, and $1.44Z_1Z_2/d$ for the Coulomb energy in MeV and the distance d in fm)

Problem 3 Compare the critical energy $E_f \sim 6\,\text{MeV}$ for ^{239}U fission to the mass-energy difference between $n + ^{238}U$ and ^{239}U (this is the minimum excitation energy gained by capturing the neutron).

Problem 4 In their 1939 paper, Bohr and Wheeler relied on phenomenological formulas to estimate the masses of fission products that were unknown at the time

(Eqs. (2)–(6) in their paper). For nuclei with the same mass number A, but different charge numbers Z_1 and Z_2, these formulas give the difference in the masses as

$$M(Z_1, A) - M(Z_2, A) = \frac{1}{2} B_A \left[(Z_1 - Z_m)^2 - (Z_2 - Z_m)^2 \right] \qquad (1.1)$$

where B_A is a free parameter to be adjusted, Z_m is the charge number with the lowest mass, and provided that either A is an odd number, or if A is even, that Z_1 and Z_2 are both either even or both odd. For $A = 100$ and $40 \leq Z_1, Z_2 \leq 48$, for example, and using modern nuclear mass values, how well does this relationship work? What is optimal value of B_A?

Problem 5 To estimate the lifetime τ_f of a nucleus against fission from its ground state (i.e., spontaneous fission), Bohr and Wheeler [7] used a rough approximation derived from Gamow's theory of α decay,

$$\tau_f \sim \frac{2\pi}{5\omega_f} \exp \left[\sqrt{2ME_f} d_f / \hbar \right] \qquad (1.2)$$

where ω_f is the vibration frequency of the parent nucleus in its ground state (they used $\hbar\omega_f = 0.8\,\text{MeV}$), M is the mass of the parent nucleus, E_f is the critical fission energy (they used $E_f = 6\,\text{MeV}$), and d_f is the distance between the centers of the nascent fission fragments at scission (they used $d_f = 13\,\text{fm}$). Using this formula and the numerical values given here, estimate the spontaneous fission lifetime for ^{236}U and compare to the known present-day value of the partial lifetime $\tau_f = 3.6 \times 10^{16}\text{y}$.

Problem 6 Bohr and Wheeler used measured fission cross sections σ_f to deduce the energy spacing d between levels in the compound nucleus. They used the Breit-Wigner formalism to obtain the following expression (Eq. (55) in [7])

$$\sigma_f = \frac{\pi}{2} \left(\frac{\lambda}{2\pi} \right)^2 \Gamma_n \frac{2\pi}{d} \qquad (1.3)$$

where λ is the de Broglie wavelength of the incident neutron ($\lambda = h/p$ where p is the momentum of the neutron) and the quantity Γ_n is the neutron capture width (we will discuss widths in greater detail in Sects. 3.6 and 7.1.2). Taking $\Gamma_n = 10^{-4}$ eV, as in the Bohr and Wheeler paper, estimate the energy level spacing if the thermal ($E_n = 25$ meV) cross section is 3 barn.

Problem 7 Use the Heisenberg uncertainty principle to show that an electron confined inside a nucleus of dimension ~ 10 fm would have to be treated relativistically. (Use $p \sim \Delta p$ where $p = \left(1 - v^2/c^2 \right)^{-1/2} mv$ is the relativistic momentum).

Problem 8 The radius of a nucleus with mass number A is given approximately by

$$R = r_0 A^{1/3} \qquad (1.4)$$

Table 1.1 List of actinides with decay fractions (as a percent of all decays) for α, β^-, and spontaneous fission (SF) processes. Data taken from the ENSDF database

	α (%)	β^- (%)	SF (%)
^{238}U	100.0	0	5.45×10^{-5}
^{238}Np	0	100.0	0
^{238}Pu	100.0	0	1.9×10^{-7}
^{235}U	100.0	0	7×10^{-9}
^{234}Th	0	100.0	0
^{234}Pa	0	100.0	0
^{234}U	100.0	0	1.64×10^{-9}
^{230}Th	100.0	0	$\leq 4 \times 10^{-12}$

where $r_0 \approx 1.2$ fm. Estimate the radius of ^{236}U. Estimate the density (number of nucleons per fm^3) of this nucleus. Compare this value to the nucleon density of ^{56}Fe.

Problem 9 Table 1.1 shows some actinides and their decay branches (i.e., the fraction of time a particular mode of decay occurs). Starting from ^{238}U, and using the data in the table to follow the most likely branch for each decay, what is the highest Z element you can form?

Problem 10 The average range of alpha particles in aluminum (density = 2.70 g/cm^3) is given in units of g/cm^2 by [9, 30, 41]

$$R_\alpha = 0.00165 \left(\frac{g}{cm^2} \right) \begin{cases} 0.56E_\alpha & E_\alpha < 4 \text{ MeV} \\ 1.24E_\alpha - 2.62 & 4 \text{ MeV} \leq E_\alpha < 8 \text{ MeV} \end{cases} \quad (1.5)$$

as a function of the kinetic energy E_α of the α particle. This average range can be converted into an average distance by dividing by the density of the material. By contrast, the range of the ^{140}Ba fission fragment in aluminum following the irradiation of ^{235}U has been measured at [2]

$$R_{140\text{Ba}} = 2.98 \text{ mg/cm}^2 \quad (1.6)$$

Compare the average distances traveled in an aluminum absorber by the α particle emitted by ^{235}U ($E_\alpha = 4.7$ MeV) and by the recoiling ^{140}Ba fission fragment, and comment on the implication for the use of absorbers in early fission experiments.

References

1. L. Alvarez-Gaume, M.A. Vazquez-Mozo, *An Invitation to Quantum Field Theory* (Springer-Verlag GmbH, 2011)
2. N.K. Aras, M.P. Menon, G.E. Gordon, Nucl. Phys. **69**, 337 (1965)

3. S.T. Belyaev, V.G. Zelevinskiĭ, Sovi. Phys. Uspekhi **28**, 854 (1985)
4. H.A. Bethe, Rev. Mod. Phys. **9**, 69 (1937)
5. N. Bohr, Nature **137**, 351 (1936)
6. N. Bohr, F. Kalckar, Dan. Mat. Fys. Medd. **14**, 1 (1937)
7. N. Bohr, J.A. Wheeler, Phys. Rev. **56**, 426 (1939)
8. N. Bohr, Nature **137**, 344 (1936)
9. W.H. Bragg, R. Kleeman, Lond. Edinburgh Dublin Philos. Mag. J. Sci. **10**, 318 (1905)
10. G. Breit, E. Wigner, Phys. Rev. **49**, 519 (1936)
11. A. Calogeracos, N. Dombey, Contemp. Phys. **40**, 313 (1999)
12. J. Chadwick, Nature **129**, 312 (1932)
13. J.J. Cowan, C. Sneden, J.E. Lawler, A. Aprahamian, M. Wiescher, K. Langanke, G. Martínez-Pinedo, F.-K. Thielemann, Rev. Mod. Phys. **93**, 015002 (2021)
14. E. Crawford, R.L. Sime, M. Walker, Physics Today **50**, 26 (1997)
15. A.J. Dempster, Phys. Rev. **53**, 869 (1938)
16. P.A.M. Dirac, Proc. R. Soc. A Math. Phys. Eng. Sci. **117**, 610 (1928)
17. A. Einstein, Annalen der Physik **18**, 639 (1905)
18. H. Eyring, J. Chem. Phys. **3**, 107 (1935)
19. E. Feenberg, Phys. Rev. **55**, 504 (1939)
20. G. Gamow, Zeit. für Phys. **52**, 510 (1928)
21. G. Gamow, Proc. R. Soc. A Math. Phys. Eng. Sci. **126**, 632 (1930)
22. O. Hahn, F. Strassmann, Naturwiss **27**, 11 (1939)
23. W. Heisenberg, in *Institut International de Physique Solvay, Structure et Propriétés des Noyaux Atomiques: Rapports et Discussions du Septième Conseil de Physique tenu à Bruxelles du 22 au 29 Octobre 1933* (1934), p. 316
24. G. Herrmann, Nucl. Phys. A **502**, 141 (1989)
25. N.E. Holden, tech. rep. BNL-NC-43163 (BNL, 1989)
26. O. Klein, Zeit. für Phys. **51**, 204 (1928)
27. L. Meitner, O. Hahn, F. Strassmann, Zeit. für Phys. **106**, 249 (1937)
28. L. Meitner, O.R. Frisch, Nature **143**, 471 (1939)
29. H. Moseley, Lond. Edinburgh Dublin Philos. Mag. J. Sci. **26**, 1024 (1913)
30. R. Murray, *Nuclear Energy* (Elsevier LTD, Oxford, 2019)
31. I. Noddack, Z. Angew. Chem. **47**, 653 (1934)
32. B.E.J. Pagel, *Nucleosynthesis and Chemical Evolution of Galaxies* (Cambridge University Press, 2009)
33. J.M. Pearson, Physics Today **68**, 40 (2015)
34. R.B. Roberts, R.C. Meyer, P. Wang, Phys. Rev. **55**, 510 (1939)
35. C. Rolfs, *Cauldrons in the Cosmos: Nuclear Astrophysics* (University of Chicago Press, Chicago, 1988)
36. E. Rutherford, Proc. R. Soc. A Math. Phys. Eng. Sci. **97**, 374 (1920)
37. E. Rutherford, F.W. Aston, J. Chadwick, C.D. Ellis, G. Gamow, R.H. Fowler, O.W. Richardson, D.R. Hartree, Proc. R. Soc. A Math. Phys. Eng. Sci. **123**, 373 (1929)
38. E. Rutherford, Lond. Edinburgh Dublin Philos. Mag. J. Sci. **4**, 580 (1927)
39. E. Schrödinger, Phys. Rev. **28**, 1049 (1926)
40. G. Shaviv, *The Synthesis of the Elements* (Springer, 2012)
41. H. Sodak, ed., *Reactor Handbook*, 2nd edn. (Interscience, New York, 1962)
42. J.L. Spradley, Phys. Teacher **27**, 656 (1989)
43. K. Starke, J. Chem. Ed. **56**, 771 (1979)
44. R.H. Stuewer, in *George Gamow Symposium; ASP Conference Series*, vol. 129 (1997), p. 29
45. F.-K. Thielemann, M. Eichler, I. Panov, B. Wehmeyer, Annu. Rev. Nucl. Part. Sci. **67**, 253 (2017)
46. C.F.v. Weizsäcker, Zeitschrift für Physik **96**, 431 (1935)
47. E. Wigner, Trans. Faraday Soc. **34**, 29 (1938)
48. H.T. Wolff, Ann. Phys. **60**, 45 (1919)

Chapter 2
General Characteristics

Abstract This chapter gives an overview of some general characteristics of fission. The topics covered include the distinction between spontaneous fission and fission induced by the capture of an incident particle, the sequence of events that occur during the fission process, the features of fission cross sections, the observed distributions in charge and mass of the fragments and how we can explain them, the calculation of the energy released in fission and how it compares to other energy sources, the concepts of fission barriers and fission isomers and how the two are related, and the properties of neutrons and gamma rays produced during fission. Fission is inherently a collective motion of the protons and neutrons that make up the nucleus, and as such it is distinguishable from other phenomena that break up the nucleus. We discuss the difference between fission and one such fragmentation process: spallation. The chapter also provides a summary of online databases for fission data.

Various characteristic features of fission have emerged from over 80 years of study. These features are important in formulating a coherent understanding of the fission process. That being said, it is important to keep in mind that, despite overall systematic trends, observed fission properties can differ significantly between neighboring nuclei. In this chapter, we survey various observables and characteristic features of fission and their overall behavior.

2.1 Spontaneous and Induced Fission

Nuclear fission is an extreme example of large-amplitude collective motion [4] that results in the division of a parent nucleus into two or more fragment nuclei. Binary fission (fission into two fragments) is far more common than ternary fission, which typically occurs in fewer than about one in a thousand events in thermal fission

© The Author(s), under exclusive license to Springer Nature Switzerland AG 2021
W. Younes, W. D. Loveland, *An Introduction to Nuclear Fission*, Graduate Texts
in Physics, https://doi.org/10.1007/978-3-030-84592-6_2

[35]. The third fragment in ternary fission is most often an α particle that tends to be emitted in a direction roughly perpendicular to the axis connecting the other two fragments [35].

The fission process can occur spontaneously, or it can be induced by an incident particle. Historically, induced fission was discovered first, and spontaneous fission (SF) was first observed in uranium nuclei some time later by G. N. Flerov and K. A. Petrjak [20, 48, 49, 55]. The SF process in actinides typically competes with α decay and is often dwarfed by the α branch but not always, as in the case of ^{250}Cm where SF dominates α decay. When viewed as a barrier penetration mechanism, the SF half-life can be calculated using the semi-classical Wentzel-Kramers-Brillouin (WKB) approximation [41] in terms of the action integral for the system [2, 35, 52]. The SF half-lives, when plotted as a function of the Z^2/A ratio of the parent nucleus (Fig. 2.1), display two important features: the SF half-life tends to decrease with increasing Z^2/A, and nuclei with an odd number of either protons or neutrons have systematically larger half-lives than the neighboring even-even nuclei [35].

The number of parent nuclei at time t that have not yet undergone SF is given by the exponential decay law

$$N(t) = N(0)\, e^{-t/\tau_{SF}} \tag{2.1}$$

where $N(0)$ is the number at time $t = 0$ and τ_{SF} is the SF lifetime, which is related to the SF half-life $T_{1/2}$ (SF) via $\tau_{SF} = T_{1/2}$ (SF) $/ \ln 2$. When several decay modes are possible (e.g., SF and alpha decay), the frequency at which each occurs

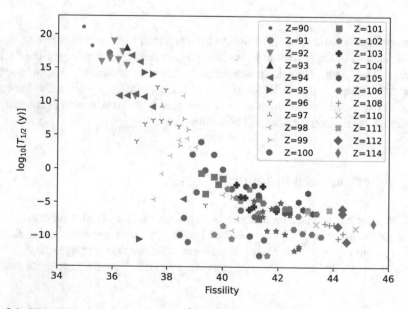

Fig. 2.1 SF half-lives as a function of the Z^2/A ratio of the parent nucleus. Data taken from the ENSDF database

is characterized by a branching ratio ρ. For example, if a nucleus decays by SF a fraction ρ_{SF} of the time and by α emission a fraction ρ_α of the time (with $\rho_{SF} + \rho_\alpha = 1$), the lifetime τ of the nucleus can be decomposed as

$$\frac{1}{\tau} = \frac{\rho_{SF}}{\tau} + \frac{\rho_\alpha}{\tau} \tag{2.2}$$

The term "partial lifetime" is sometimes used to denote the lifetime τ/ρ_i associated with a particular decay mode i, but this is something of misnomer since these partial lifetimes cannot be directly measured but are instead deduced from the lifetime τ for all modes of decay and the branching ratio ρ_i [58].

2.2 Chronology of the Fission Process

In induced fission, an incident particle fuses with a target to form a compound system.[1] If the excitation energy of the compound system is sufficiently large, the parent nucleus will likely emit neutrons and may undergo multiple-chance fission (fission without prior neutron emission is called "first-chance" fission; if one neutron is emitted, the process is referred to as "second-chance" fission, etc.). Once there is no longer enough excitation energy left to emit additional neutrons, the parent nucleus can continue to de-excite through gamma-ray emission or proceed toward scission (the breaking point of the nucleus). Various models predict a transition time to scission (also known as the saddle-to-scission time) of $10^{-20} - 10^{-19}$ s for low-energy fission [3, 8, 44]. This is a relatively long transition time compared to the orbital time of 10^{-22} s of a nucleon in the nucleus, as expected for a collective process involving the coherent motion of many or all nucleons. There are various experimental techniques that attempt to measure the time scale of the fission process. Some of these techniques probe the time scale of the entire fission process from compound-nucleus formation to scission. For example, one such approach takes advantage of the rearrangement of the electronic structure that occurs when the fragments formed and the corresponding change in the x-rays produced [76]. Another approach, the crystal blocking technique, relies on the ordered structure of a crystal to gauge how far the parent nucleus has recoiled before the fragments are produced and fly apart, and to deduce the time delay since compound-nucleus formation [43]. Other experimental methods, looking at the properties of neutrons and gamma rays emitted before scission, can give a better sense of the saddle-to-scission time scale [27, 28].

The fragments produced at scission are called "*primary fragments*" (or more simply, pre-neutron fragments) until they emit prompt neutrons and become "*secondary fragments*" (or post-neutron fragments). These secondary fragments can then beta decay into more stable nuclei, which are then referred to as "fission products" [16].

---

[1] Direct reaction mechanisms can also lead to fission but without initially going through compound-nucleus formation [18].

The gamma rays emitted by the fragments can be prompt, or "late prompt" [65] if they are issued from an isomeric state (i.e., a long-lived, metastable state), and can help identify the fragments thanks to their precisely known energies. The sequence of events that occurs during induced fission is summarized in Fig. 2.2 (see also [23, 37, 71]).

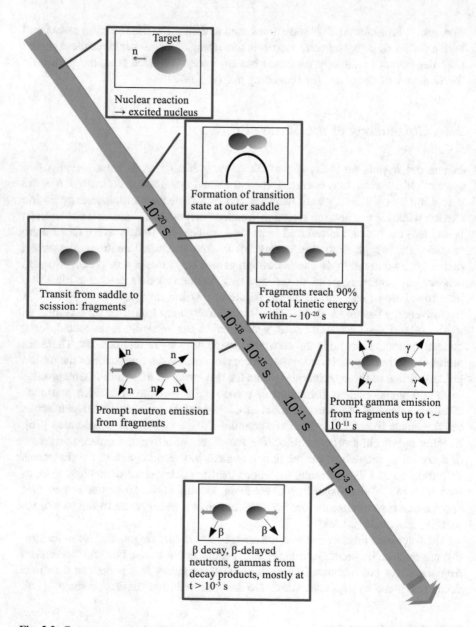

Fig. 2.2 Cartoon representing the various events that occur during induced fission. Quoted times are rough estimates. See also [37]

2.3 Fission Cross Sections

When discussing induced fission and other nuclear reactions, it will be convenient to use the standard notation [54]

$$\text{Target (projectile, ejectile) residual} \tag{2.3}$$

where the ejectile is the lighter particle (or particles) produced by the reaction of the projectile incident on the target, leaving a residual nucleus behind. The notation for a fission reaction induced, for example, by a neutron incident on a ^{235}U target is ^{235}U (n, f). For the spontaneous fission of ^{252}Cf, for example, we will write: ^{252}Cf (SF).

Fission cross sections are a measure of the probability that fission will occur in a nuclear reaction. We will discuss cross sections in greater detail in Sect. 3.6 and Chap. 7 of this book. Here, we consider them as another measurable property of induced fission reactions. Figure 2.3 shows the ^{235}U (n, f) and ^{238}U (n, f) reaction cross sections as a function of incident neutron energy E. The ^{235}U (n, f) cross section can be divided into three regions of interest: at very low energies ($E \lesssim 0.1$ eV), the cross section has a characteristic $1/v$ (inverse velocity) behavior [5]; in the range 0.1 eV $\lesssim E \lesssim 1000$ eV, individual peaks or resonances (see Sect. 7.1.2) can be seen; and for $E \gtrsim 1000$ eV, the resonances overlap and form a relatively smooth cross-section curve.

Note that the lower-energy features of the ^{235}U (n, f) cross section are absent from the ^{238}U (n, f) cross section. This is due to the fact that neutron absorption on a ^{235}U target forms a ^{236}U nucleus with excitation energy greater than the critical fission energy, whereas in the case of $n + ^{238}$U, the resulting ^{239}U nucleus has an excitation energy below the critical fission energy. Actinides like ^{235}U that fission easily following the absorption of a thermal (0.25 meV) neutron are called *fissile*, whereas those like ^{238}U that do not easily fission when they absorb a thermal neutron are called *fissionable*.

2.4 Fragment Mass and Charge Distributions

The number of fragments with a given mass (A), charge (Z), and isomeric state (I), produced in each fission event after prompt neutron emission but before any delayed decays, is called the "independent yield" $Y(A, Z, I)$ and is normalized to 2 (or 200%) when summed over all values of A, Z, and I [16]. The number of nuclei produced by each fission over all time is called the "cumulative yield" $Y_{cu}(A, Z, I)$ and is also normalized to 2 (or 200%) [16].[2] The sum yield $Y(A)$ is the sum of

---

[2] The choice of 2 or 200% for the normalization is due to the fact that the parent nucleus is most likely to fission into two fragments.

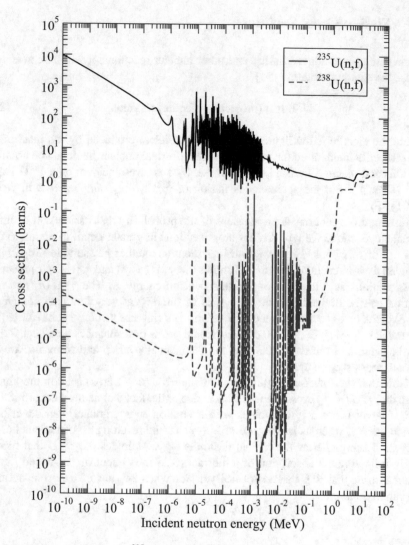

Fig. 2.3 Cross sections for the ^{235}U (n, f) and ^{238}U (n, f) reactions as a function of incident neutron energy. Adapted from [34] using data from the ENDF database

independent yields over all Z and I for a given mass A. The chain yield $Y_{ch}(A)$ is the cumulative yield of the last (i.e., stable) member of a decay chain of given mass A. The sum and chain yields usually have nearly identical values; they differ by a small amount due to the fact that sum yields are calculated before—and chain yields after—delayed neutron emission.

The mass and charge distributions of fission products have been measured for many actinides and for both SF and induced fission at various energies. Some characteristic features can be gleaned from these studies. For thermal fission of

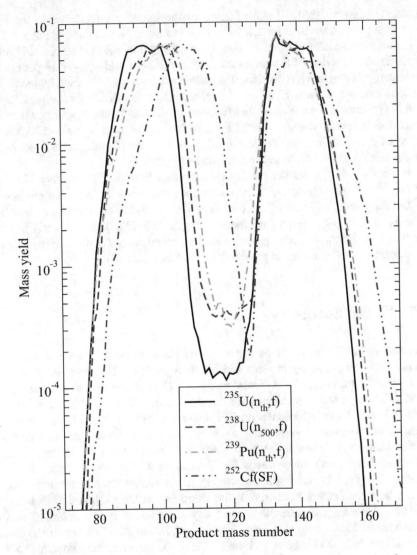

Fig. 2.4 Fragment mass distribution as a function of mass number. The data are taken from the ENDF library and correspond to incident neutrons with thermal (*th*) and 500-keV (*500*) energies and for spontaneous fission (*SF*)

actinides, the distributions as a function of A tend to be bimodal, with one peak near the doubly magic ^{132}Sn due to quantum shell effects and another corresponding to the lighter complementary nuclei, as shown in Fig. 2.4.

The two peaks are fairly broad and are separated by a dip near symmetric fission for thermal and low-energy fission [21, 72]. As the energy of the incident neutron is increased, the central dip at symmetric fission tends to fill in, and the mass peaks broaden somewhat [25]. Another feature of the fission products is that the ratio Z/A of a product tends to have a value close to the Z/A ratio of the parent. This is known as the unchanged charge distribution (UCD) rule [73]. The distribution of products as a function of Z sometimes displays an even-odd staggering pattern for low-energy fission that has been used to estimate the energy dissipated by the parent nucleus into non-collective modes of excitation before scission (see, e.g., Chapter 8 in [71]). Looking at the mass distributions of the fragments as a function of the mass A of the parent nucleus, a notable transition occurs for $A \approx 100 - 130$, known as the Businaro-Gallone region [9, 10, 57], where the distribution shifts from purely symmetric about $\approx A/2$, to asymmetric as in Fig. 2.4.

In the late 1980s, studies of the spontaneous fission of the heavy elements ^{258}Fm, ^{259,260}Md, ^{258}No, and ^{260}Rf showed a strong symmetric fission mode with higher fragment kinetic energies coexisting with a weaker broad mass distribution associated with lower fragment kinetic energies [30, 31]. This "bimodal fission" process revealed the underlying complexity potential energy surface of the nucleus (a concept we will examine in greater detail in Chaps. 3 and 6).

2.5   Fission Energetics

The total energy released in a fission event is defined as the difference in rest-mass energy between the parent nucleus and the final products. This energy can also be written as the sum of the total kinetic energy (TKE) and the total excitation energy (TXE) of the primary fragments, minus any energy contributed by the incident particle in the case of induced fission. The energy contributed by the incident particle consists of its kinetic energy plus any excitation energy gained in the formation of the compound nucleus. These two different formulations of the total energy released in fission are summarized in Eq. (2.5) below for the case of neutron-induced fission. The TKE of the fragments can be estimated as the Coulomb repulsion energy between centers of charge of the fragments at their separation distance at scission. However, this estimate ignores any pre-scission energy acquired during the transition from saddle to scission. Precisely how the initial energy of the parent nucleus is partitioned between TKE and TXE of the fragments and how the TXE is divided among the fragments remain open questions and topics of ongoing research [39, 42, 45, 56].

The total energy released in fission is given by energy conservation as the difference between the rest-mass energy of the parent and the sum of the rest-mass energies of the fragments (before prompt neutron emission). Let us look, for example, at the fission ^{239}U $\rightarrow$ ^{119}Pd $+$ ^{120}Pd, considered by Bohr and Wheeler in their 1939 paper [6]. Using the online tool at http://t2.lanl.gov/nis/data/astro/molnix96/massd.html, the mass excess retrieval tool at https://www-nds.iaea.org/RIPL-3/, or published mass tables [1], we can retrieve the required nuclear masses and calculate the total energy released,

$$E_r = M\left(^{239}\text{U}\right)c^2 - \left[M\left(^{119}\text{Pd}\right)c^2 + M\left(^{120}\text{Pd}\right)c^2\right]$$

$$= 222.631690\,\text{GeV} - [110.753618\,\text{GeV} + 111.686217\,\text{GeV}] \qquad (2.4)$$

$$= 191.855\,\text{MeV}$$

For neutron-induced fission, this same energy can also be written (again using energy conservation) as the total fragment kinetic energy (TKE) and total fragment excitation energy (TXE) minus the sum of the incident neutron kinetic energy (E_n) and neutron separation energy (S_n) in the compound nucleus,

$$E_r = M\,(\text{parent})\,c^2 - \sum_{\text{fragment}} M\,(\text{fragment})\,c^2$$

$$\qquad (2.5)$$

$$= \text{TKE} + \text{TXE} - (E_n + S_n)$$

So far, we have only considered one specific division of the nucleus at a time (e.g., ^{239}U into ^{119}Pd and ^{120}Pd), but in reality the fission of a large number of parent nuclei will produce over time a full distribution of fragments and products (see Sect. 2.4). Therefore it will be useful to consider an average total prompt energy release, $\langle E_r \rangle$, obtained by weighting Eq. (2.5) with the probability of each division (which is itself given by the independent yields of the fragments, as described in Sect. 2.4). Madland [38] has obtained empirical formulas for $\langle E_r \rangle$ as a function of incident neutron energy for three target nuclei, ^{235}U, ^{238}U, and ^{239}Pu, and valid for neutron energies from thermal to 15 MeV:

$$\langle E_r \rangle = \begin{cases} 185.6 - 0.0995E_n & n +^{235}\text{U} \\ 187.7 - 0.1318E_n + 0.0034E_n^2 & n +^{238}\text{U} \\ 197.2 - 0.1617E_n - 0.0017E_n^2 & n +^{239}\text{Pu} \end{cases} \qquad (2.6)$$

where both $\langle E_r \rangle$ and E_n are in MeV. These formulas account for the prompt energy release in fission, i.e., from the kinetic energy of the fragments and the neutrons and gamma rays they emit as they lose their initial excitation energy. The formulas do not include contributions from subsequent β decays and β-delayed processes.

To appreciate the impressive amount of energy released by fission, we can compare it to the energy obtained from chemical processes such as the combustion of methane and the oxidation of hydrogen. The energy released by chemical reactions (or change in enthalpy, ΔH) is due to the breaking of bonds in the reactants and the formation of new bonds in the products. For this exercise, we will use average bond dissociation energies taken from Table 4.11 in [17]. Thus, for the combustion of methane,

$$CH_4 + 2O_2 \longrightarrow CO_2 + 2H_2O$$

We will need dissociation energies for the bonds C-H (= 337.2 kJ/mol), O-O (= 498.3 kJ/mol), C=O (= 749 kJ/mol), and H-O (= 428.0 kJ/mol). Using these values, we calculate the change in enthalpy

$$\Delta H = [4 \times 337.2 + 2 \times 498.3] - [2 \times 749 + 2 \times (2 \times 428.0)]$$
$$= -864.6 \text{ kJ/mol} \tag{2.7}$$

Thus, 1 mol of methane (16 g) releases 864.6 kJ of energy. To put it in different terms, the complete combustion of 1 kg of CH_4 would release $\approx 5.4 \times 10^7$ J of energy.

Next, we consider the oxidation of hydrogen,

$$2H_2 + O_2 \longrightarrow 2H_2O$$

which is used in hydrogen fuel cells to produce energy. In addition to the bond dissociation energies already given above, we will also need the one for H-H (= 436 kJ/mol). The change in enthalpy is

$$\Delta H = [2 \times 436 + 498.3] - [2 \times (2 \times 428.0)]$$
$$= -341.7 \text{ kJ/mol} \tag{2.8}$$

Thus, for 1 mol of hydrogen (1 g), we obtain $341.7/4 = 85.4$ kJ/mol, or 8.5×10^7 J of energy for 1 kg of H. This is $\approx 57\%$ more energy than is obtained from the combustion of methane for an equivalent mass of CH_4 (with the added benefit that the by-product of the reaction is simply water).

Finally, we consider the thermal-induced fission of ^{235}U. One kilogram of ^{235}U contains

$$\frac{1000 \text{ g}}{235.0 \text{ g/mol}} \times 6.02 \times 10^{23} \text{ atoms/mol} = 2.56 \times 10^{24} \text{ atoms}$$

If we assume that all those ^{235}U atoms fission (in order to compare fairly with the chemical reactions above where we assumed complete combustion/oxydation), and if each fission releases 185.6 MeV on average (see Eq. (2.6) above), then the total prompt energy release is $\approx 7.6 \times 10^{13}$ J, or about a million times more than the hydrogen reaction.

Most of the energy produced by fission ($\sim 85\%$) is in the form of fragment kinetic energies, with the rest in the form of prompt neutrons and gamma rays ($\sim 6\%$), neutrons and gamma rays emitted after β decay ($\sim 6\%$), and neutrinos produced by the β decay ($\sim 3\%$) [59].

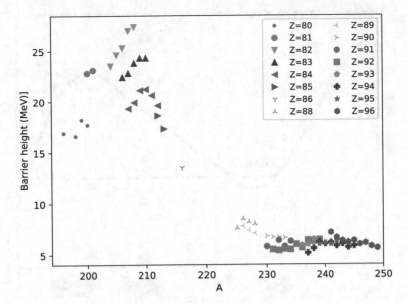

Fig. 2.5 Plot of barrier heights as a function mass number A of the parent nucleus. Data taken from the RIPL database

2.6 Fission Barriers

Bohr and Wheeler invoked the concept of a fission barrier to explain (1) the stability of uranium against fission from its ground state and (2) the existence of a critical energy (the barrier height) that has to be supplied in order to trigger fission. In their paper [6], they used the liquid-drop model to explain and estimate the barrier heights. Some estimated barrier heights [11] are shown in Fig. 2.5.

The calculation of barrier heights follows from the calculation of the energy of the nucleus as a function of its deformation, which we will discuss in Chaps. 3 and 6. Barriers also play an important role in modeling fission cross sections (see Chap. 7).

2.7 Fission Isomers

We have so far referred to *the* fission barrier of a nucleus, perhaps leaving the reader with the impression that each fissioning nucleus has a single fission barrier. In reality the situation is often more complicated. From Bohr and Wheeler's original work in 1939 [6] through the 1950s, the fission barrier imagined by physicists consisted of a single barrier, and spontaneous fission could be understood as tunneling through this barrier starting from the nucleus in its ground state.

Then in 1962, Polikanov et al. [51] identified the nucleus $^{242}_{95}$Am, formed in the bombardment of ^{238}U with ^{16}O and ^{22}Ne projectiles, and measured its half-life of

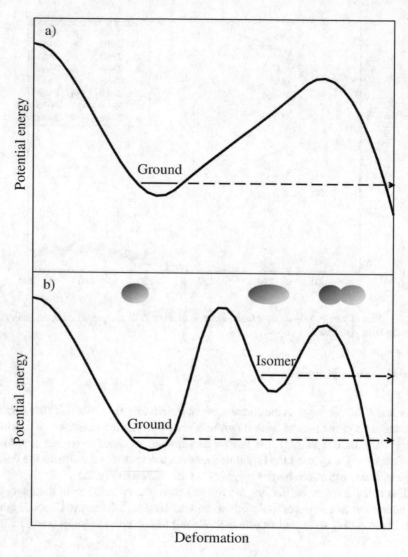

Fig. 2.6 Evolution in the interpretation of spontaneous fission from (**a**) a single barrier to (**b**) a double-humped fission barrier

13.5 ± 1.5 ms. From overall trends, this half-life is very much shorter than what can be expected from spontaneous fission of $^{242}_{95}$Am in its ground state. Since then, about 40 such cases have been identified with 6 ps $\lesssim T_{1/2}$ (SF) $\lesssim$ 14 ms [14, 35].

The explanation for this unexpected behavior was given by Strutinsky in 1966 [61–63]. He showed that the shape of the classical barrier should be modified to account for the influence of quantum mechanics and could acquire a "dip" that turns one barrier into two, as depicted in Fig. 2.6.

We will explore Strutinsky's approach in greater detail in Chap. 3. For now, we simply note that this dip acts as a local potential well several MeV above the ground state that can support a metastable state between the barriers on either side of it. Because this state exists at a larger deformation than the ground states and because it can fission by tunneling through the outer barrier, it is referred to alternately as a *shape isomer* or a *fission isomer*. The discovery of these fission isomers was a major clue in establishing the "double-humped" shape of the fission barrier, which is common throughout the actinide region. In fact, nuclei with $Z \leq 89$ tend to have a single barrier while those with $Z \geq 90$ tend to have two.

For actinide nuclei, it has long been known [36] that the shape of the fissioning nucleus breaks axial symmetry at deformations corresponding to the first barrier; that is to say near the first (or inner) barrier, the shape can no longer simply be described as a surface of revolution generated by rotating a curve about an axis of symmetry. Near the second (or outer) barrier, the shape of the nucleus is axially symmetric again but, at least in low-energy fission, prefers to break reflection symmetry about a plane going through its center of mass and perpendicular to its symmetry axis; or in plain terms, the left and right parts of the nucleus "look" different and will evolve into a light and a heavy fragment with different masses.

2.8 Neutron and Gamma-Ray Distributions

In addition to the information that can be gathered from the fragments and products (e.g., mass distributions and TKE), the neutrons and gamma rays that they emit can also provide useful data that shed light on the fission process. In principle, the excitation energy and initial angular momentum imparted to the fragments could be reconstructed by measuring the neutrons and gamma rays they emit. The average number of prompt neutrons (multiplicity, $\bar{\nu}$) emitted as a function of primary fragment mass (A) typically follows a characteristic "sawtooth" shape shown in Fig. 2.7 and can be used to estimate the average fragment excitation energy, removed by neutron emission [38].

Figure 2.8 shows the average neutron multiplicity as a function of the incident neutron energy, summed over the contributions from all fragment masses, for neutrons incident on a ^{232}Th target. The energy thresholds for the different multiple-chance fission reactions are also shown, and the effect on the neutron multiplicity can be seen, at least for first-chance fission.

The energy spectrum of the neutrons emitted by the fragments is expected to have a Maxwellian shape in the center-of-mass frame of the fragment which, when transformed to the laboratory frame, takes on a Watt functional form [74], plotted in Fig. 2.9.

In addition to the prompt neutrons emitted by the fragments, and multi-chance neutrons emitted by the parent nucleus, there are two additional sources of neutrons associated with fission that must be mentioned: delayed neutrons and scission neutrons. Delayed neutrons are produced long after scission when, in a few special cases, a fission fragment undergoes a beta decay that leaves the daughter nucleus

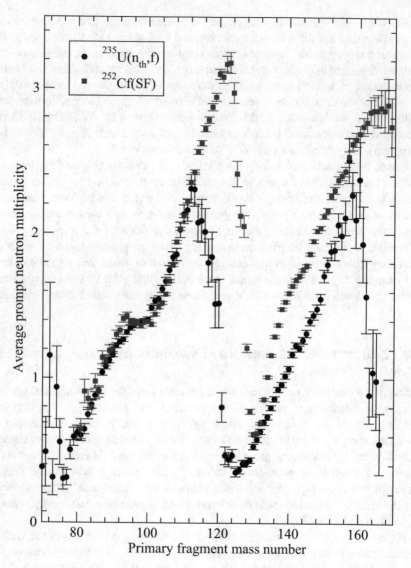

Fig. 2.7 Average neutron multiplicity as a function of primary fragment mass [69]. Data taken from the EXFOR database

with enough excitation energy to emit a neutron. These neutrons, which can be observed up to minutes after scission in some cases, are referred to as delayed neutrons because they are delayed compared to prompt emissions by the relatively slower beta decay process [64].

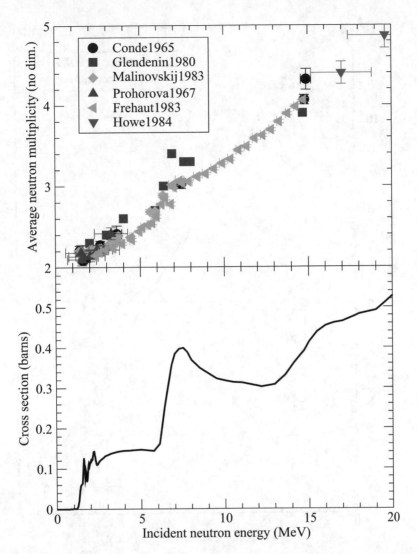

Fig. 2.8 The top panel shows the average neutron multiplicity for the ^{232}Th (n, f) reaction as a function of incident neutron energy. The bottom panel shows the ^{232}Th (n, f) cross section to highlight the onset of second-chance fission near 6.5 MeV, and of third-chance fission near 15 MeV, and the effect on the neutron multiplicity. The data sets labeled Conde1965 [15], Glendenin1980 [24], Malinovskij1983 [40], Prohorova1967 [60], Frehaut1982 [22], and Howe1984 [29] are taken from the EXFOR database. See also [32]

In contrast to delayed neutrons, for which there is ample experimental evidence, scission neutrons remain a controversial topic [50, 68]. Scission neutrons are thought to be emitted from the neck region of the parent nucleus as it breaks apart. Because they originate from the slowly recoiling parent nucleus (in contrast

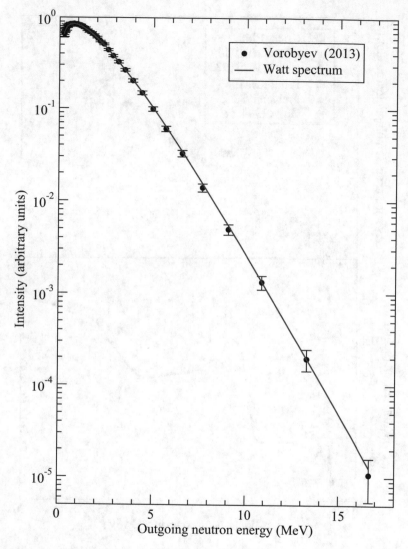

Fig. 2.9 Watt spectrum, fitted to experimental data from the ^{235}U (n_{th}, f) reaction [67]. Data taken from the ENDF database

to neutrons emitted by the rapidly moving fragments), scission neutrons were originally expected to exhibit a telltale isotropic angular distribution [7]. It was also hypothesized that the neutron energy spectrum would be affected by the presence of scission neutrons [33]. However, subsequent studies have shown that a portion of the scission neutrons should be absorbed by the moving fragments [12] and that simplifying assumptions about the angular distribution of the neutrons emitted by the fragments may have led to their mischaracterization as scission neutrons

[69, 70]. Thus, the observation of an additional isotropic source of neutrons in the data should not be automatically taken as evidence of the presence of scission neutrons.

The detection of gamma rays emitted by the fragments can provide additional information about them. The sum of the energies of neutrons and gamma rays emitted by a fragment provides an estimate of its initial excitation energy. Furthermore, gamma rays can also be used to reconstruct the initial angular momentum of the fragments [75]. As a result, there is ongoing interest in measuring these gamma rays, and this has recently been done for several fissioning nuclei [13, 46, 47].

2.9   Spallation Versus Fission

The collective character of the fission process distinguishes it from other processes that break up the nucleus, such as spallation [19]. In a spallation reaction,[3] a high-energy incident nucleon (e.g., an 800-MeV proton [53]) interacts with individual protons and neutrons in the nucleus causing the emission of secondary particles via an intranuclear cascade mechanism [19]. Once the nucleus loses enough energy through this process, particle (mostly neutron) emission continues via statistical evaporation, which can eventually result in fission. The nature of the spallation process is revealed by the de Broglie wavelength of the incident particle. For an incident proton of kinetic energy $E_p = 800$ MeV, we can calculate the de Broglie wavelength

$$\lambda = \frac{h}{\sqrt{2m_p E_p}} \approx 1 \text{ fm} \tag{2.9}$$

which is on the order of the size of an individual nucleon. Therefore, unlike fission which is a collective process involving many or all nucleons, the initial stage of spallation involves a few individual nucleons. The different mechanisms underlying fission and spallation (provided it does not end in fission) lead to differences in their observable properties: spallation tends to produce fragments that are close to the target in mass, whereas (low-energy) fission more often results in a pair of fragments significantly lighter than the target nucleus.

2.10   Online Resources for Fission Data

The following list provides links to useful databases for fission (and other nuclear) data:

- EXFOR, for experimental data: https://www-nds.iaea.org/exfor/
- ENSDF, for evaluated properties of nuclear levels and decays: https://www.nndc.bnl.gov/ensdf/

---

[3] The term "spallation" was coined by Glenn Seaborg in 1947 (www2.lbl.gov/Publications/Seaborg/hits.htm).

- ENDF, for evaluated cross-section data: https://www.nndc.bnl.gov/exfor/endf00.
  jsp
- RIPL, the Reference Input Parameter Library for various quantities: https://
  www-nds.iaea.org/RIPL-3/
- QCALC for reaction Q values: https://www.nndc.bnl.gov/qcalc
- LANL Qtool reaction Q value calculator: https://t2.lanl.gov/nis/data/qtool.html
- LANL lookup tool for nuclear masses and deformations: https://t2.lanl.gov/nis/
  data/astro/molnix96/massd.html
- NUDAT interactive chart of nuclides to retrieve various nuclear properties:
  https://www.nndc.bnl.gov/nudat2/
- AMEDEE database of calculated nuclear properties: http://www-phynu.cea.
  fr/science_en_ligne/carte_potentiels_microscopiques/carte_potentiel_nucleaire_
  eng.htm
- WIMS database of reactor fission product yields: https://www-nds.iaea.org/
  wimsd/fpyield.htm
- IAEA database of cumulative yields at thermal, fast, and 14-MeV neutron
  energies https://www-nds.iaea.org/sgnucdat/c3.htm

2.11   Questions

11. What is the general trend of spontaneous fission half-lives as a function of the charge number squared over mass of the parent nucleus (Z^2/A)?

12. How many neutrons are emitted in third-chance fission, before the fission process begins?

13. What are some experimental techniques that can be used to estimate fission times?

14. Define and contrast the terms: independent, cumulative, sum, and chain yield.

15. What is the general shape of the distribution of the number of fragments per fission, as a function of fragment mass for low-energy fission?

16. How are the mass and charge of the parent nucleus and daughter fragments related?

17. How do the energies released by methane combustion, hydrogen oxidation, and fission compare?

18. If you could (hypothetically) suppress the Coulomb repulsion between protons, what would happen to the fission barrier?

19. What was one of Strutinsky's important contributions to the fission problem in the 1960s?

20. Why do fission isomers usually have half-lives that are very different from spontaneous fission in the same nucleus?

21. What is the typical shape of the average neutron multiplicity as a function of fragment mass?

22. What is the typical shape of the neutron energy spectrum in the frame of reference of the emitting nucleus?

23. When a fragment de-excites to its ground state, what are the particles most likely to be emitted?

24. What are some observable differences between spallation and fission?

2.12 Exercises

Problem 11 How many SF events per second do you expect on average from a 1 μg ^{248}Cm sample ($T_{1/2} = 3.48 \times 10^5$ y with 91.61% α decay and 8.39% SF and a molecular weight of 248.072 g/mol).

Problem 12 The spontaneous fission half-lives of ^{252}Fm, ^{254}Fm, and ^{256}Fm are 3.97×10^9 s, 1.97×10^7 s, and 1.03×10^4 s, respectively. Use this information to estimate the spontaneous fission half-life of ^{258}Fm. How does your estimate compare to the known value of 370 μs?

Problem 13 List five fission products from the ^{235}U (n, f) reaction produced in a reactor that have a mass yield of at least 1%. You can use the WIMS database for this exercise, for example.

Problem 14 Plot Z as a function of A for the isotopes of I, Xe, Cs, and Nd with the largest yields from each of the ^{235}U (n, f), ^{238}U (n, f), and ^{239}Pu (n, f) reactions with reactor neutrons. How well do they follow the UCD rule? What is the largest deviation in Z/A from that of the parent nucleus? (e.g., you can use look up fission product yields at https://www-nds.iaea.org/wimsd/fpyield.htm)

Problem 15 Figure 2.10 shows relative barrier heights as a function of fissility calculated using the LDM and given by $0.02Z^2/A$. According to this model, what is the barrier height for ^{209}Pb relative to ^{240}Pu.

Problem 16 Consider the thermal neutron induced fission of a ^{235}U target (i.e., $n + ^{235}$ U). How much energy is released by (1) symmetric fission and (2) fission that produces ^{134}Te as the heavy fragment?

Problem 17 What is the maximum number of neutrons that can be emitted (excluding neutrons emitted by the fragments) following the $n + ^{235}$ U reaction with a 20-MeV incident neutron?

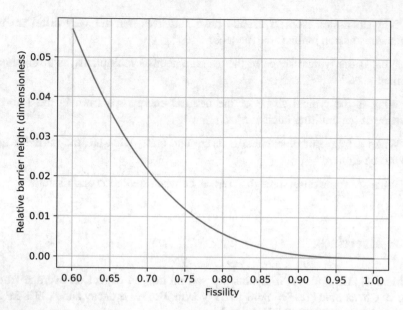

Fig. 2.10 Relative fission barrier heights predicted by the LDM for fissilities $0.6 \leq x \leq 1$. Adapted from [26]

Problem 18 The following is a list of secondary fragments produced by the fission of ^{236}U : ^{142}Ba, ^{100}Nb, ^{95}Sr, ^{99}Zr, ^{139}Xe, ^{133}Sb, ^{91}Kr, and ^{135}Te. Assuming each fission produces two fragments, match them by pairs and note the number of neutrons emitted in each case.

Problem 19 ^{235}U has a half-life $T_{1/2} = 7.04 \times 10^8$ y. Only $\approx 7 \times 10^{-9}$% of those decays are spontaneous fission, with the remainder being essentially α decays. What is the rate of ^{235}U (SF) for a 1-g sample? (Molecular weight of ^{235}U = 235.04393 g/mol).

Problem 20 A natural uranium sample today consists of 99.2739% ^{238}U ($T_{1/2} = 4.47 \times 10^9$ y), 0.7205% ^{235}U ($T_{1/2} = 7.04 \times 10^8$ y), and trace amounts of ^{234}U which we will ignore. What was its isotopic composition 4.5 billion years ago, when the earth was formed?

Problem 21 The nucleus ^{252}Cf decays with a half-life $T_{1/2} = 2.645$ years with 3.09% of those decays by SF. Each fission releases 186.25 MeV on average in kinetic energy of the fragments. What is the power in watts produced by the kinetic energy of these fragments for 1 g of ^{252}Cf. (Molecular weight of ^{252}Cf = 252.08163 g/mol).

Problem 22 What is the Q value for the (unlikely) spontaneous fission of ^{237}Np into three identical fragments.

Fig. 2.11 Toy model of the ^{236}U nucleus at scission represented by two spheres (corresponding to doubly magic ^{78}Ni and ^{132}Sn) connected by a cylindrical neck of length L. The neck breaks at a position x, leaving a light fragment of mass number A_L and a heavy one of mass number A_H

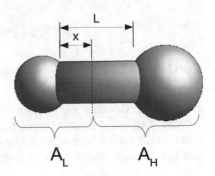

Problem 23 If the yearly power consumption of a typical US household is 10,000 kWh, and if each ^{235}U fission releases $\sim$ 200 MeV in energy, for how long could a household be powered by 1 g of ^{235}U, assuming the entire sample fissions and all the energy is recovered?

Problem 24 Suppose a fission event produces the fragments ^{102}Zr and ^{134}Te. How much energy is released by their subsequent β decays?

Problem 25 This exercise is adapted from [71] pp. 523–524. Consider the schematic model depicted in Fig. 2.11.

Each fragment consists of a doubly magic core plus a part of the neck determined by the position x of the break along its length L. The number of particles A_L and A_H in each fragment is proportional to its volume. Let $A_L^{(0)} \equiv A_L (x = 0)$ and $A_H^{(0)} \equiv A_H (x = L)$, with $A_L + A_H = A$, the mass number of the parent nucleus.

1. Write formulas for A_L and A_H as a function of x, L, A, $A_L^{(0)}$, and $A_H^{(0)}$.
2. If the neck is equally likely to break anywhere along its length, plot the expected mass distribution as a function of fragment mass for $A = 236$, $A_L^{(0)} = 78$, and $A_H^{(0)} = 132$, and compare to Fig. 2.4.

Problem 26 Flynn et al. [21] fit the mass number for the peak of the light mass product yield (A_L) as a function of the mass number of the fissioning nucleus (A_F) for SF and low-energy fission. They found the following linear relationship

$$A_L \approx 0.8593 A_F - 108.177 \tag{2.10}$$

Ignoring neutron emission, plot A_L and the corresponding value A_H for the heavy mass product peak as a function of A_F for $225 \leq A_F \leq 260$ and comment on the result.

Problem 27 Viola et al. [66] found the following empirical relation between the total kinetic energy (TKE) of the fragments averaged over all fragments and the charge (Z) and mass (A) numbers of the fissioning nucleus for low-energy fission,

$$\text{TKE} \approx 0.1189 \frac{Z^2}{A^{1/3}} + 7.3 \text{ (MeV)} \tag{2.11}$$

Calculate the TKE for the low-energy ^{235}U (n, f), ^{238}U (n, f), ^{239}Pu (n, f), and ^{252}Cf (SF) processes.

Problem 28 Consider the symmetric fission of nucleus with A nucleons and Z protons. Suppose that, at scission, the fissioning nucleus looks like two touching spheres with each sphere representing one of the fragments. Calculate the Coulomb repulsion energy assuming the fragments can be treated as point charges:

$$E_C = e^2 \frac{Z_1 Z_2}{d} \tag{2.12}$$

where $e^2 \approx 1.44$ MeV $\cdot$ fm, Z_1 and Z_2 are the charge numbers of the fragments, and d is the distance between their centers. Compare your answer with the empirical formula for TKE in Exercise 27.

References

1. G. Audi, A. Wapstra, C. Thibault, Nucl. Phys. A **729**, 337 (2003)
2. A. Baran, Phys. Lett. B **76**, 8 (1978)
3. J. Berger, M. Girod, D. Gogny, Nucl. Phys. A **502**, 85 (1989)
4. G. Bertsch, Nucl. Phys. A **574**, 169 (1994)
5. J.M. Blatt, V.F. Weisskopf, *Theoretical Nuclear Physics* (Dover Publications, 1991)
6. N. Bohr, J.A. Wheeler, Phys. Rev. **56**, 426 (1939)
7. H.R. Bowman, S.G. Thompson, J.C.D. Milton, W.J. Swiatecki, Phys. Rev. **126**, 2120 (1962)
8. A. Bulgac, P. Magierski, K.J. Roche, I. Stetcu, Phys. Rev. Lett. **116**, 122504 (2016)
9. U.L. Businaro, S. Gallone, Il Nuovo Cimento **1**, 629 (1955)
10. U.L. Businaro, S. Gallone, Il Nuovo Cimento **1**, 1277 (1955)
11. R. Capote et al., Nucl. Data Sheets **110**, 3107 (2009)
12. N. Carjan, M. Rizea, Phys. Rev. C **82**, 014617 (2010)
13. A. Chyzh et al., Phys. Rev. C **90**, 014602 (2014)
14. D.D. Clark, Physics Today **24**, 23 (1971)
15. H. Conde, M. Holmberg, in *Physics and Chemistry of Fission. Proceedings of a Symposium*, Vol. 2 (1965), pp. 57–65
16. E.A.C. Crouch, Atomic Data Nucl. Data Tables **19**, 417 (1977)
17. J. Dean, *Lange's Handbook of Chemistry* (McGraw-Hill, New York, 1999)
18. J.E. Escher, J.T. Burke, F.S. Dietrich, N.D. Scielzo, I.J. Thompson, W. Younes, Rev. Mod. Phys. **84**, 353 (2012)
19. D. Filges, F. Goldenbaum, *Handbook of Spallation Research* (Wiley, 2009)
20. G.N. Flerov, K.A. Petrjak, Phys. Rev. **58**, 89 (1940)
21. K.F. Flynn, E.P. Horwitz, C.A.A. Bloomquist, R.F. Barnes, R.K. Sjoblom, P.R. Fields, L.E. Glendenin, Phys. Rev. C **5**, 1725 (1972)
22. J. Frehaut, R. Bertin, R. Bois, in *International Conference on Nuclear Data for Science and Technology (CEA, 1982)*, pp. 6–10
23. J.E. Gindler, J.R. Huizenga, in *Nuclear Chemistry: v. 2*, ed. by L. Yaffe (Academic Press, 1968)

24. L.E. Glendenin, J.E. Gindler, I. Ahmad, D.J. Henderson, J.W. Meadows, Phys. Rev. C **22**, 152 (1980)
25. L.E. Glendenin, J.E. Gindler, D.J. Henderson, J.W. Meadows, Phys. Rev. C **24**, 2600 (1981)
26. R.W. Hasse, W.D. Myers, *Geometrical Relationships of Macroscopic Nuclear Physics* (Springer, 1988)
27. D.J. Hinde, D. Hilscher, H. Rossner, Nucl. Phys. A **538**, 243 (1992)
28. D.J. Hofman, B.B. Back, I. Diószegi, C.P. Montoya, S. Schadmand, R. Varma, P. Paul, Phys. Rev. Lett. **72**, 470 (1994)
29. R.E. Howe, Nucl. Sci. Eng. **86**, 157 (1984)
30. E.K. Hulet, J.F. Wild, R.J. Dougan, R.W. Lougheed, J.H. Landrum, A.D. Dougan, P.A. Baisden, C.M. Henderson, R.J. Dupzyk, R.L. Hahn, M. Schädel, K. Sümmerer, G.R. Bethune, Phys. Rev. C **40**, 770 (1989)
31. E.K. Hulet, J.F. Wild, R.J. Dougan, R.W. Lougheed, J.H. Landrum, A.D. Dougan, M. Schadel, R.L. Hahn, P.A. Baisden, C.M. Henderson, R.J. Dupzyk, K. Sümmerer, G.R. Bethune, Phys. Rev. Lett. **56**, 313 (1986)
32. N.V. Kornilov, A.B. Kagalenko, V.M. Maslov, Y.V. Porodzinskij, tech. rep. INDC(CCP)-437 (IAEA, 2003)
33. N. Kornilov, *Fission Neutrons* (Springer-Verlag GmbH, 2014)
34. K.S. Krane, *Introductory Nuclear Physics* (Wiley, 1987)
35. H.J. Krappe, K. Pomorski, *Theory of Nuclear Fission* (Springer-Verlag GmbH, 2012)
36. S.E. Larsson, G. Leander, in *Symposium on the Physics and Chemistry of Fission*, Vol. 1 (IAEA, 1974), pp. 177–201
37. W.D. Loveland, D.J. Morrissey, G.T. Seaborg, *Modern Nuclear Chemistry 2nd edition* (Wiley, Hoboken, 2017), 780 pp.
38. D. Madland, Nucl. Phys. A **772**, 113 (2006)
39. D. Madland, A. Kahler, Nucl. Phys. A **957**, 289 (2017)
40. V.V. Malinovskij, V.G. Vorob'eva, B.D. Kuz'minov, V.M. Piksaikin, N.N. Semenova, P.S. Soloshenkov, Atomnaya Energia **54**, 209 (1983)
41. A. Messiah, *Quantum Mechanics* (Dover Publications, Mineola, NY, 1999)
42. M. Mirea, Phys. Rev. C **83**, 054608 (2011)
43. M. Morjean et al., Eur. Phys. J. D **45**, 27 (2007)
44. J.W. Negele, S.E. Koonin, P. Möller, J.R. Nix, A.J. Sierk, Phys. Rev. C **17**, 1098 (1978)
45. D. Neudecker, T. Taddeucci, R. Haight, H. Lee, M. White, M. Rising, Nucl. Data Sheets **131**, 289 (2016)
46. A. Oberstedt, R. Billnert, F.-J. Hambsch, S. Oberstedt, Phys. Rev. C **92** (2015). https://doi.org/10.1103/physrevc.92.014618.
47. S. Oberstedt, R. Billnert, T. Belgya, T. Bryś, W. Geerts, C. Guerrero, F.-J. Hambsch, Z. Kis, A. Moens, A. Oberstedt, G. Sibbens, L. Szentmiklosi, D. Vanleeuw, M. Vidali, Phys. Rev. C **90**, 024618 (2014)
48. K.A. Petrjak, G.N. Flerov, J. Phys. **3**, 275 (1940)
49. K.A. Petrjak, G.N. Flerov, Dokl. Akad. Nauk SSSR **28**, 500 (1940)
50. G.A. Petrov et al., in *AIP Conference Proceedings*, vol. 1175 (2009) p. 289.
51. S.M. Polikanov et al., Sov. Phys. JETP **15**, 1016 (1962)
52. R. Rodríguez-Guzmán, L.M. Robledo, Phys. Rev. C **89**, 054310 (2014)
53. G.J. Russell, in *Proceedings of the Eleventh Meeting of the International Collaboration on Advanced Neutron Sources (ICANS-XI)*, vol. 23 (1991), pp. 291–299
54. G.R. Satchler, *Introduction to Nuclear Reactions* (Macmillan, Houndmills, Basingstoke, Hampshire, 1990)
55. G. Scharff-Goldhaber, G.S. Klaiber, Phys. Rev. **70**, 229 (1946)
56. K.-H. Schmidt, B. Jurado, Rep. Progress Phys. **81**, 106301 (2018)
57. C. Schmitt, K. Mazurek, P. Nadtochy, Acta Phys. Pol. B Proc. Suppl. **8**, 685 (2015)
58. P. Marmier, E. Sheldon, *Physics of Nuclei and Particles: v. 1* (Academic Press, 1969)
59. R. Sher, C. Beck, tech. rep. NP-1771 (Stanford University, 1981)
60. G.N. Smirenkin, L.I. Prohorova, tech. rep. FEI-107 (1967)

61. V.M. Strutinsky, Sov. J. Nucl. Phys. **3**, 449 (1966)
62. V.M. Strutinsky, Nucl. Phys. **A95**, 420 (1967)
63. V.M. Strutinsky, Nucl. Phys. **A122**, 1 (1968)
64. K.D. Talley, Beta-delayed neutron data and models for scale, Ph.D. thesis (University of Tennessee, Knoxville, 2016)
65. P. Talou, T. Kawano, I. Stetcu, J.P. Lestone, E. McKigney, M.B. Chadwick, Phys. Rev. C **94**, 064613 (2016)
66. V.E. Viola, K. Kwiatkowski, M. Walker, Phys. Rev. C **31**, 1550 (1985)
67. A.S. Vorobyev, O.A. Shcherbakov, tech. rep. INDC(NDS)-0655 (International Atomic Energy Agency (IAEA), 2014)
68. A.S. Vorobyev, O.A. Shcherbakov, Y.S. Pleva, A.M. Gagarski, G.V. Val'ski, G.A. Petrov, V.I. Petrova, T. A.Zavarukhina, in *ISINN-17* (2009)
69. A. Vorobyev, O. Shcherbakov, A. Gagarski, G. Val'ski, G. Petrov, EPJ Web Confer. **8**, 03004 (2010)
70. A. Vorobyev, O. Shcherbakov, Y. Pleva, A. Gagarski, G. Val'ski, G. Petrov, V. Petrova, T. Zavarukhina, Nucl. Instrum. Methods Phys. Res. A **598**, 795 (2009)
71. C. Wagemans, *The Nuclear Fission Process* (CRC Press, 1991)
72. A.C. Wahl, in *Physics and Chemistry of Fission*, Vol. 1 (1965), p. 317
73. A.C. Wahl, Atomic Data Nucl. Data Tables **39**, 1 (1988)
74. B.E. Watt, Phys. Rev. **87**, 1037 (1952)
75. J.B. Wilhelmy, E. Cheifetz, R.C. Jared, S.G. Thompson, H.R. Bowman, J.O. Rasmussen, Phys. Rev. C **5**, 2041 (1972)
76. H. Wilschut, V. Kravchuk, Nucl. Phys. A **734**, 156 (2004)

Chapter 3
Models

Abstract This chapter introduces a few critical models of fission. These models include (a) the liquid-drop model (LDM), one of the earliest successful models of the nucleus, and which provides information about the nuclear shape, its changes during fission, and the associated energetic consequences (such as the origin of the widely used semi-empirical mass formula); (b) the Strutinsky shell correction method, which allows us to include quantum effects in the LDM; and (c) various scission point models designed to predict properties of the fission fragments, without taking into account the evolution of the parent nucleus from formation to scission. The kinematics of the fission process resulting in the division into a light and a heavy fragment is presented as a consequence of energy and momentum conservation laws. We also give a qualitative discussion of the transition state model, used to calculate fission cross sections, in preparation for the more detailed presentation in Chap. 7.

The goal of this chapter is to introduce a few crucial models of fission. The chapter focuses on the concepts and ideas behind the models, while the detailed math and formalism are deferred to later chapters. Structure properties of the nucleus relevant to fission (e.g., energy, barrier heights) will be discussed in the context of the liquid-drop model, while the reaction aspects of fission will be addressed with the goal of obtaining fission cross-section estimates.

3.1 The Liquid-Drop Model

The liquid-drop model (LDM) was the first approach used to describe the fission process [5, 35] (see also Chap. 1). The original LDM has been significantly improved, and its modern successor, the macroscopic-microscopic model, is widely used today [25, 36, 37]. The remarkable success of the LDM, even in its original form, can be attributed to the fact that it is well suited to describe those properties of the nucleus that vary smoothly with the number of its nucleons (i.e., the bulk behavior of the nucleus). Within this approach, any deviations from the bulk behavior are expected to be caused primarily by contributions from nucleons in the thin surface region of the nucleus [23]. Near the surface, nucleons have fewer neighbors to interact with, and a more careful treatment of those interactions may be necessary (see, e.g., Section 4.3 in [39]).

3.1.1 The Nuclear Shape

The formulation of the basic LDM begins with a curve characterizing the "surface" of the nucleus, which is taken as a geometrical object. For example, a general shape can be described as a surface using an expansion in terms of spherical harmonics [15, 26], or if the shape is symmetric about an axis, in terms of Legendre polynomials [15, 26]. Thus, a general shape can be described in spherical coordinates by the expansion [3, 23]

$$R(\theta, \varphi) = R_0 \left[1 + \sum_{\lambda=0}^{\infty} \sum_{\mu=-\lambda}^{\lambda} \alpha_{\lambda\mu} Y_{\lambda\mu}(\theta, \varphi) \right] \tag{3.1}$$

where $R(\theta, \varphi)$ gives the radial position of the nuclear surface in the direction (θ, φ), the quantity

$$R_0 = r_0 A^{1/3} \tag{3.2}$$

with $r_0 \approx 1.2$ fm is the usual nuclear radius, the $\alpha_{\lambda\mu}$ are parameters that determine the particular shape of the nucleus, and the $Y_{\lambda\mu}(\theta, \varphi)$ functions are spherical harmonics, which can be written in terms of Legendre polynomials P_λ [1],

$$Y_{\lambda\mu}(\theta, \varphi) = \sqrt{\frac{2\lambda+1}{4\pi} \frac{(\lambda-\mu)!}{(\lambda+\mu)!}} P_\lambda(\cos\theta) e^{i\mu\varphi} \tag{3.3}$$

If the shape is axially symmetric (i.e., if it does not depend on the azimuthal angle φ), then Eq. (3.1) reduces to

$$R(\theta) = R_0 \left[1 + \sum_{\ell=0}^{\infty} a_\ell P_\ell(\cos\theta) \right] \tag{3.4}$$

The geometric description of the nuclear surface is usually subject to two constraints as the shape changes: (1) the volume inside the surface is conserved and (2) the center-of-mass position is conserved (and usually held at the origin of the coordinate system). Figure 3.1 shows examples of surfaces of revolution generated by Eq. (3.4).

The expansion in terms of spherical harmonics or Legendre polynomials is not necessarily the most convenient choice for the variety of shapes explored during

Fig. 3.1 Surfaces of
revolution generated by
Eq. (3.4)

$a_2 = 2.0$

$a_2 = 0.5$

$a_2 = 0.0$

fission, however. Other choices that have been used in the literature include triaxial
ellipsoids [23], which might be used to describe the nucleus around its ground-state
shape, Cassinian ovaloids [8, 23], which can describe a shape with two nascent
fragments connected by a neck, and smoothly joined parabolas rotated around the z
axis to form a versatile family of shapes for fission [37, 41].

3.1.2   From Shape to Energy

Having established a shape for the liquid drop, the next step is to associate an energy
with that shape. In one of its earliest applications to nuclear physics, the LDM was
used to calculate nuclear masses. This work led to the standard semi-empirical mass
formula [23, 39, 52], which expresses the rest-mass energy of the nucleus in terms of
the rest-mass energies of the nucleons minus the binding energy of the nucleus. The
binding energy in turn is expressed as the sum of several contributions, including
a volume, surface, and Coulomb energy. There are additional terms that can be

added to improve the estimate of the binding energy, such as the contributions from symmetry and pairing [23, 39] (see also Sect. 6.3). The finite-range liquid-drop model (FRLDM), for example, includes terms that account for the finite range of the nuclear force and the diffuseness of the charge distribution in nuclei [25, 36]. In its most basic form, the standard semi-empirical mass formula for a nucleus with N neutrons and Z protons can be written as [23, 52]

$$M(N, Z) = Nm_n + Zm_p \underbrace{-a_V A}_{\text{volume}} \underbrace{+a_S A^{2/3} B_{\text{surf}}}_{\text{surface}} \underbrace{+a_c \frac{Z^2}{A^{1/3}} B_{\text{Coul}}}_{\text{Coulomb}} \tag{3.5}$$

where $m_n \approx 939.573 \, \text{MeV}/c^2$ is the neutron mass, $m_p \approx 938.279 \, \text{MeV}/c^2$ is the proton mass, and $Nm_n + Zm_p$ is simply the sum of the masses of the nucleons. The remaining terms in Eq. (3.5) subtract the binding energy of the nucleus from the simple sum of masses of its constituents (the binding energy BE is positive by definition). The coefficients that appear in the formula have empirically determined values [43]

$$a_V \approx 16 \, \text{MeV}/c^2$$
$$a_S \approx 18 \, \text{MeV}/c^2 \tag{3.6}$$
$$a_C \approx 0.711 \, \text{MeV}/c^2$$

The factors B_{surf} and B_{Coul} contain the information about the shape of the nucleus, and they represent the ratio of an energy for the deformed nucleus relative to its spherical form,

$$B_{\text{surf}} \equiv \frac{E_S}{E_S^{(0)}}$$
$$B_{\text{Coul}} \equiv \frac{E_C}{E_C^{(0)}} \tag{3.7}$$

where the superscript "(0)" indicates an energy calculated for the nucleus in a spherical form. The pairing term has been neglected in Eq. (3.5). We will return to the role of pairing in fission in Chap. 6 (see also Problem 39).

In order to understand how the energy of the nucleus changes with its shape on the way to scission, it will be sufficient to focus on the surface and Coulomb energies for now. The surface energy is proportional to the surface area $\mathcal{A}$ of the nuclear shape [21, 23],

$$E_S = \sigma \mathcal{A} \tag{3.8}$$

where the constant σ is the surface tension, and the Coulomb energy is calculated by integrating the inverse-distance Coulomb potential over the volume inside the nuclear surface, assuming a uniform charge distribution [6, 9, 14, 23]. Recall that the Coulomb potential energy between Z_1 and Z_2 positive point charges separated by a distance d is given by

$$V_C = \frac{e^2 Z_1 Z_2}{d} \tag{3.9}$$

where $e^2 \approx 1.44\, \text{MeV} \cdot \text{fm}$. In particular, for a spherical shape,

$$E_S^{(0)} = 4\pi R_0^2 \sigma$$
$$E_C^{(0)} = \frac{3}{5} \frac{(Ze)^2}{R_0} \tag{3.10}$$

where $R_0 = 1.2\, A^{1/3}$ fm is the nuclear radius.

Given the Coulomb energy $E_C^{(0)}$ and the surface energy $E_S^{(0)}$ for the spherical shape of the nucleus, it is useful to define the fissility parameter,

$$x \equiv \frac{E_C^{(0)}}{2 E_S^{(0)}} \approx \frac{1}{50} \frac{Z^2}{A} \tag{3.11}$$

which has been estimated here using the semi-empirical mass formula. In broad terms, the larger the fissility x, the more unstable the nucleus is against fission.

We define the deformation energy of the nucleus relative to its spherical shape by [23]

$$E_{\text{def}} = E_S + E_C - \left[E_S^{(0)} + E_C^{(0)} \right] \tag{3.12}$$

or, in terms of dimensionless quantities,

$$B_{\text{def}} \equiv \frac{E_{\text{def}}}{E_S^{(0)}} = (B_{\text{surf}} - 1) + 2x (B_{\text{Coul}} - 1) \tag{3.13}$$

with x the fissility parameter defined in Eq. (3.11).

Example: consider a shape described by the harmonic expansion in spherical coordinates

$$R(\theta) = \frac{R_0}{\lambda} [1 + \alpha_2 P_2 (\cos \theta)] \tag{3.14}$$

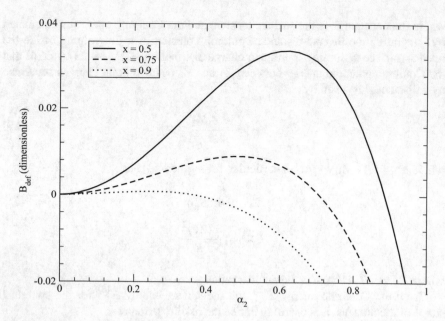

Fig. 3.2 Deformation energy ratio B_{def} for the shape in Eq. (3.14), plotted for values of the fissility parameter $x = 0.5, 0.75, 0.9$

where $R_0 = 1.2\, A^{1/3}$ fm is the nuclear radius and λ is a scale factor adjusted to enforce volume conservation as the shape deforms. It can be shown that [23]

$$\lambda^3 = 1 + \frac{3}{5}\alpha_2^2 + \frac{2}{35}\alpha_2^3 \tag{3.15}$$

and the surface and Coulomb energy ratios are

$$B_{\text{surf}} = 1 + \frac{2}{5}\alpha_2^2 - \frac{4}{105}\alpha_2^3 - \frac{66}{175}\alpha_2^4 + \dots \tag{3.16}$$

and

$$B_{\text{Coul}} = 1 - \frac{2}{5}\alpha_2^2 - \frac{4}{105}\alpha_2^3 + \frac{51}{245}\alpha_2^4 + \dots \tag{3.17}$$

Using Eq. (3.13), we plot the (dimensionless) deformation energy for different values of the fissility in Fig. 3.2. The plots show a fission barrier that disappears with increasing values of the fissility parameter.

3.2 Introduction to the Strutinsky Shell Correction Method

The extension of the LDM to include quantum effects has led to the development of the macroscopic-microscopic model. The macroscopic-microscopic model starts from a family of curves that describe the surface of the nucleus as a geometrical object. The binding energy calculated using the basic LDM represents a "macroscopic" contribution. The macroscopic contribution often consists primarily of a surface energy, proportional to the surface integral of the nuclear shape, and a Coulomb energy calculated as the volume integral of the inverse-distance potential inside the nucleus, assuming a uniform charge distribution, as discussed in Sect. 3.1.2 [6, 23]. There are additional terms that can be added to improve the accuracy of the macroscopic-energy term, and various liquid-drop prescriptions have been developed depending on which terms are included, such as the generalized liquid-drop model [2, 19], the Lublin-Strasbourg liquid drop [42], and the finite-range liquid-drop model [25, 36].

Regardless of the prescription used for the macroscopic part of the energy, a further, "microscopic" correction is needed to account for quantum effects due to the individual nucleons and their interactions. V. M. Strutinsky introduced a prescription to calculate this microscopic contribution [48, 49], motivated by the observation of fission isomers (see Sect. 2.7). The macroscopic quantities (viz., energy) given by the LDM vary smoothly as more particle are added to the nucleus, whereas quantum mechanics predicts irregular jumps instead. This is because (in a simplified picture) the nucleons fill irregularly spaced energy levels as they are added. These energy levels form shells separated by energy gaps, in analogy with the atomic orbital structure filled by electrons. Thus we can calculate both an exact and "smoothed" energy for the nucleus corresponding to the filling of energy levels by nucleons, and the difference those energies is a microscopic shell correction that can be added to the LDM binding energy to take into account the impact of quantum shell effects. The mathematical details of this approach will be discussed in greater detail in Sect. 6.2.

3.3 Potential Energy Surfaces

To each configuration, or shape, of the nucleus, there corresponds a potential energy that can be calculated using the LDM estimate, modified by the Strutinsky correction. Varying the shape of the nucleus with respect to one parameter (e.g., its elongation or a related quantity, the quadrupole moment) generates a potential energy curve, such as those plotted in Fig. 3.2. With two parameters controlling the shape (e.g., the elongation and the "lopsidedness" of the shape or a related quantity: the octupole moment), a potential energy surface (PES) is generated. The PES is a very useful concept in the description of fission. There may be more than two relevant shape parameters needed to properly describe the fission process, but

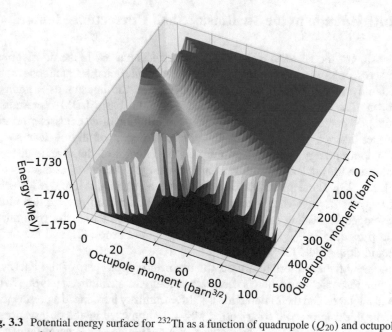

Fig. 3.3 Potential energy surface for ^{232}Th as a function of quadrupole (Q_{20}) and octupole (Q_{30}) moments of the nucleus. The cliff-like features near the front of the figure define the boundary where the neck between the fragments breaks and the energy drops sharply

visualizing more than two becomes a challenge, and for the purposes of this book, two will suffice. We will assume that the additional shape parameters (or *degrees of freedom*) are relaxed rather than fixed (and therefore that the potential energy has been minimized with respect to all degrees of freedom not held explicitly fixed). Note that the PES does not account for any kinetic energy associated with changes in the nuclear shape, and it therefore provides a necessarily incomplete picture of the problem. Figure 3.3 shows an example of a PES, plotted as a function of the quadrupole and octupole moments of the nucleus.

In a classical picture, we expect the fissioning nucleus to evolve along a path that minimizes the potential energy on the PES. Thus, we expect the fission process to begin in a local minimum (the ground state) and evolve over one or more saddles (points on the PES that are a maximum along the fission path and a minimum in the direction perpendicular to it). These saddles correspond to the fission barriers, when viewed in one dimension, along the fission path.

3.4 Scission Point Models

Scission point models were developed starting in the 1950s [17, 18] and represent a conceptually different approach to fission modeling. These models largely ignore the dynamical evolution of the parent nucleus up to scission and rely instead on statistical arguments formulated at scission to predict fission-fragment properties. Scission point models depend on two critical assumptions: (1) the fission-fragment properties can be determined entirely from an analysis of the system at scission (i.e., the history of the nucleus up to that point is largely irrelevant), and (2) a thermodynamic equilibrium exists at scission between the fragments. Based on these assumptions and a calculated energy of the system at scission, a probability distribution is constructed as a function of all possible divisions of the parent nucleus. Average properties of the fragments can then be obtained by integrating the property of interest (e.g., the deformation of the fragments) with this probability distribution. These types of models have continued to evolve to the present day [31, 54], and in recent calculations by Lemaître et al. [31], for example, they have been used to obtain chain yields for a wide range of fissioning actinide nuclei.

3.5 Kinematics of Fission Reactions

We summarize some useful non-relativistic formulas for induced binary fission, obtained from conservation of energy and momentum. Standard formulas for reaction kinematics can be found in [27, 33, 34] (with the correction pointed out in [32]) and also in [28, 47].

Consider the reaction depicted in Fig. 3.4, where a projectile P (e.g., a neutron) hits a target T (e.g., an actinide nucleus) and two ejectiles (e.g., fission fragments) are produced: one light (L) and one heavy (H). The velocity vector of the incident projectile defines the z axis going through both target and projectile. We define the

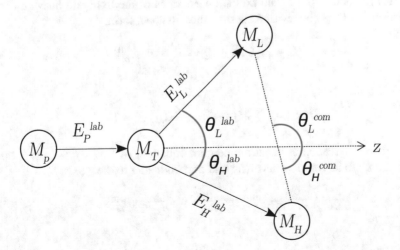

Fig. 3.4 Schematic representation of a projectile incident on a target resulting in two reaction products

rest masses

$$M_P = \text{projectile mass}$$
$$M_T = \text{target mass}$$
$$M_L = \text{light ejectile mass}$$
$$M_H = \text{heavy ejectile mass}$$

(3.18)

and the kinetic energies in the lab frame (the target is assumed to be at rest in the lab frame)

$$E_P^{\text{lab}} = \text{projectile energy}$$
$$E_L^{\text{lab}} = \text{light ejectile energy}$$
$$E_H^{\text{lab}} = \text{heavy ejectile energy}$$

(3.19)

along with the angles with respect to the z axis in the lab frame

$$\theta_L^{\text{lab}} = \text{light ejectile angle}$$
$$\theta_H^{\text{lab}} = \text{heavy ejectile angle}$$

(3.20)

We can write the conservation of energy as

$$M_P c^2 + E_P^{\text{lab}} + M_T c^2 = M_L c^2 + E_L^{\text{lab}} + M_H c^2 + E_H^{\text{lab}} + \text{TXE} \tag{3.21}$$

where TXE is the sum of the excitation energies of the light and heavy ejectiles. Equation (3.21) can be rewritten in the more compact form,

$$E_{\text{tot}} = E_P^{\text{lab}} + Q = E_L^{\text{lab}} + E_H^{\text{lab}} \tag{3.22}$$

with

$$Q \equiv (M_P + M_T - M_L - M_H)\, c^2 - \text{TXE} \tag{3.23}$$

Next, the conservation of linear momentum equation, $\mathbf{p}_P^{\text{lab}} = \mathbf{p}_L^{\text{lab}} + \mathbf{p}_H^{\text{lab}}$, can be broken up into its components along and perpendicular to the z axis:

$$p_P^{\text{lab}} = p_L^{\text{lab}} \cos \theta_L^{\text{lab}} + p_H^{\text{lab}} \cos \theta_H^{\text{lab}} \tag{3.24}$$

and

$$0 = p_L^{\text{lab}} \sin \theta_L^{\text{lab}} - p_H^{\text{lab}} \sin \theta_H^{\text{lab}} \tag{3.25}$$

with the classical relation $E_i^{\text{lab}} = \left(p_i^{\text{lab}}\right)^2 / (2M_i)$ between kinetic energy and momentum of particle i. From Eq. (3.25), we obtain a relation between the angles of the ejectiles:

$$\sin\theta_H^{\text{lab}} = \sqrt{\frac{M_L E_L^{\text{lab}}}{M_H E_H^{\text{lab}}}} \sin\theta_L^{\text{lab}} \tag{3.26}$$

and from Eqs. (3.22), (3.24), and (3.26), we get the lab kinetic energy of the light ejectile [32]

$$E_L^{\text{lab}} = \left\{ \frac{\sqrt{M_L M_P E_P^{\text{lab}}}}{M_L + M_H} \cos\theta_L^{\text{lab}} \pm \left[\left(\frac{M_H - M_P}{M_L + M_H} \right. \right.\right.$$
$$\left.\left.\left. + \frac{M_L M_P}{(M_L + M_H)^2} \cos^2\theta_L^{\text{lab}} \right) E_P^{\text{lab}} + \frac{M_H Q}{M_L + M_H} \right]^{1/2} \right\}^2 \tag{3.27}$$

and similarly for the heavy ejectile:

$$E_H^{\text{lab}} = \left\{ \frac{\sqrt{M_H M_P E_P^{\text{lab}}}}{M_L + M_H} \cos\theta_H^{\text{lab}} \pm \left[\left(\frac{M_L - M_P}{M_L + M_H} \right. \right.\right.$$
$$\left.\left.\left. + \frac{M_H M_P}{(M_L + M_H)^2} \cos^2\theta_H^{\text{lab}} \right) E_P^{\text{lab}} + \frac{M_L Q}{M_L + M_H} \right]^{1/2} \right\}^2 \tag{3.28}$$

The velocity of the center of mass (com) of the projectile-target system is (see, e.g., Appendix B in [44])

$$\begin{aligned} \mathbf{v}_{\text{com}} &= \frac{M_P \mathbf{v}_P^{\text{lab}} + M_T \mathbf{v}_T^{\text{lab}}}{M_P + M_T} \\ &= \frac{M_P \mathbf{v}_P^{\text{lab}}}{M_P + M_T} \end{aligned} \tag{3.29}$$

where $\mathbf{v}_P^{\text{lab}}$ and $\mathbf{v}_T^{\text{lab}}$ are the projectile and target velocities in the lab frame, and where the second line follows because we assume the target is at rest in the lab. Assuming non-relativistic velocities, the velocity of the projectile in the com frame is

$$\mathbf{v}_P^{\text{com}} = \mathbf{v}_P^{\text{lab}} - \mathbf{v}_{\text{com}} \tag{3.30}$$

In the center-of-mass frame, the energies of the ejectiles in Fig. 3.4 are given by

$$E_L^{\text{com}} = R_{TH} E_{\text{tot}} + R_{PH} Q$$
$$E_H^{\text{com}} = R_{TL} E_{\text{tot}} + R_{PL} Q$$

$$(3.31)$$

where we have used the notation

$$R_{ij} \equiv \frac{M_i M_j}{(M_P + M_T)(M_L + M_H)}$$

$$(3.32)$$

and the angles with respect to the z axis are given by

$$\sin \theta_L^{\text{com}} = \frac{E_L^{\text{lab}}}{E_L^{\text{com}}} \sin \theta_L^{\text{lab}}$$
$$\theta_H^{\text{com}} = \pi - \theta_L^{\text{com}}$$

$$(3.33)$$

3.6 Fission Cross-Section Models

We will defer the more formal definition of cross sections until Chap. 7. For the present discussion, it is useful to think of a fission cross section in the following way: when an incident beam impinges on a target, the fission cross section represents an effective area around each target nucleus and perpendicular to the beam such that, if an incident particle falls within that area, a fission event will occur.

The fission cross section is therefore related to the probability that a fission will be induced by the incident particle, and as such it depends on the physics of the fission process itself. Thus, fission cross sections are important quantities to measure, both for applications which need to know how often fission occurs (compared to other competing outcomes), and to improve our understanding of the fission process.

When a nucleus is excited by an incident particle, fission is often only one of many possible outcomes. Neutron emission and gamma decay are processes that typically compete with fission. We refer to these outcomes as *exit channels*. The decay of a nuclear state—be it through fission, neutron emission, gamma radiation, or some other process—implies a finite lifetime for that state. The Heisenberg uncertainty principle, in turn, implies a corresponding finite energy width for the state. The width of a state with lifetime τ is defined as

$$\Gamma \equiv \frac{\hbar}{\tau}$$

$$(3.34)$$

We will return to the concept of widths in Chap. 7, but from its definition in Eq. (3.34), we can see that for a collection of nuclei occupying the same unstable state with lifetime τ, the width is proportional to the rate of decay and therefore to the probability of decay. The widths for different modes of decay (e.g., fission, neutron, gamma) will therefore enter into the calculation of fission cross sections, as we will see in great detail in Chap. 7. The fission widths are usually calculated using the *transition state theory* first introduced by Eyring to describe chemical reaction rates [16] and later adapted to nuclear reaction rates by Wigner [53].

In the transition state theory, the saddle (or saddles) represents a bottleneck along the fission path. Whatever trajectory the system follows from the initial well to a final state, the system must cross the saddle. In that theory, the transition state is taken as a dividing surface at the saddle, perpendicular to the unstable mode (i.e., perpendicular to the fission path). There is no possibility of return once the system passes the transition state at the saddle, and the transition rate across the saddle is proportional to the total flux of classical trajectories going through the dividing surface [22].

In the case of fission, and taking into account the quantum aspects of the fission process, passage across the saddle is mediated by a set of transition states (also called *fission channels* [3]). At energies near the top of the barrier, these fission channels are discrete levels with well-defined quantum numbers (viz., spin, parity, and projection on the symmetry axis if the the system is axially symmetric). At higher energies above the barrier, the fission channels are more aptly described by a continuous *level density* function (i.e., the average number of nuclear levels per excitation-energy interval). Transmission through individual transition states is calculated using a formula suggested by Hill and Wheeler [24] (see also [3] p. 373).

We will discuss the calculation of level densities starting from more basic principles in Sect. 6.4, but for now it will be sufficient to introduce some empirical formulas that are commonly used in the literature. At low excitation energies E_x, a simple exponential formula, the *constant temperature model*, usually gives a good description of the average density of nuclear levels [7]

$$\rho\left(E_x\right) = \exp\left[-\frac{(E_x - E_0)}{T}\right] \tag{3.35}$$

where E_0 and T are adjustable parameters (T can be thought of as an "effective" nuclear temperature). At higher excitation energies, the *back-shifted Fermi gas model* gives a better description of level density. Its analytical form is [4, 7, 29]

$$\rho\left(E_x\right) = \frac{\sqrt{\pi}}{12} a^{-1/4} \left(E_x - \Delta\right)^{-5/4} \exp\left[2\sqrt{a\left(E_x - \Delta\right)}\right] \tag{3.36}$$

where a is a level density parameter and Δ is a pairing-gap parameter. The average trend of a as a function of the mass number A of the nucleus is given approximately by [7]

$$\bar{a} = 0.0722 + 0.1953 A^{2/3} \tag{3.37}$$

in units of MeV^{-1}. More sophisticated treatments can include a dependence on energy and shell corrections [7],

$$a = \bar{a}\left[1 + \frac{\delta W}{U}\left(1 + e^{-\gamma U}\right)\right] \tag{3.38}$$

where $U \equiv E_x - \Delta$, δW is a shell correction energy and γ is a positive-valued damping parameter. Gilbert and Cameron [20] proposed a composite-level density formulation that uses Eq. (3.35) at lower excitation energies and Eq. (3.36) at higher excitation energies, matching both functions and their first derivative at some energy E_M.

There are also experimental techniques used to measure (or at least to constrain) level densities. If a complete set of levels can be measured up to some excitation energy, then the level density can be deduced by directly counting levels in energy bins. In practice, it is difficult to measure all levels up to some reasonably high excitation energy, but it has been done in a few cases (e.g., for ^{168}Er [13]). Near the neutron or proton separation energy (the minimum energy needed to remove a neutron or proton from the nucleus), the energy spacing between resonances can be measured to deduce a level density at that energy [7, 38]. Particle evaporation spectra can also be fit by reaction models to extract a level density [10, 51] (see also Sect. 7.2). Recently, the *Oslo method* has been developed to extract level densities from sets of γ-ray spectra measured as a function of excitation energy [30, 45].

3.7 Questions

25. What was the first model used to describe fission?

26. Name a functional form that could be used to describe the shape of the parent nucleus near scission.

27. What are the two most important LDM contributions to the nuclear binding energy when describing fission?

28. What useful quantity in fission modeling is proportional to the charge squared divided by the mass of the parent nucleus? What does this quantity tell us?

29. What important physics missing from the basic LDM does the Strutinsky procedure attempt to recover?

30. Name some important degrees of freedom that can be used to construct a PES for fission.

31. What role do saddle points play on the PES?

32. What critical assumptions do scission point models usually depend on?

33. What "area" does a fission cross section represent?

34. What are the most likely exit channels in a fission reaction?

35. What model is often used to calculate fission cross sections?

3.8 Exercises

Problem 29 What is the volume (in fm^3) of ^{240}Pu at scission?

Problem 30 Consider the division of ^{240}Pu into ^{134}Te and ^{106}Mo fragments at scission. Assume that the nucleus at scission can be depicted as two touching spheres. Estimate the surface energy of ^{240}Pu at scission. Use $\sigma = 0.72$ MeV/fm^2 for the surface tension [21].

Problem 31 Estimate the Coulomb energy of ^{240}Pu at scission in Problem 30, by assuming the system can be approximated by two point charges at the centers of the fragments.

Problem 32 Using the results from Problems 30 and 31, calculate the deformation energy of ^{240}Pu at scission in the LDM.

Problem 33 The surface of a spheroid can be described in cylindrical coordinates by the equation

$$\frac{r^2}{a^2} + \frac{z^2}{c^2} = 1 \tag{3.39}$$

where we will assume a prolate shape ($c \geq a$). We can define the eccentricity parameter ε by

$$\varepsilon^2 \equiv 1 - \frac{a^2}{c^2} \tag{3.40}$$

The volume of a spheroid is

$$V = \frac{4}{3}\pi a^2 c \tag{3.41}$$

its surface energy is [23]

$$E_S(\varepsilon) = \frac{E_S(0)}{2}\left(1 - \varepsilon^2\right)^{1/3}\left(1 + \frac{\arcsin \varepsilon}{\varepsilon\sqrt{1 - \varepsilon^2}}\right)$$

$$= E_S(0)\left(1 + \frac{2}{45}\varepsilon^4 + \dots\right) \tag{3.42}$$

and its Coulomb energy is [23]

$$E_C(\varepsilon) = E_C(0) \frac{(1-\varepsilon^2)^{1/3}}{2\varepsilon} \ln \frac{1+\varepsilon}{1-\varepsilon}$$

$$\approx E_C(0) \left(1 - \frac{1}{45}\varepsilon^4 + \dots\right) \tag{3.43}$$

1. Use volume conservation to express a and c in terms of ε and the nuclear radius R_0.
2. Write the deformation energy E_{def} for a prolate spheroid in terms of ε and the fissility x, and show that for small values of ε

$$E_{\text{def}} \approx \alpha (c - R_0)^2 \tag{3.44}$$

and express α in terms of the surface tension σ and the fissility. The quantity $c - R_0$ measures the "stretching" of the prolate spheroid relative to the spherical shape.
3. Using $\sigma = 1.03 \, \text{MeV/fm}^2$, calculate α for ^{134}Te and ^{102}Zr.

Problem 34 Consider the single-particle levels in Fig. 3.5 below. The levels are equidistant with a spacing of 100 keV, and the energy of the lowest level is $\varepsilon_0 = 0 \, \text{keV}$. You can build up a nucleus by placing nucleons in those levels, starting from the lowest. Assume each level can hold at most two nucleons. You can estimate the energy of the nucleus by adding up the single-particle energies of the states occupied by each nucleon (we will call this the shell model energy in Sect. 6.2). Plot this energy as a function of the number of nucleons for $A = 0$ to 20 nucleons.

Problem 35 Consider the single-particle levels in Fig. 3.6 below. As in Problem 34, plot the shell model energy as a function of the number of nucleons for $A = 0$ to 20

Fig. 3.5 Single-particle states of an equidistant-energy model

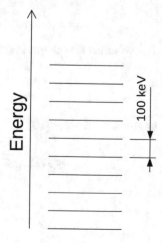

Fig. 3.6 Single-particle states

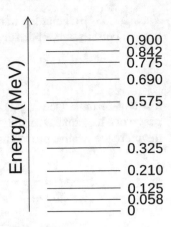

nucleons, assuming each level can hold at most 2 particles. Next, plot the difference between the energy as a function of number of nucleons in this problem and the one in Problem 34. The graph in Problem 34 represents more or less the energy predicted by the LDM, while the graph in this problem is closer to a realistic nuclear energy (obtained from single-particle energies with a shell gap). The difference between the two represents the Strutinsky shell correction in the toy model of this problem.

Problem 36 Consider the following PES:

$$V(x, y) = y^2 - x^2 \tag{3.45}$$

where x represents the elongation of the nucleus and y some other shape parameter (e.g., the mass asymmetry of the nucleus). Where is the top of the barrier located? Write down an equation that describes the fission path on this simplified PES.

Problem 37 Consider Problems 30 and 31, and plot the available energy $E_{avail}(A_H)$ (which we will take as the sum of the surface and Coulomb contributions) at scission as a function of the mass number of the heavy fragment, A_H, for $A_H = 120$ to 140 assuming the UCD rule holds exactly and that no neutrons are emitted so that $A_L + A_H = A_F$ with A_F the mass number of the fissioning nucleus. If the probability of scission into a particular mass division is given by

$$\text{Prob}(A_H) \propto \exp\left[-\frac{E_{avail}(A_H)}{T}\right] \tag{3.46}$$

where $T > 0$ is a parameter (which we don't need to set for this exercise), what is the most probable division predicted by this very simplified scission point model?

Problem 38 By counting transition states and states populated by neutron decay using a Fermi gas-level density formula,

$$\rho(E) \propto \exp\left[2\sqrt{aE}\right] \tag{3.47}$$

a useful formula (see, e.g., Eq. (VII-7) in [50]) has been obtained and used in past work to estimate roughly the ratio of neutron emission to fission width (and therefore the relative probability of those two processes),

$$\frac{\Gamma_n}{\Gamma_f} = \frac{4A^{2/3}a_f(E - B_n)}{K_0 a_n \left[2\sqrt{a_f(E - E_f)} - 1\right]} \times \exp\left[2\sqrt{a_n(E - B_n)} - 2\sqrt{a_f(E - E_f)}\right] \tag{3.48}$$

where A is the mass number of the fissioning nucleus; E is its excitation energy; B_n is the neutron separation energy; E_f is the critical fission energy (barrier height); a_n and a_f are level density parameters for neutron emission and fission, respectively; and $K_0 \approx 10$ MeV is a constant.

Taking $A = 240$, $B_n = 6$ MeV, and $a_n = a_f = 25$ MeV^{-1}, plot Γ_n/Γ_f from $E = E_f + 1$ MeV to $E = 20$ MeV for a fission barrier 0.5 MeV below and 0.5 MeV above the neutron separation energy. Comment on any differences in shapes between the two curves.

Problem 39 Using the semi-empirical mass formula, estimate the neutron separation energy of ^{236}U in the spherical limit. Compare your result to a calculation using experimental masses. For this problem, you should include the pairing contribution to the binding energy in Eq. (3.5),

$$BE_{pair}(Z, N) \approx \frac{(12\,\text{MeV})}{\sqrt{Z + N}} \times \begin{cases} +1 & Z \text{ and } N \text{ are even} \\ 0 & Z + N \text{ is odd} \\ -1 & Z \text{ and } N \text{ are odd} \end{cases} \tag{3.49}$$

Remember that in Eq. (3.5), $M = nM_n + Zm_p - BE/c^2$.

Problem 40 The Coulomb energy of Z charges uniformly distributed in inside a cube of side length L is given by [11]

$$E_C \approx 0.941156 \frac{Z^2 e^2}{L} \tag{3.50}$$

with $e^2 \approx 1.44$ MeV $\cdot$ fm. Using the LDM with surface and Coulomb contributions, calculate the binding energy of a ^{240}Pu nucleus with a cubic shape. Note that the *positive* binding energy is given approximately by $E_V - E_S - E_C$ where E_V is the volume energy, and since E_V is the same for all shapes by volume conservation, we will ignore it in this exercise which aims to compare binding energies of different shapes. Compare this to the binding energy of the spherical ^{240}Pu nucleus having the same volume. Which of these shapes is more likely to be relevant to low-energy fission and why? Use the surface tension parameter value $\sigma = 1.03\ \mathrm{MeV/fm^2}$.

Problem 41 A particular LDM surface near asymmetric scission is generated by rotating the following set of smoothly joined functions about the z axis (see [6, 23, 40]):

$$
f(z) = \begin{cases}
\sqrt{11.56 - 0.597107\,(z - 4.4)^2} & 0 \le z < 6.748527 \\
\sqrt{4.333589 + 0.5\,(z - 9.553172)^2} & 6.748527 \le z < 13.17244 \\
\sqrt{16 - 0.64\,(z - 16)^2} & 13.17244 \le z < 21
\end{cases}
\tag{3.51}
$$

where z is in fm.

1. Plot the function defined by Eq. (3.51).
2. At what position z is the neck between the nascent fragments thinnest?

Problem 42 Adapt Eqs. (3.26), (3.27), and (3.28) to spontaneous fission, and verify that the fragments in that case are emitted back to back in the lab frame.

Problem 43 What is the width of a nuclear state with a femtosecond lifetime?

Problem 44 The cross section for the absorption of a thermal neutron by a ^{235}U target is 681 barns [46]. Compare this cross section to the cross-sectional area of the ^{235}U nucleus, taken as a sphere of radius $R = 1.2A^{1/3}$ (fm).

Problem 45 Use the semi-empirical mass formula in Eq. (3.5) to estimate the energy required to break up ^{239}U into its individual protons and neutrons, and compare to the energy of the ^{239}U $\rightarrow$ ^{119}Pd $+$ ^{120}Pd division considered by Bohr and Wheeler [5].

Problem 46 Show that the summed kinetic energy E_K^{com} of the projectile and target in the center-of-mass frame can be written in terms of the kinetic energy E_P^{lab} of the projectile in the lab frame as

$$
E_K^{\mathrm{com}} = \frac{M_T}{M_P + M_T} E_P^{\mathrm{lab}}
\tag{3.52}
$$

Problem 47 For a reaction to be energetically allowed, the total kinetic energy in the center-of-mass frame (see Exercise 46) must equal the absolute value of the Q value of the reaction. In that case, the kinetic energy of the projectile in the lab frame is called the *threshold energy* for the reaction. What is the threshold energy for the ^{235}U $(n, 2n)$ reaction?

Problem 48 Taking $E_f = 6$ MeV as the critical energy for ^{239}U fission, what is the minimum kinetic energy needed for the incident neutron in the $n +^{238}$ U reaction to cause fission?

Problem 49 A ^{239}Pu nucleus at rest in the lab frame decays by emitting an α particle. What is the maximum recoil velocity in the lab frame of the residual ^{235}U nucleus? (Hint: for maximum recoil, assume the entire Q value goes into kinetic energy.)

Problem 50 The ^{236}U (t, p) reaction has been used in the past as a "substitute" for neutron absorption on a ^{237}U target, since they both form a ^{238}U nucleus [12] (see also Sect. 7.3.2). Suppose the incident triton energy is 18 MeV and that the outgoing proton is detected at an angle of 140 degrees with respect to the beam direction and with a kinetic energy of 9.38 MeV. What are the possible values of the initial excitation energy of the ^{238}U nucleus?

Problem 51 A 5-MeV neutron incident on a ^{235}U nucleus is absorbed, resulting in an excited ^{236}U compound nucleus. What is the recoil velocity of the compound nucleus?

Problem 52 Consider a ^{235}U (n, f) reaction with 14-MeV incident neutrons. The reaction produces the fragments ^{134}Te and ^{102}Zr. The light fragment is observed in a detector placed at 60 degrees with respect to the beam direction. Assume that the fragments are produced with no excitation energy for this exercise.

1. What are the lab energies of the light and heavy fragment?
2. At what angle in the lab frame did the heavy fragment fly off?

References

1. M. Abramowitz, I. Stegun, *Handbook of Mathematical Functions: With Formulas, Graphs, and Mathematical Tables* (Dover Publications, 1965)
2. X. Bao, H. Zhang, G. Royer, J. Li, Nucl. Phys. A **906**, 1 (2013)
3. A. Bohr, B.R. Mottelson, *Nuclear Structure vol. II: Nuclear Deformations* (World Scientific, 1998)
4. A. Bohr, B.R. Mottelson, *Nuclear Structure vol. I: Single-Particle Motion* (World Scientific, 1998)
5. N. Bohr, J.A. Wheeler, Phys. Rev. **56**, 426 (1939)
6. M. Bolsterli, E.O. Fiset, J.R. Nix, J.L. Norton, Phys. Rev. C **5**, 1050 (1972)
7. R. Capote et al., Nucl. Data Sheets **110**, 3107 (2009)
8. N. Carjan, F. Ivanyuk, V. Pashkevich, Physics Procedia **31**, 66 (2012)
9. B.C. Carlson, G.L. Morley, Am. J. Phys. **31**, 209 (1963)
10. R.J. Charity, Phys. Rev. C **82**, 014610 (2010)
11. O. Ciftja, Phys. Lett. A **375**, 766 (2011)
12. J.D. Cramer, H.C. Britt, Phys. Rev. C **2**, 2350 (1970)
13. W.F. Davidson et al., J. Phys. G Nucl. Phys. **7**, 455 (1981)
14. K. Davies, A. Sierk, J. Comput. Phys. **18**, 311 (1975)
15. A.R. Edmonds, *Angular Momentum in Quantum Mechanics* (Princeton University Press, 1996)

16. H. Eyring, J. Chem. Phys. **3**, 107 (1935)
17. P. Fong, Phys. Rev. **89**, 332 (1953)
18. P. Fong, Phys. Rev. **102**, 434 (1956)
19. J. Gao, X. Bao, H. Zhang, J. Li, H. Zhang, Nucl. Phys. A **929**, 246 (2014)
20. A. Gilbert, A.G.W. Cameron, Can. J. Phys. **43**, 1446 (1965)
21. W. Greiner, J.A. Maruhn, *Nuclear Models* (Springer, 1996)
22. P. Hänggi, P. Talkner, M. Borkovec, Rev. Mod. Phys. **62**, 251 (1990)
23. R.W. Hasse, W.D. Myers, *Geometrical Relationships of Macroscopic Nuclear Physics* (Springer, 1988)
24. D.L. Hill, J.A. Wheeler, Phys. Rev. C **89**, 1102 (1953)
25. T. Ichikawa, A. Iwamoto, P. Möller, A.J. Sierk, Phys. Rev. C **86**, 024610 (2012)
26. J.D. Jackson, *Classical Electrodynamics* (Wiley, 1998)
27. N. Jarmie, J.D. Seagrave, tech. rep. LA-2014 (Los Alamos Scientific Laboratory, 1957)
28. A. Kamal, *Nuclear Physics* (Graduate Texts in Physics) (Springer, 2014)
29. H.J. Krappe, K. Pomorski, *Theory of Nuclear Fission* (Springer-Verlag GmbH, 2012)
30. A.C. Larsen et al., Phys. Rev. C **83**, 034315 (2011)
31. J.-F. Lemaître, S. Panebianco, J.-L. Sida, S. Hilaire, S. Heinrich, Phys. Rev. C **92**, 034617 (2015)
32. X.-T. Lu, Nucl. Instrum. Methods Phys. Res. **225**, 283 (1984)
33. J.B. Marion, F.C. Young, *Nuclear Reaction Analysis: Graphs and Tables* (North-Holland, 1968)
34. J.W. Mayer, E. Rimini, *Ion Beam Handbook for Material Analysis (Casebound)* (Academic Press, 2012)
35. L. Meitner, O.R. Frisch, Nature **143**, 471 (1939)
36. P. Möller, A.J. Sierk, T. Ichikawa, A. Iwamoto, M. Mumpower, Phys. Rev. C **91**, 024310 (2015)
37. P. Möller, T. Ichikawa, Eur. Phys. J. A **51**, 173 (2015)
38. S. Mughabghab, *Atlas of Neutron Resonances* (Elsevier Science & Technology, 2018)
39. S.G. Nilsson, I. Ragnarsson, *Shapes and Shells: Nuclear Structure* (Cambridge University Press, 1995)
40. J.R. Nix, tech. rep. UCRL-17958 (Lawrence Radiation Laboratory, 1968)
41. J.R. Nix, Nucl. Phys. A **130**, 241 (1969)
42. K. Pomorski, J. Dudek, Phys. Rev. C **67**, 044316 (2003)
43. J.W. Rohlf, *Modern Physics from Alpha to Z0* (Wiley, 1994)
44. G.R. Satchler, *Introduction to Nuclear Reactions* (Macmillan, Houndmills, Basingstoke, Hampshire, 1990)
45. A. Schiller, L. Bergholt, M. Guttormsen, E. Melby, J. Rekstad, S. Siem, Nucl. Instrum. Methods Phys. Res. A **447**, 498 (2000)
46. V.F. Sears, Neutron News **3**, 26 (1992)
47. P. Marmier, E. Sheldon, *Physics of Nuclei and Particles: v. 1* (Academic Press, 1969)
48. V.M. Strutinsky, Nucl. Phys. **A95**, 420 (1967)
49. V.M. Strutinsky, Nucl. Phys. **A122**, 1 (1968)
50. R. Vandenbosch, J.R. Huizenga, *Nuclear Fission* (Academic Press, 1973)
51. A.V. Voinov, S.M. Grimes, C.R. Brune, M.J. Hornish, T.N. Massey, A. Salas, Phys. Rev. C **76**, 044602 (2007)
52. C.F. von Weizsäcker, Zeit. für Phys. **96**, 431 (1935)
53. E. Wigner, Trans. Faraday Soc. **34**, 29 (1938)
54. B.D. Wilkins, E.P. Steinberg, R.R. Chasman, Phys. Rev. C **14**, 1832 (1976)

Chapter 4
Fission Fragments and Products

Abstract This chapter focuses on the properties of the fission fragments. A brief overview of some of the experimental techniques used to study fission is given, and the properties of fission-fragment detectors are presented. The double-energy technique, used to extract fission-fragment properties, is examined in some detail. The functional forms commonly used to model the distribution of the fission fragments as function of both their mass and their charge numbers are discussed. The origin and distribution of excitation energy and angular momentum in the fragments are also discussed and related to experimentally observed quantities. The independent yields of the fission fragments, taken to be the yields of the fragments prior to beta decay, and the cumulative yields after beta decay are discussed. The activities of the beta decaying species can be understood using the Bateman equations, and the chapter concludes with an extensive discussion of the solution of these equations and how they are used to calculate both independent and cumulative yields.

The fission process releases many different types of particles (in particular, fragments, neutrons, gamma rays). In this chapter, we focus on some of the characteristic of the fission fragments. We focus on the detection and analysis of the properties of fission fragments and products.

4.1 Fission-Fragment Detectors

The fission-fragment properties can be measured using a variety of detectors. The choice of detector is determined by the quantity of interest and cost, among other considerations. We will briefly mention some of the more commonly used fission-fragment detectors.

Fission track detectors are typically made from inorganic crystals, glasses, or plastics (e.g., CR-39 is a plastic track detector which has been used for decades [5]). In a track detector, charged particles (e.g., the fission fragments) leave damage

© The Author(s), under exclusive license to Springer Nature Switzerland AG 2021
W. Younes, W. D. Loveland, *An Introduction to Nuclear Fission*, Graduate Texts
in Physics, https://doi.org/10.1007/978-3-030-84592-6_4

trails (typically $\sim 50 \overset{\circ}{A}$ in diameter) that are visible under a microscope, possibly after chemical etching. The number of tracks is proportional to the number of target atoms, the incident beam flux, the reaction cross section, and the duration of irradiation.

In a *gas ionization chamber*, the recoiling fission fragments create ion pairs (positive ion + negative electron) in the gas. A potential difference then gives rise to a current proportional to the number of pairs and therefore to the numbers of fission fragments. An example of a gas ionization chamber is the fission chamber used at LANL's LANSCE/WNR facility to measure the neutron beam flux via the $^{235}U(n, f)$ and $^{238}U(n, f)$ reactions [36].

A *proportional gas chamber* is a variant of the ionization chamber described above. The proportional chamber uses an avalanche effect to produce a pulse that is proportional to the primary ionization, with greater voltage applied leading to proportionately more ions collected.

The *multi-wire proportional counter* (MWPC) was developed by G. Charpak in 1968 [6]. The detector consists essentially of a plane of parallel independent sense wires (anode) between two plane electrodes (cathodes). The anode wires act as sensors for particles traveling between the cathode planes. Positive ions incident on the anode wires induce a negative signal on the nearest wire and positive signals on neighboring wires, leaving little ambiguity as to where the ionization event occurred.

In a *time projection chamber* (TPC), a particle such as a fission fragment leaves an ionizing track in a gas. The resulting electrons drift toward a segmented plate where the charges are collected. A full 3D track can then be reconstructed using both the cathode pad information and drift times of the electrons. In addition to position information, a TPC can also provide excellent energy resolution as long as attention is paid to fragment energy losses in the target and target backing [8].

Surface barrier detectors (e.g., solar cells) offer a low-cost alternative to the detectors listed above. They provide position information with usually good energy resolution [24].

4.2 Fragment Mass, Charge, and Kinetic Energy

4.2.1 Double-Energy Measurements

Some of the quantities that can be extracted from fission fragment detectors include yield (or the amount of a particular element produced), mass, charge, and kinetic energy of the fragments. Other properties, such as those of the neutrons and gamma rays emitted by the fragments, will be discussed in later chapters.

As an illustrative example, we describe one particular type of fission-fragment experiment: double-energy (2E) measurements [3, 7, 21, 28]. One goal of these experiments is to extract the number of fission events for each given kinetic energy and primary fragment mass. The basic principle of these measurements is to detect two fission fragments in coincidence (e.g., using two separate silicon detectors, or back-to-back ionization chambers with a common cathode). The quantities of interest can be reconstructed from the measured kinetic energies, conservation of number of protons and neutrons and of linear momentum, and a knowledge (e.g., from other experiments or additional detectors in the current experimental setup) of the number of neutrons emitted by the primary fragment.

The conservation laws of momentum and energy for the two fragments in the center-of-mass (com) frame can be combined into a single equation,

$$A_1^* E_1^* = A_2^* E_2^* \tag{4.1}$$

where the "*" superscript refers to the quantities before prompt neutron emission, A_i^* is the mass number of fragment i, and E_i^* is its kinetic energy. We also have

$$A_1^* + A_2^* = A \tag{4.2}$$

where A is the mass number of the parent nucleus. The pre-neutron (A_i^*) and post-neutron (A_i) fragment masses are related through

$$A_i = A_i^* - \nu_i \approx A_i^* - \bar{\nu}_i \tag{4.3}$$

where ν_i is the neutron multiplicity for that fragment in a given fission event and $\bar{\nu}_i$ its value averaged over all fission events. Assuming isotropic emission of the neutrons, we can then write an expression for the post-neutron energy E_i of the fragment in the com frame

$$E_i = \frac{A_i}{A_i^*} E_i^* + E_i^R$$
$$\approx \frac{A_i}{A_i^*} E_i^* \tag{4.4}$$

where E_i^R is the recoil energy of the fragment due to the emission of a neutron, which is usually very small and can be neglected. Combining Eqs. (4.1), (4.2), (4.3), and (4.4) gives a new relation between pre- and post-neutron masses, in terms of the measured energies. For fragment 1, for example,

$$A_1^* \approx \frac{A E_2}{E_2 + E_1 (1 + \xi_1)} \tag{4.5}$$

where

$$\xi_1 \equiv \frac{1 + \nu_1/A_1}{1 + \nu_2/A_2} - 1 \tag{4.6}$$

with a similar equation for A_2^*. The primary fragment mass A_i^* is not directly measured and must be deduced; it is therefore useful to introduce the concept of a *provisional mass* μ_i, which is defined as the fragment mass assuming no neutron emission. For the provisional mass, Eq. (4.5) reduces to

$$\mu_1 = A \frac{E_2}{E_1 + E_2} \tag{4.7}$$

and for fragment 2,

$$\mu_2 = A \frac{E_1}{E_1 + E_2} \tag{4.8}$$

Furthermore, not all electrons produced in the detector by the recoiling fragments are recorded; there is usually a loss of electrons due to non-ionizing collisions and incomplete charge collection leading to a so-called pulse-height defect (PHD) which must be corrected for. The corrections for neutron emission from the primary fragment and PHD can be applied in an iterative procedure, which starts with the measured energies and provisional masses for the fragments and converges to the primary fragment kinetic energies and masses. More explicitly, the algorithm is [7]:

1. Measure fragment energies $E_1^{(0)}$ and $E_2^{(0)}$, given by the post-neutron energies *not* corrected for PHD. Initialize ε_i, E_i, μ_i as follows:

$$\varepsilon_1 = E_1 = E_1^{(0)}$$
$$\varepsilon_2 = E_2 = E_2^{(0)} \tag{4.9}$$
$$\mu_1 = \mu_2 = 0$$

2. Calculate provisional masses using

$$\mu_{1,2} = A \frac{\varepsilon_{2,1}}{\varepsilon_1 + \varepsilon_2} \tag{4.10}$$

3. Apply the PHD correction to get new post-neutron fragment energies:

$$E_i = E_i^{(0)} + \text{PHD}\left(\mu_i, E_1 + E_2\right) \tag{4.11}$$

4. Correct for average neutron multiplicity to obtain pre-neutron energies:

$$E_i^* = E_i \frac{\mu_i}{\mu_i - \bar{\nu}\left(\mu_i, E_1 + E_2\right)} \tag{4.12}$$

5. Compare the μ_1 and μ_2 values to those from the previous iteration:

   a. If the values have changed significantly, then substitute $\varepsilon_1 = E_1^*$ and $\varepsilon_2 = E_2^*$, and return to step 2.
   b. Otherwise stop: μ_1 and μ_2 have converged to the pre-neutron (primary) fragment masses, and E_1^* and E_2^* have converged to the pre-neutron fragment energies.

Figure 4.1 shows the type of data that can be extracted from a double-energy measurement, in the case of ^{252}Cf spontaneous fission. The "lung" shape in the figure is characteristic of two-dimensional plots of yields as a function of primary fragment mass and TKE for both induced and spontaneous fission.

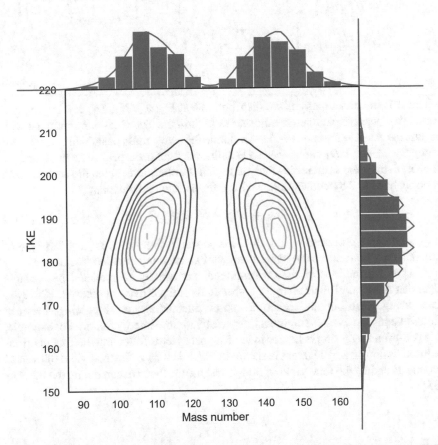

Fig. 4.1 Yields of fission fragments as a function of primary mass and total kinetic energy (TKE) for ^{252}Cf (SF). The data are from measurements in [7], retrieved from the EXFOR database

4.2.2 Modeling Fragments Mass and Charge Distributions

The double-humped shape of the fragment mass distribution as a function of primary fragment mass number can be fairly well fitted using a sum of five Gaussian functions: two Gaussians for each of the light and heavy mass peaks and one for the symmetric-division region. The explicit form of this function is [9]

$$
\begin{aligned}
Y(A) = \frac{N_1}{\sqrt{2\pi}\sigma_1} &\left\{ \exp\left[-\frac{\left(A - \bar{A} - D_1\right)^2}{2\sigma_1^2} \right] + \exp\left[-\frac{\left(A - \bar{A} + D_1\right)^2}{2\sigma_1^2} \right] \right\} \\
+ \frac{N_2}{\sqrt{2\pi}\sigma_2} &\left\{ \exp\left[-\frac{\left(A - \bar{A} - D_2\right)^2}{2\sigma_2^2} \right] + \exp\left[-\frac{\left(A - \bar{A} + D_2\right)^2}{2\sigma_2^2} \right] \right\} \\
+ \frac{N_3}{\sqrt{2\pi}\sigma_3} &\exp\left[-\frac{\left(A - \bar{A}\right)^2}{2\sigma_3^2} \right]
\end{aligned}
$$

$$(4.13)$$

where $\bar{A}$ is the mean mass of the distribution and N_1, N_2, N_3, D_1, D_2, σ_1, σ_2, and σ_3 are free positive parameters adjusted to fit the data. The Gaussian functions with arguments $A - \bar{A} - D_1$ and $A - \bar{A} - D_2$ define the heavy mass peak, while those with arguments $A - \bar{A} + D_1$ and $A - \bar{A} + D_2$ define the light mass peak. The final term, in $A - \bar{A}$, defines the symmetric-division region. Furthermore, since the overall yield is normalized to 2 by convention, we have the additional constraint

$$
2N_1 + 2N_2 + N_3 = 2
$$

$$(4.14)$$

A generic curve produced by Eq. (4.13) is shown in Fig. 4.2. Note that this type of function only reproduces the overall behavior and not any detailed fluctuations.

Charge distributions of fission fragments can usually be modeled by a single Gaussian function of their charge number Z for a fixed mass number A. However, these charge distributions sometimes show fluctuations, or staggering, between yields of odd- and even-Z fragments, especially for low-energy fission. For example, this odd-even staggering behavior in the fragment yields following thermal neutron-induced fission on a ^{235}U target is shown in Fig. 4.3. If Y_e is the total yield of even-Z fragments and Y_o the total yield of odd-Z fragments, then we can define the number [35]

$$
\delta \equiv \frac{Y_e - Y_o}{Y_e + Y_o}
$$

$$(4.15)$$

to quantify the amount of staggering in the charge distributions.

The independent yields $Y(A, Z)$ as a function of both fragment mass and charge can be modeled through the use of fractional independent yields $f(A, Z)$ [9],

$$
Y(A, Z) = Y(A) \times f(A, Z)
$$

$$(4.16)$$

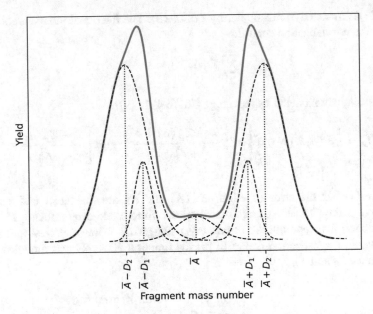

Fig. 4.2 Five-Gaussian function of Eq. (4.13) (after Ref. [37]). The individual components are also plotted and identified by their centroid position labeled on the x axis

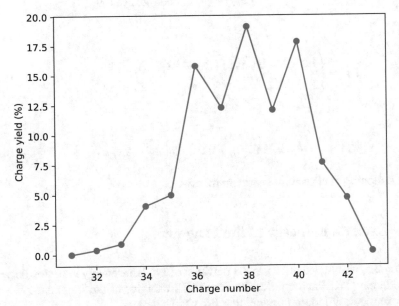

Fig. 4.3 Fragment charge yields as a function of heavy fragment charge number following the $^{235}U\left(n_{th}, f\right)$ reaction. The data are taken from the EXFOR database

with $Y(A)$ modeled, for example, by Eq. (4.13). The fractional independent yields satisfy the normalization condition

$$\sum_Z f(A, Z) = 1 \tag{4.17}$$

for all A and are modeled by (see, e.g., Eq. (6.4) in [9])

$$f(A, Z) = \frac{1}{2} F(A, Z) N(A) \left\{ \mathrm{erf} \left[\frac{Z - Z_p(A) + \frac{1}{2}}{\sqrt{2}\sigma_Z} \right] - \mathrm{erf} \left[\frac{Z - Z_p(A) - \frac{1}{2}}{\sqrt{2}\sigma_Z} \right] \right\} \tag{4.18}$$

where $\mathrm{erf}(x)$ is the error function, $Z_p(A)$ and σ_Z are the mean and standard deviation of the Gaussian charge distribution (without odd-even effects), $F(A, Z)$ is a function that reproduces the odd-even staggering at low energies, and $N(A)$ is an overall normalization factor. In [9] the function $F(A, Z)$ was modeled using parameters $\bar{F}_Z$ and $\bar{F}_N$,

$$F(A, Z) \equiv \begin{cases} \bar{F}_Z \bar{F}_N & \text{even } Z, \text{ even } N \\ 1/\left(\bar{F}_Z \bar{F}_N\right) & \text{odd } Z, \text{ odd } N \\ \bar{F}_Z / \bar{F}_N & \text{even } Z, \text{ odd } N \\ \bar{F}_N / \bar{F}_Z & \text{odd } Z, \text{ even } N \end{cases} \tag{4.19}$$

with

$$\bar{F}_Z = \begin{cases} 1.2723 - 0.06415\left(Z_f - 92\right) & Z_f \le 96 \\ 1 & Z_f \ge 97 \end{cases} \tag{4.20}$$

and

$$\bar{F}_N = 1.0758 - 0.01245\left(Z_f - 92\right) \tag{4.21}$$

for a fissioning nucleus with charge number Z_f.

4.3 Excitation Energy of the Fragments

In Sect. 2.5, we alluded briefly to the partition of available energy in fission between total kinetic (TKE) and total excitation (TXE) energies of the fragments. We can show, for neutron-induced fission, that Eq. (2.5) becomes

$$E_r + E_n + S_n = \underbrace{E_{\mathrm{pre}} + E_{\mathrm{Coul}}}_{\mathrm{TKE}} + \underbrace{E_x^{(L)} + E_x^{(H)}}_{\mathrm{TXE}} \tag{4.22}$$

where E_r is the total energy released (difference in mass-energy between fissioning nucleus and primary fragments), E_n is the kinetic energy of the incident neutron, and S_n is the neutron separation energy in the fissioning nucleus. On the right-hand side of the equation, we have introduced E_{pre} the pre-scission kinetic energy acquired by the pre-fragments during the transition from saddle to scission, E_{Coul} the Coulomb repulsion energy between the nascent fragments at scission, and the excitation energies of the light ($E_x^{(L)}$) and heavy ($E_x^{(H)}$) fragment at scission. The fragment excitation energies are often themselves subdivided into intrinsic (E_{int}) and deformation (E_{def}) energy contributions [16, 26],

$$E_x^{(i)} = E_{int}^{(i)} + E_{def}^{(i)}, \quad i = L, H \tag{4.23}$$

The deformation energy $E_{def}^{(i)}$ is due to the deformation of the fragment at scission and eventually appears as intrinsic excitation as the fragment relaxes into its ground-state shape. The intrinsic excitation energy $E_{int}^{(i)}$ can be dissipated by particle emission (e.g., neutrons and gamma rays). The partition of the intrinsic energy at scission between light and heavy fragments remains an active area of research. One common prescription for this partition is to take [16, 34]

$$\frac{E_x^{(L)}}{E_x^{(H)}} = \frac{a_L}{a_H} \tag{4.24}$$

where a_i ($i = L, H$) is the level density parameter (see Sect. 3.6). Experimentally, the TXE can be estimated by adding up the energies carried away by the emitted neutrons and gammas [21], although this is usually done as an average over the duration of the experiment rather than on an event-by-event basis.

4.4 Angular Momentum and Angular Distribution of Fragments

Measurements of prompt γ rays emitted by the fragments show that they can be formed with several units of angular momentum at scission (see Sect. 5.11). It may seem counterintuitive at first that fission fragments can have a fair amount of (or indeed any) angular momentum. For example, ^{252}Cf fissions spontaneously from a ground state with zero angular momentum: Where does the angular momentum of the fragments come from then?

In fact, it is possible to generate angular momentum in the fragments without adding any net angular momentum to the system as a whole. The angular momentum bearing modes of the nucleus responsible for this process have been described in the literature [17–19, 22, 23]. There are five modes of motion, shown in Fig. 4.4, that can impart angular momentum to the fragments without adding any additional angular

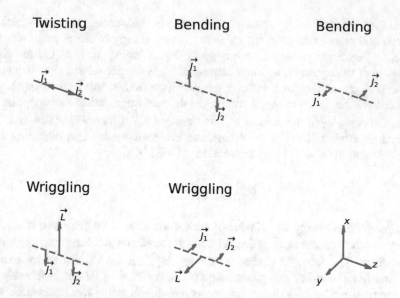

Fig. 4.4 Angular momentum bearing modes at scission. The vectors $\mathbf{J}_1$ and $\mathbf{J}_2$ represent the angular momenta of the fragments, and $\mathbf{L}$ is their relative orbital angular momentum. The dashed line represents the symmetry axis of the nucleus, taken as the z axis. See also [19]

momentum to the system. If $\mathbf{J}_1$ and $\mathbf{J}_2$ are the angular momenta of the fragments and $\mathbf{L}$ their relative *orbital* angular momentum, and taking the z axis as the symmetry axis of the nucleus with x and y axes perpendicular to it, then the modes can be described as

- One *twisting* mode with $\mathbf{J}_1 + \mathbf{J}_2 = 0$, $\mathbf{L} = 0$, and where

  - $\mathbf{J}_1$ and $\mathbf{J}_2$ are along the z axis

- Two *bending* modes with $\mathbf{J}_1 + \mathbf{J}_2 = 0$, $\mathbf{L} = 0$, and where

  - $\mathbf{J}_1$ and $\mathbf{J}_2$ are parallel to the x axis
  - $\mathbf{J}_1$ and $\mathbf{J}_2$ are parallel to the y axis

- Two *wriggling* modes with $\mathbf{J}_1 + \mathbf{J}_2 + \mathbf{L} = 0$, $\mathbf{L} \neq 0$, and where

  - $\mathbf{J}_1$ and $\mathbf{J}_2$ and $\mathbf{L}$ are parallel to the x axis
  - $\mathbf{J}_1$ and $\mathbf{J}_2$ and $\mathbf{L}$ are parallel to the y axis

In all five cases, the total sum of individual and relative angular momenta is

$$\mathbf{J}_1 + \mathbf{J}_2 + \mathbf{L} = 0 \tag{4.25}$$

Thus, by playing these angular momentum vectors against one another, angular momentum can be imparted to the fragments *ex nihilo*.

Fig. 4.5 Projections of the total angular momentum on the intrinsic and laboratory reference axes

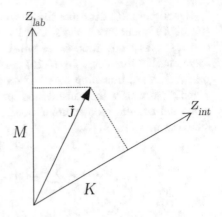

For simplicity, we assume a deformed nucleus with an axially symmetric shape in the following discussion. The nascent fragments fly apart along the axis of symmetry of the parent nucleus. If the parent nucleus is formed in a reaction, its total angular momentum **J** is the sum of the target and projectile intrinsic spins and their relative orbital angular momentum in the reaction (assuming for simplicity that no particles are emitted before fission). We can define a projection M of **J** on an arbitrary axis in the laboratory frame (typically the beam direction) and its projection K on the symmetry axis of the nucleus (and therefore in the intrinsic frame of the nucleus). The modulus J of the vector **J**, along with the projections M and K, determine the angular distribution of orientations of the symmetry axis in the laboratory frame and therefore the angular distribution of the emitted fission fragments themselves. Figure 4.5 shows these various definitions.

Assuming that the fragments separate and fly apart along the symmetry axis of the nucleus, the angular distribution of the fragments is given by [31]

$$W_{M,K}^J (\theta) = \frac{2J + 1}{2} \left| d_{M,K}^J (\theta) \right|^2 \tag{4.26}$$

where the $d_{M,K}^J (\theta)$ are Wigner small-d functions [12, 31], and θ is the angle with respect to the lab axis. The $d_{M,K}^J (\theta)$ functions have a complicated form, which simplifies somewhat when **J** is aligned in a perpendicular direction to the beam (taken as the lab frame axis) by the reaction. In that case, $M = 0$ and

$$d_{0,K}^J (\theta) = (J!) \sqrt{(J + K)! (J - K)!}$$

$$\times \sum_i \frac{(-1)^i \left(\sin \frac{\theta}{2}\right)^{K+2i} \left(\cos \frac{\theta}{2}\right)^{2J-K-2i}}{(J - K - i)! (J - i)! (K + i)! i!} \tag{4.27}$$

where $i = 0, 1, 2, \ldots$ spanning all integer values that do not give a negative argument for the factorials in the denominator.

Equation (4.26) describes fission from a single transition state with well-defined (J, M, K) quantum numbers. Usually, and especially at higher energies above the barriers, many transition states will contribute to the angular distribution of the fragments. In that case, Eq. (4.26) must be generalized to account for the density $\rho(E_x, J, K)$ of transition states at excitation energy E_x and with quantum numbers J and K, as well as the transmission coefficient T_J for populating the parent nucleus with total angular momentum J. Assuming again $M = 0$, the angular distribution then becomes [10, 31]

$$
W(\theta) \propto \sum_{J=0}^{\infty} (2J + 1)\, T_J \sum_{K=-J}^{J} \frac{2J + 1}{2}
$$
$$
\times \left| d_{M,K}^J(\theta) \right|^2 \frac{\rho(E_x, J, K)}{\sum_{K'=-J}^{J} \rho(E_x, J, K')}
$$
(4.28)

At sufficiently high excitation energy (where statistical arguments can be made), the transition state density $\rho(E_x, J, K)$ is often taken to have a Gaussian form [10, 31],

$$
\rho(E_x, J, K) \propto \exp\left(-\frac{K^2}{2 K_0^2} \right)
$$
(4.29)

where K_0^2 is related to the nuclear temperature T and an "effective" moment of inertia $\Im_{\text{eff}}$ of the nucleus,

$$
K_0^2 = \frac{T \Im_{\text{eff}}}{\hbar^2}
$$
(4.30)

The total angular momentum of the parent nucleus is partitioned at scission into the sum of angular momenta of the fragments and their relative orbital angular momentum. The study of gamma rays emitted by the fragments can be used to gauge angular momentum of the fragments, which will be discussed further in Sect. 5.11.

4.5 Decay of Fragments

Independent yields are the yields after prompt neutron emission, but before any delayed decays (see Sect. 2.4). In principle, cumulative, sum, and chain yields can all be deduced from the independent yields. In practice, it is not always possible to directly measure independent yields if the decays occur too quickly. Therefore, it is often necessary to deduce independent yields from measurements performed over a finite time interval and in many cases long after all induced fission processes have ended.

For each element, the decay rate is given by a simple exponential law. Typically, however, many elements observed soon enough after fission has occurred may be both depleted by beta decay into a daughter nucleus and replenished by beta decay from a precursor parent element. These interdependent decays lead to a system of coupled time-differential equations that has to be solved in order to obtain the number of elements of each species at any given time. A general solution for this type of problem was worked out by Bateman in 1910 [1]. The Bateman formula can readily be generalized to account for a finite time during which the fissioning material is irradiated, followed by a cooling time before measurements begin (see, e.g., pp. 40–42 in [2]). Using these formulas, the independent and cumulative yields can be related to the actual measured number of elements.

The rate of β decay (or *activity*) for fragments follows a standard exponential decay law for the number $N(t)$ of nuclei of a given species at time t,

$$\frac{dN}{dt} = -\lambda N \Rightarrow N(t) = N(0) e^{-\lambda t} \tag{4.31}$$

where λ is the *decay constant*. The half-life $T_{1/2}$, lifetime τ, and decay constant λ are related through

$$\lambda = \frac{1}{\tau} = \frac{\ln 2}{T_{1/2}} \tag{4.32}$$

If a fragment is part of a *decay chain*, it is written symbolically as

$$N_1 \xrightarrow{\lambda_1} N_2 \xrightarrow{\lambda_2} N_3 \xrightarrow{\lambda_3} \dots$$

so that fragment 2, for example, is simultaneously populated by the decay of fragment 1 and depleted by its own decay to fragment 3. In that case, the number of nuclei of each species (1, 2, 3, etc.) in this chain is obtained by solving the system of equations

$$\frac{dN_1}{dt} = -\lambda_1 N_1$$

$$\frac{dN_2}{dt} = \lambda_1 N_1 - \lambda_2 N_2 \tag{4.33}$$

$$\frac{dN_3}{dt} = \lambda_2 N_2 - \lambda_3 N_3$$

$$\vdots$$

We can solve these equations by proceeding systematically from one to the next, starting with the first one. We illustrate the process for N_1 and N_2. The solution for N_1 is given by Eq. (4.31),

$$N_1(t) = N_1(0) e^{-\lambda_1 t} \tag{4.34}$$

For $N_2(t)$, we multiply both sides of the equation by dt and plug in the solution for $N_1(t)$,

$$dN_2 + \lambda_2 N_2 dt = \lambda_1 N_1 dt = \lambda_1 N_1(0) e^{-\lambda_1 t} dt \tag{4.35}$$

Next, we multiply both sides of the equation by $e^{\lambda_2 t}$,

$$e^{\lambda_2 t} dN_2 + e^{\lambda_2 t} \lambda_2 N_2 dt = e^{\lambda_2 t} \lambda_1 N_1(0) e^{-\lambda_1 t} dt \tag{4.36}$$

and use the rule for the differential/derivative of a product to rewrite the left-hand side of the equation,

$$d\left(e^{\lambda_2 t} N_2\right) = \lambda_1 N_1(0) e^{(\lambda_2 - \lambda_1)t} dt \tag{4.37}$$

We now integrate both sides with respect to t, from $t = 0$ to some later time T, and re-arrange the terms to isolate $N_2(T)$ on the left-hand side of the equation,

$$N_2(T) = \frac{\lambda_1}{\lambda_2 - \lambda_1} N_1(0) \left(e^{-\lambda_1 T} - e^{-\lambda_2 T}\right) \\ + N_2(0) e^{-\lambda_2 T} \tag{4.38}$$

An interesting situation is that of a sample that is being irradiated, so that fragments are continuously generated as they also decay. Assuming that the production rate of a species of fragment is a constant P, its decay equation is

$$\frac{dN}{dt} = P - \lambda N \tag{4.39}$$

and the solution assuming no initial nuclei (i.e., $N(0) = 0$) is

$$N(t) = \frac{P}{\lambda}\left(1 - e^{-\lambda t}\right) \tag{4.40}$$

which can be checked by direct substitution into Eq. (4.39). In a realistic experiment, the irradiation lasts until some time T, and we may ask: What is the number of nuclei at time $T + \Delta t$? In other words, after a cooling period Δt. In this case, the number of nuclei follows Eq. (4.40) until $t = T$, and then decays normally according to Eq. (4.34), but with $N(0)$ replaced by $N(T)$:

$$N(T, \Delta t) = N(T) e^{-\lambda \Delta t} \\ = \frac{P}{\lambda}\left(1 - e^{-\lambda T}\right) e^{-\lambda \Delta t} \tag{4.41}$$

So far, we have only considered a single nuclear species produced by the irradiation, but in reality we likely need to consider an entire decay chain. The formula for this more general case can be found by solving the corresponding system of differential equations, as we did for Eq. (4.33), which can be quite tedious. Fortunately, general formulas for various decay and irradiation schemes have already been derived in the literature [1, 2]. For the ith element in the chain [2],

$$
N_i\,(T, \Delta t) = \sum_{\ell=1}^{i-1} \left\{ P_\ell \lambda_1 \lambda_2 \dots \lambda_{i-1} \sum_{j=\ell}^{i} \frac{\left(1 - e^{-\lambda_j T}\right) e^{-\lambda_j \Delta t}}{\lambda_j \prod_{\substack{k = \ell \\ k \neq j}}^{i} \left(\lambda_k - \lambda_j\right)} \right\}
$$
$$
+ \frac{P_i}{\lambda_i} \left(1 - e^{-\lambda_i T}\right) e^{-\lambda_i \Delta t}
$$

(4.42)

where the P_ℓ are constant production rates and the λ_j are decay constants. Equation (4.42) can be derived in various ways, for example, using Laplace transforms [25] or matrix exponentiation [13]. The saturation activity $\lambda_i N_i^*$ of the ith element is obtained from Eq. (4.42) in the limit of very long irradiation time ($T \to \infty$) and very short cooling time ($\Delta t \to 0$),

$$
\lambda_i N_i^* = \lim_{\substack{T \to \infty \\ \Delta t \to 0}} N_i\,(T, \Delta t)
$$

$$
= \sum_{\ell=1}^{i} P_\ell
$$

(4.43)

Given the fission rate R_F, we can then relate the various quantities above to the independent ($Y\,(i)$) and cumulative ($Y_c\,(i)$) yields for the ith element in the chain:

$$
Y\,(i) \equiv \frac{P_i}{R_F}
$$

$$
Y_c\,(i) \equiv \frac{\lambda_i N_i^*}{R_F} = \frac{1}{R_F} \sum_{\ell=1}^{i} P_\ell
$$

(4.44)

Finally, note that independent yields of the fission fragments are not usually measured directly but must be deduced instead from cumulative yields. In certain cases, however, when the β-decay parent of a fragment is very long-lived compared to the measurement time, the daughter nucleus is said to be "shielded," and its independent yield can be measured directly. This is the case, for example, of ^{134}Cs, which is unlikely to be populated by β^- decay of ^{134}Xe whose half-life is $> 5.8 \times 10^{22}$ years.

4.6   Computer Models of Fission-Fragment Properties

Various codes have been developed to model fission-fragment properties. For example, the code FREYA [32, 33] simulates the complete decay of the fission fragments on an event-by-event basis, keeping track of all kinematic information of the fragments and emitted particles. Other codes that similarly model the fission fragments and their ejecta after scission include FIFRELIN [14], CGMF [11], and GEF [27]. The FIER code [15] calculates the time dependence of delayed γ ray spectra following fission by using the solutions of the Bateman equation discussed in Sect. 4.5. For an overview of fission simulation codes and their application to the evaluation of nuclear data, see [4, 30].

4.7   Questions

36. What type of detector could you use if you wanted to reconstruct the path of ternary fission fragments in individual fission events?

37. What types of detectors could you use in a double-energy measurement?

38. What is the provisional mass of a fragment?

39. What causes pulse-height defects in the measurement of fission-fragment properties in a gas ionization chamber?

40. Which quantum numbers determine the angular distribution of the fragments?

41. Using the nomenclature of Sect. 4.4, if $K = M = 0$, how is the nucleus oriented relative to the laboratory axis?

42. Which lengths of time should be made "as long as possible" and which "as short as possible" for the irradiation, cooling, and counting phases of a cumulative yield measurement?

4.8   Exercises

Problem 53 For a parent nucleus of mass number A spontaneously fissioning into fragments of mass numbers A_1 and A_2 with a given value of the TKE, what are the individual kinetic energies of each fragment?

Problem 54 Consider the decay chain

$$A \xrightarrow{\lambda_A} B \xrightarrow{\lambda_B} C$$

where the final element (C) is stable, and where at $t = 0$

$$N_B(0) = N_C(0) = 0$$

What is the total activity of this chain at a given time t?

Problem 55 In Prob. 54, at what time $t = t_M$ does the activity reach a maximum?

Problem 56 Verify that Eq. (4.41) is a special case of Eq. (4.42).

Problem 57 Write the explicit form of $N_2(T, \Delta t)$ from Eq. (4.42).

Problem 58 Write the explicit form of Eq. (4.26) in the case where $K = J$ and $M = 0$.

Problem 59 To lowest order, neutron emission results in an overall shift of the fragment mass distribution before neutron emission. For a parent nucleus with mass A_f and average neutron multiplicity $\bar{v}$, how would you modify Eq. (4.13) to fit post-neutron emission mass yields?

Problem 60 Carry out three iterations of the double-energy measurement analysis in Sect. 4.2.1 with the following inputs:

$$E_1^{(0)} = 50 \text{ MeV}$$

$$E_2^{(0)} = 100 \text{ MeV}$$

$$A = 240 \tag{4.45}$$

$$\bar{v} = 2.3 \text{ (regardless of mass or energy)}$$

$$\text{PHD} = 1 \text{ MeV (regardless of mass or energy)}$$

What values do you obtain for the pre-neutron masses and kinetic energies of the fragments?

Problem 61 Table 4.1 gives recommended independent yields for mass 140 fission products following the thermal fission of a ^{235}U target. Find the most probable charge number Z_p, assuming a Gaussian distribution of the yields.

Table 4.1 Recommended independent yields for $A = 140$ fission products following the ^{235}U (n, f) reaction with thermal incident neutrons. The data are from the JEFF-3.3 library in the ENDF database

Z	Yield
52	2.4771×10^{-5}
53	1.9313×10^{-3}
54	3.8343×10^{-2}
55	1.8190×10^{-2}
56	5.0254×10^{-3}
57	6.4628×10^{-6}

Problem 62 Repeat Problem 61 using the fractional yield formula (Eq. (4.18)) with $F(A, Z) = 1$ to fit the yields, instead of a Gaussian. Then, repeat your fit using

$$F(A = 140, Z) = 1 + \varepsilon \cos(\pi Z) \tag{4.46}$$

and compare the quality of the fit to those obtained in the first part of this problem and in Problem 61.

Problem 63 Using the charge yields for the ^{235}U (n_{th}, f) reaction in Table 4.2, calculate the even-odd staggering parameter δ for these yields.

Problem 64 Table 4.3 gives the parameters for the mass distribution model in Eq. (4.13) for the ^{235}U (n, f) reaction at thermal and 6-MeV incident kinetic energies. Plot the corresponding mass yields on a log scale, and describe their qualitative differences.

Problem 65 For ^{238}U fission with a 14-MeV incident neutron beam, the parameters of the five-Gaussian model are [9] $\bar{A} = 117.42$, $N_1 = 0.6605$, $D_1 = 22.32$, $\sigma_1 = 5.962$, $N_2 = 0.1921$, $D_2 = 15.81$, $\sigma_2 = 2.938$, and $\sigma_3 = 11.80$. What is the ratio of the highest yield divided by the yield in the symmetric region at $A = \bar{A}$ (peak-to-valley ratio)?

Table 4.2 Charge yields for the ^{235}U (n_{th}, f) reaction, as plotted in Fig. 4.3. Data taken from the EXFOR database

Z	Y(Z)
31	0.017
32	0.396
33	0.900
34	4.075
35	4.983
36	15.777
37	12.273
38	19.074
39	12.059
40	17.783
41	7.651
42	4.718
43	0.254

Table 4.3 Parameter values for the five-Gaussian model for thermal and 6-MeV neutrons incident on a ^{235}U target [29] (data for fit available on the EXFOR database)

	$E_n = 25$ meV	$E_n = 6$ MeV
N_1	0.186	0.062
N_2	0.813	0.928
D_1	15.64	14.79
D_2	22.76	21.29
σ_1	2.58	2.54
σ_2	4.89	6.21
σ_3	3.01	5.47

Problem 66 The TKE for ^{235}U (n_{th}, f) leading to light and heavy fragment mass numbers $A_L = 106$ and $A_H = 130$, respectively, has been measured as 182.90 ± 0.26 MeV [20]. Using the UCD rule, and assuming the TKE is due entirely to Coulomb repulsion between the fragments at scission (i.e., no pre-scission energy) treated as point charges, estimate the separation distance between the centers of charge of the fragments at scission.

Problem 67 Repeat Exercise 66 for the case of symmetric fission following the ^{235}U (n_{th}, f) reaction, given that the TKE in this case has been measured at 146.30 ± 2.62 MeV [20].

Problem 68 Consider a simple model of the nucleus at scission as two co-linear touching spheroids representing the nascent fragments, as shown in Fig. 4.6. The surface of each spheroid is described by the equation in cylindrical coordinates

$$\frac{r^2}{a_i^2} + \frac{z^2}{c_i^2} = 1 \tag{4.47}$$

and we will assume a prolate shape ($c_i \geq a_i$). In this problem, we will assume the TKE consists only of the Coulomb energy between the fragments (i.e., $E_{\text{pre}} = 0$ in Eq. (4.22)) and the TXE consists entirely of deformation energies (i.e., $E_{\text{int}}^{(i)} = 0$ in Eq. (4.23)).

1. Using the form of the deformation energy in Problem 33 in Chap. 3, write the total energy $E_{\text{tot}} = \text{TKE} + \text{TXE}$ for this system in terms of c_1 and c_2 (you may approximate the Coulomb energy by assuming point charges $Z_1 e$ and $Z_2 e$ at the centers of the fragments).
2. Minimize this total energy with respect to c_1 and c_2, and obtain expressions for the excitation energies $E_x^{(1)}$ and $E_x^{(2)}$ of the fragments in terms of the TKE and that do not depend explicitly on c_1 and c_2.
3. Consider the fission reaction $n + ^{235}$ U $\rightarrow ^{134}$ Te $+ ^{102}$ Zr at an incident neutron energy of $E_n = 25$meV. The Q value of the reaction is 203.112 MeV; the neutron separation energy is $S_n = 6.545$ MeV. Use the results above to solve Eq. (4.22) numerically for the TKE. (Use $\alpha = 3.09$ MeV/fm^2 for ^{134}Te and 3.55 MeV/fm^2 for ^{102}Zr in Eq. (3.44).)

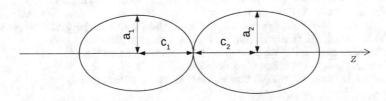

Fig. 4.6 Two touching spheroids representing the nascent fragments at scission

References

1. H. Bateman, Proc. Camb. Philos. Soc. **15**, 423 (1910)
2. M. Benedict, *Nuclear Chemical Engineering* (McGraw-Hill, New York, 1981)
3. H.R. Bowman, J.C.D. Milton, S.G. Thompson, W.J. Swiatecki, Phys. Rev. **129**, 2133 (1963)
4. R. Capote, Y.-J. Chen, F.-J. Hambsch, N. Kornilov, J. Lestone, O. Litaize, B. Morillon, D. Neudecker, S. Oberstedt, T. Ohsawa, N. Otuka, V. Pronyaev, A. Saxena, O. Serot, O. Shcherbakov, N.-C. Shu, D. Smith, P. Talou, A. Trkov, A. Tudora, R. Vogt, A. Vorobyev, Nuclear Data Sheets **131**, 1 (2016)
5. R.M. Cassou, E.V. Benton, Nucl. Track Det. **10**, 173 (1978)
6. G. Charpak, R. Bouclier, T. Bressani, J. Favier, Č. Zupančič, Nucl. Instr. Methods **62**, 262 (1968)
7. A. Göök, F.-J. Hambsch, M. Vidali, Phys. Rev. C **90**, 064611 (2014)
8. M. Heffner et al., Nuclear Instruments and Methods in Physics Research Section A: Accelerators, Spectrometers, Detectors and Associated Equipment **759**, 50 (2014)
9. M. James, R. Mills, D. Weaver, Progr. Nuclear Energy **26**, 1 (1991)
10. B. John, S.K. Kataria, Phys. Rev. C **57**, 1337 (1998)
11. T. Kawano, P. Talou, M.B. Chadwick, T. Watanabe, J. Nucl. Sci. Technol. **47**, 462 (2010)
12. R. Lamphere, Nucl. Phys. **38**, 561 (1962)
13. E. Levy, Am. J. Phys. **86**, 909 (2018)
14. O. Litaize, O. Serot, L. Berge, Eur. Phys. J. A **51** (2015) https://doi.org/10.1140/epja/i2015-15177-9
15. E. Matthews, B. Goldblum, L. Bernstein, B. Quiter, J. Brown, W. Younes, J. Burke, S. Padgett, J. Ressler, A. Tonchev, Nucl. Instr. Meth. Phys. Res. Sect. A: Accel. Spectrom. Detect. Assoc. Equip. **891**, 111 (2018)
16. C. Morariu, A. Tudora, F.-J. Hambsch, S. Oberstedt, C. Manailescu, J. Phys. G: Nucl. Part. Phys. **39**, 055103 (2012)
17. L.G. Moretto, R.P. Schmitt, Phys. Rev. C **21**, 204 (1980)
18. L.G. Moretto, G.J. Wozniak, Annu. Rev. Nucl. Part. Sci. **34**, 189 (1984)
19. L.G. Moretto, G.F. Peaslee, G.J. Wozniak, Nucl. Phys. A **502**, 453 (1989)
20. R. Müller, A.A. Naqvi, F. Käppeler, F. Dickmann, Phys. Rev. C **29**, 885 (1984)
21. J.N. Neiler, F.J. Walter, H.W. Schmitt, Phys. Rev. **149**, 894 (1966)
22. J.R. Nix, Nucl. Phys. A **130**, 241 (1969)
23. J.R. Nix, W.J. Swiatecki, Nucl. Phys. **71**, 1 (1965)
24. M. Petit, T. Ethvignot, T. Granier, R. Haight, J. O'Donnell, D. Rochman, S. Wender, E. Bond, T. Bredeweg, D. Vieira, J. Wilhelmy, Y. Danon, Nucl. Instrum. Methods Phys. Res. Sect. A: Accel. Spectrom. Detect. Assoc. Equip. **554**, 340 (2005)
25. D.S. Pressyanov, Am. J. Phys. **70**, 444 (2002)
26. A. Ruben, H. Märten, D. Seeliger, Zeitschrift für Physik A Hadrons Nuclei **338**, 67 (1991)
27. K.-H. Schmidt, B. Jurado, C. Amouroux, C. Schmitt, Nuclear Data Sheets **131**, 107 (2016)
28. H.W. Schmitt, J.H. Neiler, F.J. Walter, Phys. Rev. **141**, 1146 (1966)
29. C. Straede, C. Budtz-Jørgensen, and H.-H. Knitter, Nucl. Phys. A **462**, 85 (1987)
30. P. Talou, R. Vogt, J. Randrup, M.E. Rising, S.A. Pozzi, J. Verbeke, M.T. Andrews, S.D. Clarke, P. Jaffke, M. Jandel, T. Kawano, M.J. Marcath, K. Meierbachtol, L. Nakae, G. Rusev, A. Sood, I. Stetcu, C. Walker, Eur. Phys. J. A **54** (2018). https://doi.org/10.1140/epja/i2018-12455-0
31. R. Vandenbosch, J.R. Huizenga, *Nuclear Fission* (Academic, New York, 1973)
32. J. Verbeke, J. Randrup, R. Vogt, Comput. Phys. Commun. **222**, 263 (2018)
33. R. Vogt, J. Randrup, Phys. Rev. C **96**, 064620 (2017)
34. R. Vogt, J. Randrup, J. Pruet, W. Younes, Phys. Rev. C **80**, 044611 (2009)

35. C. Wagemans, *The Nuclear Fission Process* (CRC Press, Boca Raton, 1991)
36. S. Wender, S. Balestrini, A. Brown, R. Haight, C. Laymon, T. Lee, P. Lisowski, W. McCorkle, R. Nelson, W. Parker, N. Hill, Nucl. Instrum. Methods Phys. Res. Sect. A: Accel. Spectrom. Detect. Assoc. Equip. **336**, 226 (1993)
37. W. Younes, J.A. Becker, L.A. Bernstein, P.E. Garrett, C.A. McGrath, D.P. McNabb, R.O. Nelson, G.D. Johns, W.S. Wilburn, D.M. Drake, Phys. Rev. C **64**, 054613 (2001)

Chapter 5
Fission Neutrons and Gamma Rays

Abstract Chapter 5 deals with the neutrons and photons associated with the fission process. Fission can be induced by capture of neutrons and/or photons. The fission process also produces neutrons and photons that can be used to gain knowledge about various aspects of fission. The various sources of these neutrons and photons are discussed along with their characteristics and how to detect them. The energy spectrum and the number of neutrons (the multiplicity) emitted during fission are presented. The phenomenon of delayed neutrons, neutrons emitted from a few ms to minutes after fission has occurred, is discussed. Next, the chapter focuses on photons (or gamma rays in this context), starting with a discussion of photons as a means of inducing fission. A brief discussion of the detection of photons is followed by an analysis of the energy and multiplicity distributions of gamma rays in fission. The chapter also describes how gamma rays can be used to deduce fragment yields and estimate the amount of angular momentum imparted to the fragments by fission.

Neutrons can come from a variety of sources in a fission experiment. If a neutron is used to induce fission, some of these neutrons can reach the detectors either directly or after being scattered by the target, as well as surrounding equipment and walls (known as "room return"). Neutrons are also usually produced after the incident particle is absorbed by the target. Neutrons can be emitted by the excited nucleus before the fission process begins, by the de-excitation of the fragments, and in the form of delayed neutrons following some beta decays of the products. As in the case of neutrons, the fission process can also produce gamma rays (photons ranging from a few keV to several MeV) emitted by the fragments. The study of these γ rays provides critical information about nuclear levels such as their angular momentum, parity, excitation energy, and the transition probabilities between levels. As with neutrons, gamma rays can also be used to trigger fission.

© The Author(s), under exclusive license to Springer Nature Switzerland AG 2021   89
W. Younes, W. D. Loveland, *An Introduction to Nuclear Fission*, Graduate Texts
in Physics, https://doi.org/10.1007/978-3-030-84592-6_5

5.1   Neutron Beams

Various facilities around the world can produce neutron beams with different energy profiles and overall intensities [6]. Rather than provide an exhaustive list of those facilities, we discuss in broad terms three common types of neutron sources: reactors, monoenergetic beams, and time-of-flight facilities.

Nuclear reactors generate neutrons through fission and can provide a very high flux of neutrons (e.g., $\sim 10^{15}$ n/cm^2/s). In a typical fission experiment using a reactor as a neutron source, a sample of interest is irradiated inside the reactor for some period of time, removed from the reactor and, after it has sufficiently cooled, placed inside a detector setup to study the resulting fission products. The irradiation of the sample involves a range of neutron energies (depending on the reactor) but can span from meV to several MeV, and it is not generally possible to study the resulting fission events as a function of incident neutron energy. Reactors provide what is known as an *integral* measurement [14].

Several nuclear reactions can be used to produce nearly monoenergetic neutrons (or at least neutrons in a relatively narrow energy range), for example, ^{7}Li $(p, n)^7$ Be, ^{3}H $(p, n)^4$ He, ^{2}H $(d, n)^3$ He, and ^{3}H $(d, n)^4$ He [48]. These types of beams are especially useful for fission research, since they allow fission properties to be measured as a function of incident energy. Figure 5.1 shows the cross sections for some of these neutron-producing reactions.

Time-of-flight (TOF) facilities also provide incident neutrons with known energies, but using a white (broad) neutron spectrum from a pulsed beam, rather than a monoenergetic source. These white neutrons can, for example, be produced by a high-energy spallation reaction. The energy of an individual neutron is obtained from its flight time over a flight path of precisely known length. It is essential for the beam to be pulsed so that the start time of the neutron flight is known, its stop time is fixed by the instant of detection, and the net flight time can then be extracted [67].

Different neutron energy ranges have different uses and applications. Table 5.1 summarizes some of the labels commonly given to neutron sources, based on their energy.[1]

---

[1] These terms may be used slightly differently by various authors; therefore some caution is always warranted.

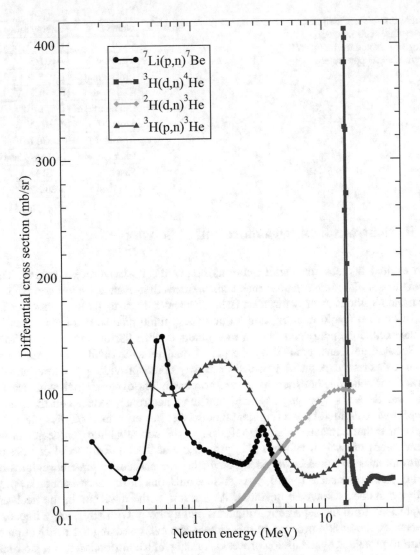

Fig. 5.1 Cross sections for neutron producing reactions. Adapted from [48] using data from the ENDF database

5.2 Neutron Detection

For fast neutrons ($E_n \gtrsim 1$ MeV), detection usually relies on measuring the proton produced by the (n, p) reaction on the detector material (e.g., plastic or organic material). For thermal neutrons, the (n, γ) and (n, α) reactions are preferred due to the high cross sections for target materials such as lithium and boron [53]. These different reactions are compared in Fig. 5.2.

Table 5.1 Names assigned
to specific neutron energy
ranges. Adapted from https://
www.nuclear-power.net

Name	Energy range
Cold	0–0.025 eV
Thermal	~ 0.025 eV
Epithermal	0.025–0.4 eV
Cadmium	0.4–0.5 eV
Epicadmium	0.5–1 eV
Slow	1–10 eV
Resonance	10–300 eV
Intermediate	300 eV–1 MeV
Fast	1–20 MeV
Ultrafast	> 20 MeV

5.3   Neutrons Emitted Before and at Scission

An excited nucleus may emit neutrons before the fission process begins. This leads to a *multi-chance fission* mechanism where first-chance fission corresponds to fission without prior neutron emission, second-chance to fission following the emission of a single neutron, and so on. These multi-chance neutrons affect the initial excitation energy and angular momentum that the fission process starts from.

Scission neutrons, emitted by the parent nucleus as it divides, are interesting from a theoretical point of view because they could provide a window into the physics of scission, but there is as yet no convincing experimental evidence for their existence. It was long assumed that most neutrons produced in the fission process were emitted by the fragment themselves; however, in 1962, R. Fuller [30] speculated that nucleons might be released by the non-adiabatic[2] processes in the latter stages of fission. Halpern [41] proposed a semi-classical model of scission neutron emission wherein the neck connecting the nascent fragments ruptures on a very short time scale ($\sim 10^{-22}$ s) so that nucleons in the neck cannot keep up with the sudden change in potential. As a result, some nucleons in the neck are "orphaned" from the separating fragments and fly away with any energy imparted to them by the fission process. In this picture, the measured angular distribution of neutrons produced by the fission process is the sum of a contribution that is isotropic in the frame of reference of the recoiling fragments and the contribution from scission neutrons that is isotropic in the frame of reference of the parent nucleus.

Thus, based on a measurement of the angular distribution of neutrons from the spontaneous fission of ^{252}Cf, Bowman et al. [11] found that 15% of those neutrons were consistent with isotropic emission from the parent nucleus at rest in the laboratory frame, which they identified as scission neutrons. Other measurements have found similarly significant fractions of scission neutrons in spontaneous and thermal neutron-induced fission [29, 65].

---

[2] Non-adiabatic motion is not purely collective and can excite individual nucleons.

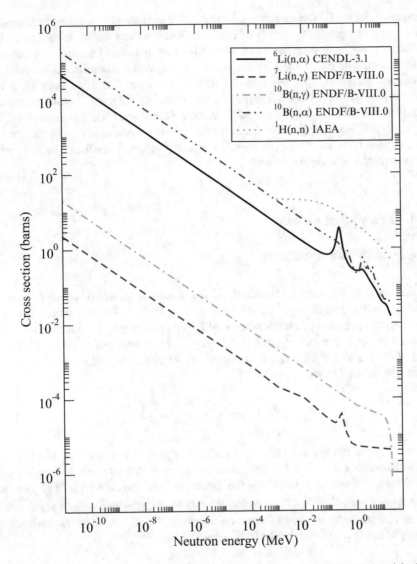

Fig. 5.2 Cross sections for the (n, n), (n, γ), and (n, α) reactions on detector materials. The $^1H\,(n, n)$ data are taken from the IAEA Neutron Data Standards repository (https://www-nds. iaea.org/standards/); the remaining data are taken from the CENDL and ENDF databases

Since then, however, various objections have been raised regarding the interpretation of the neutron angular-distribution measurements. First, some have questioned the assumption that neutrons emitted by the fragments are isotropically distributed within the fragment's reference frame [31, 40, 90, 92]. This objection is based on the expectation that the rotational angular momentum **R** of a fragment will favor the "ejection" of neutrons from the equatorial plane of the fragment (i.e., the plane perpendicular to **R**). The resulting anisotropy of the emitted neutrons in the fragment frame can complicate the interpretation of measured angular distributions in the laboratory. Furthermore, theoretical work [13] has shown that some of the scission neutrons may be re-absorbed by the recoiling fragments. Therefore those scission neutrons ejected at right angle to the fission axis (and therefore away from the recoiling fragments) have a greater chance of being detected, further confusing the interpretation of the measured angular distributions. For these reasons, the experimental evidence for scission neutrons remains weak at best. Whether or not they exist at all remains an open question, and any evidence for these particles must be interpreted with great caution.

5.4   Prompt Neutrons

5.4.1   Neutron Spectrum

In this section, we consider the shape of the energy spectrum of prompt neutrons emitted by the fragments (i.e., the average number of neutrons emitted with a given energy) and the prompt neutron multiplicity (the average number of neutrons emitted by the fragments, regardless of energy). In the center-of-mass (com) frame, and with certain simplifying assumptions, the neutron spectrum can be shown to have a Weisskopf distribution [85, 95],

$$N_n^{\text{com}}(E) = \frac{E}{T^2} \exp\left(-\frac{E}{T}\right) \tag{5.1}$$

where T is the nuclear temperature. Next, the com spectrum must be transformed to the laboratory frame, where the neutrons are actually measured. N. Feather [26, 85] derived a formula to transform the spectrum from the com ($N_n^{\text{com}}(E)$) to the laboratory frame ($N_n^{\text{lab}}(E)$) assuming (1) that the fragments are fully accelerated, (2) that all neutrons originate from the fragments, and (3) that neutron emission is isotropic in the com frame of the fragments,[3]

$$N_n^{\text{lab}}(E) = \int_{\left(\sqrt{E}-\sqrt{E_{F/N}}\right)^2}^{\left(\sqrt{E}+\sqrt{E_{F/N}}\right)^2} dE' \frac{N_n^{\text{com}}(E')}{4\sqrt{E_{F/N}E'}} \tag{5.2}$$

---

[3] A generalization of Feather's formula to the case of anisotropic neutron emission in the fragment's reference frame can be found in [85].

where $E_{F/N}$ is the fragment energy per nucleon,

$$E_{F/N} \equiv \frac{m_n}{m_F} E_F \tag{5.3}$$

with m_n and m_F the mass of the neutron and fragment, respectively, and E_F the kinetic energy of the fragment.

Applying Feather's formula to the Weisskopf distribution produces a spectrum in good agreement with experiment for the first few MeV in energy but shows significant discrepancies at higher energies [94]. This discrepancy can be attributed to an inherent limitation of the Weisskopf distribution. This limitation is overcome by allowing for a proper distribution of nuclear temperatures over the fission fragments (as opposed to a single common temperature for all of them) [85]. This improvement then leads to a Maxwellian distribution of neutron energies in the com frame of the fragment, instead of the Weisskopf distribution,

$$N_n^{\text{com}}(E) = \frac{2}{\sqrt{\pi} \left(\frac{8}{9}T\right)^{3/2}} \sqrt{E} \exp\left(-\frac{E}{\frac{8}{9}T}\right) \tag{5.4}$$

Applying Feather's formula to the Maxwellian distribution produces what is known as a Watt neutron energy spectrum [94], which fits the experimental data extremely well at all energies,

$$N_n^{\text{lab}}(E) = \frac{3}{2\sqrt{2\pi T E_{F/N}}} \exp\left[-\frac{9\left(E + E_{F/N}\right)}{8T}\right] \sinh\left(\frac{9\sqrt{E_{F/N}E}}{4T}\right) \tag{5.5}$$

Figure 5.3 shows an example of a Watt spectrum compared to a Weisskopf spectrum.

The accurate measurement of the prompt fission neutron spectrum (PFNS) is prone to experimental errors and challenges, and many PFNS measured in the past disagree among themselves by statistically significant amounts. Common sources of systematic error in these measurements include multiple scattering of neutrons in the target material and in the material around the detectors (shielding, collimators, support frame, etc.), detector efficiency, and calibrations (energy, TOF, etc.). Of these various sources of error multiple scattering phenomena are among the more serious ones, and procedures have been developed to take their effect into account [46]. Recent efforts [62] have focused on including multiple-scattering effects in the uncertainties of older PFNS data.

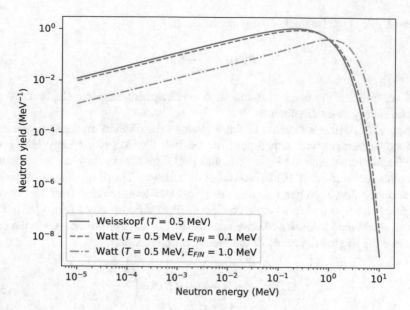

Fig. 5.3 Watt spectra for $T = 0.5$ MeV and $E_{F/N} = 0.1$ and 1.0 MeV, compared with a Weisskopf spectrum with the same nuclear temperature

5.4.2 Neutron Multiplicity

The neutron multiplicity can be estimated as the excitation energy available for neutron energy in the fragment divided by the energy removed by the emission of a single neutron. The excitation energy available for neutron emission is the total excitation energy of the fragment, minus the energy removed by gamma-ray emission (we are assuming here that only neutrons and gammas are emitted by the fragment, which is usually the case). The average energy removed by the emission of a neutron is the sum of the neutron's separation energy and its average kinetic energy (which can be obtained from the neutron spectrum and is typically $\sim$ $1-2$ MeV). For fissioning nuclei with low or no initial excitation energy, the neutron multiplicity is observed to increase roughly linearly with the mass number of the parent nucleus [87]. This effect can be understood from energy considerations: greater mass number of the parent implies more mass energy available as the excitation energy of fragments, which can therefore appear as more emitted neutrons (see also [43]). The probability distribution of the number of neutrons emitted is approximately Gaussian in shape, which has been interpreted as an indication that they are produced primarily by a statistical (evaporation) process [84].

The average neutron multiplicity as a function of primary fragment mass has a characteristic "sawtooth" shape, as shown in Fig. 5.4. This sawtooth shape can be understood using a schematic geometrical model of the nucleus as two fragments connected by a long neck which can break at random locations anywhere along

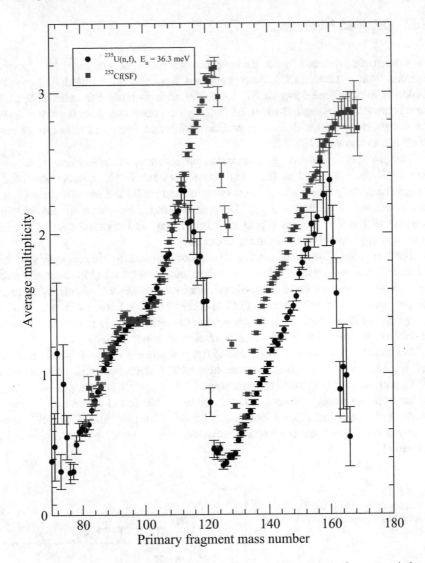

Fig. 5.4 Average neutron multiplicities as a function primary fragment mass for neutron-induced fission on ^{235}U with near thermal neutrons and for spontaneous fission of ^{252}Cf [91]. Data taken from the EXFOR database

its length (see p. 524 in [93]. If the number of neutrons emitted by a fragment is taken to be proportional to the length of the neck stub that remains attached to it (determined by the position of the random rupture), then a sawtooth shape is obtained (see Exercise 76).

5.5 Delayed Neutrons

Most neutrons produced by the fission process are emitted within $\sim 10^{-14}$ s of an induced fission reaction [93]. Some neutrons, however, are emitted $\gtrsim 1$ ms after fission is over, a time delay that is consistent with β-decay lifetimes. These late-time neutrons are generated when the β decay of a precursor fission product leaves the daughter nucleus with enough residual excitation energy to emit one or more neutrons, as shown in Fig. 5.5.

This β-delayed neutron (β-n) emission process was first identified experimentally by R. B. Roberts et al. [69]. The process is more likely to occur when the β daughter has only a few neutrons outside a closed shell (and therefore a relatively small neutron separation energy). Delayed neutrons, by virtue of the longtime constant of the β-n process (up to ~ 1 min), play an important role in the safe and controlled operation of fission reactors [60].

There are many β-n nuclei produced in fission, but to simplify their analysis, they are usually lumped together according to their half-lives into six [49] or eight groups [70]. There is some evidence for a slight dependence of the yield of delayed neutrons on the energy of incident neutrons (a few percent variation between thermal and fast energies), but more measurements are needed to confirm this [15, 16, 70]. The yield of delayed neutrons also depends on the choice of parent nucleus [70].

Any nucleus for which the Q value of the β-n process is positive is a candidate delayed neutron emitter. Thus, there are about 200 known β-n emitters from ^{8}He to ^{210}Tl, with most in the fission product range ($A \sim 70$–150). Although most delayed emitters release a single neutron, a few cases have been found where two β-delayed neutrons are produced [3], for example, ^{11}Li [4], ^{98}Rb [66], and ^{136}Sb [12]. There is even a candidate for three-neutron emission, ^{135}In, but more data are needed at present [1].

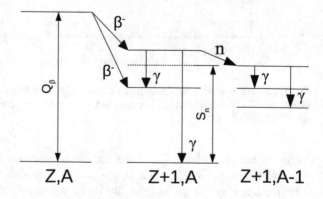

Fig. 5.5 Diagram of mechanism responsible for β-delayed neutron emission

5.6 Photon-Induced Fission

An incident photon can be absorbed by a nucleus and can initiate a reaction which results in fission, also referred to as *photofission*. Traditionally, beams used in photofission experiments have been produced by bremsstrahlung. The bremsstrahlung process, however, produces photons with a broad range of energies, and the experiments that use these beams are therefore integral measurements. More recently [2, 54], Compton backscattering of photons off of electron beams has been used to produce quasi-monoenergetic γ-ray beams.

In general, these photon beams populate nuclear states by electric dipole excitation. Thus, for an even-even nucleus with ground-state spin and parity $J^\pi = 0^+$, the photon beam will preferentially populate 1^- states. Early on, a strong resonance near 15–20 MeV in excitation energy was observed in photofission cross sections for actinide targets [5]. When an incident photon is absorbed by the nucleus, its electric field induces a coherent displacement of the protons in the nucleus and a corresponding displacement of neutrons in the opposite direction (see, e.g., Chapter 14 in [25]). This sets up a large-scale collective oscillation of neutrons and protons against one another, which corresponds to the giant resonance observed in photofission cross sections. A typical photofission cross section is shown in Fig. 5.6.

5.7 Gamma-Ray Detection

The detection of γ rays produced by fission has a long history, going back to shortly after the official discovery of fission [97]. Already in 1939, Mouzon et al. [58] had identified fission γ rays from the recoils they caused inside a cloud chamber subjected to a 1500-gauss magnetic field. From roughly 1939 to 1949, fission γ rays were detected mostly using cloud chambers, ionization chambers, and Geiger counters. From 1949 to 1965, scintillation detectors became the preferred γ-ray detection instrument [45]. In a typical sodium iodide (NaI) scintillation detector, for example, incident photons liberate electrons in a photocathode and are then amplified in a photomultiplier tube (PMT) via the emission of secondary electrons. The resulting charges are collected at the anode to produce a measurable signal. Since 1965, semiconductor detectors such as germanium (Ge) and high-purity germanium (HPGe) have become more prevalent in γ-ray detection, offering energy resolutions an order of magnitude better than scintillation detectors [81]. Despite their popularity, Ge-based detectors have a few drawbacks compared to scintillators: they are more expensive and fragile, they require more complex infrastructure (e.g., for cooling), and their timing (5–10 ns) is not as good as that of a scintillator ($\ll$ 1 ns). Therefore, although HPGe detectors have become ubiquitous nowadays, NaI and other scintillation detectors continue to be routinely used in fission and other nuclear physics experiments.

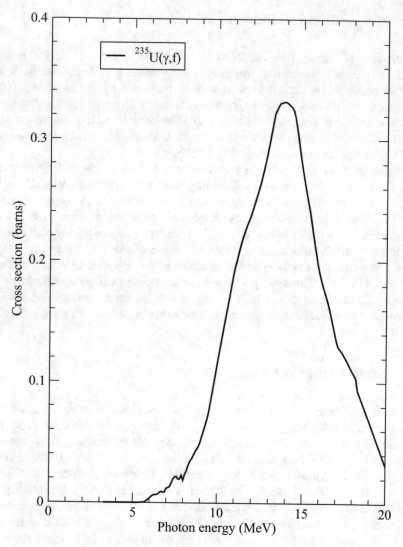

Fig. 5.6 Photofission cross section for a ^{235}U target. Data taken from the ENDF database

The decay from an excited level in a nucleus to a lower one usually entails the emission of a photon, but not always. Occasionally, an orbital electron (typically from the K or L shell) can venture inside the nucleus. The nucleus can then cause a transition between two states by transferring the energy to the electron, kicking it out of the nucleus in the process. The transition in this case occurs without the emission of a gamma ray. This process, known as *internal conversion* (IC or sometimes CE, for conversion electron) [53, 71], must be accounted for when γ rays are counted to deduce the yield of a particular fission product. If the transition between nuclear levels occurs at a rate R by both γ ray emission at rate R_γ, and internal conversion at rate R_e, then

$$R = R_\gamma + R_e \qquad (5.6)$$

and we can define an internal conversion coefficient by

$$\alpha \equiv \frac{R_e}{R_\gamma} \qquad (5.7)$$

Then, the total transition rate can be written as

$$R = R_\gamma \,(1 + \alpha) \qquad (5.8)$$

In a typical experiment where only γ rays (and not electrons) are detected, R_γ is measured, but if the quantity of interest is R—which is the case in fission product yield measurements—then Eq. (5.8) must be used. Values of IC coefficients can be found, for example, as graphs in [27] and can also be obtained using online calculators such as those at the National Nuclear Data Center (https://www.nndc.bnl.gov/hsicc/) and the Australian National University (http://bricc.anu.edu.au/).

5.8   The Use of Large Arrays of Gamma-Ray Detectors in Fission

The advent of large arrays Compton-suppressed high-purity germanium (HPGe) detectors in the 1990s [51, 72] has led to new insights into fission physics [42]. These arrays combine the advantages of the high resolution of HPGe detectors to identify fission fragments by their characteristic decay γ rays, with the ability to require the coincident detection of γ rays from complementary fragments. In this way, the exact configuration of fragments and emitted neutrons after scission could be reconstructed. Thus, for example, by identifying the fragment pair ^{108}Mo/^{144}Ba from the spontaneous fission of ^{252}Cf, Ter-Akopian et al. [83] were able to observe fission with zero neutron emission for the first time. Fission measurements with large multi-detector arrays have also been used to extract fission-fragment yields and to study the spectroscopy of neutron-rich nuclei produced in fission [42].

5.9   General Properties of Fission Gamma Rays

Prompt fission gamma rays are emitted as early as $\sim 10^{-14}$ s after scission, with most emitted within ~ 100 ns [40]. These are followed by the late prompt gamma rays produced by the fragments up to ~ 1 ms after scission [79]. Beyond that time, β

decays become more prevalent, and the resulting β-delayed gammas are observed. The emission times of fission γ rays have been measured primarily by two methods [40]: (1) using the known high recoil velocity of the fragment along with the distance traveled when the γ rays are emitted [47] or (2) by the *angular aberration method*, using the known slowing curve of the fragment as a function of time in a dense material, and the measured forward-backward anisotropy of the emitted gammas [73–75]. In the first method, the γ rays are observed through a thin collimator that can be placed at various points along the flight path of a recoiling fragment. In the second method, the angular correlation between γ rays and fragments (i.e., the angular distribution of the γ rays relative to the fission axis defined by the recoiling fragments in the case of binary fission) is analyzed to extract the emission time of the γ rays. As the fragment travels through backing material, it slows down and an anisotropy results between the count rate for γ rays emitted in the direction of motion of the fragment (forward) compared to the opposite direction (backward). This forward-backward anisotropy, or angular aberration, can be related to the velocity of the slowing fragment, which in turn is related to the emission time.

The average energy and multiplicity of fission γ rays are important quantities in the design of nuclear reactors [68]. Phenomenological models have been developed to describe them, and roughly speaking ~ 7 γ rays are produced per fission, with an average energy of ~ 1 MeV per gamma ray (but these quantities vary depending on the parent nucleus and reaction) [63]. More detailed analyses have produced simple polynomial relationships for the average total prompt γ-ray energy released per fission, as a function of incident neutron energy [55]. In addition, it has been observed that the average total prompt γ-ray energy varies smoothly as a function of the average neutron multiplicity and as a function of the TKE [63]. The γ-ray multiplicity plotted as a function of fragment mass displays a sawtooth shape, similar to the neutron multiplicity [44, 63].

The shape of the γ-ray energy spectrum is also needed in the design of shielding and cooling systems for nuclear reactors [7]. As an important step to extract the continuous energy distribution of fission gammas, the response of the detector must be unfolded [50, 64, 78, 89]. Once the energy spectrum has been obtained, certain common features can be identified. In very broad terms, below ~ 1 MeV, the spectrum consists mainly of discrete γ rays; between ~ 1 and 8 MeV, the shape of the spectrum is dominated by an exponentially decreasing yield of (mainly E1) statistical γ rays; and above ~ 8 MeV, the spectrum typically reveals a slight enhancement from the giant dipole resonance [7].

5.10   From Detector Counts to Fragment Yield

A commonly studied source of γ rays in fission experiments are those produced by the γ decay of a fission product populated in an excited state by the β decay of its precursor. The rate at which the fission product decays can be deduced from the rate of events counted in the detector with a γ ray energy from the daughter nucleus (assuming there are no contaminant γ rays from a different level or nucleus with a similar energy) after correcting for a number of factors: (1) the fraction of source decays that produce that particular γ ray (i.e., the *branching ratio*) including corrections for IC if necessary, (2) the live time fraction of the detection apparatus, (3) a correction for any anisotropy in the angular distribution of the γ rays, (4) the solid angle subtended by the detector, and (5) the intrinsic efficiency of the detector. Once these factors and their uncertainties have been determined, measured γ-ray intensities can be turned into fission product yields, thereby taking advantage of the precisely known γ-ray energies to home in on specific fission products.

We illustrate the extraction of yields from γ rays with a worked example. The ^{135}Xe fission product undergoes a β^- decay from its ground state to several excited states in ^{135}Cs with a half-life $T_{1/2} = 9.14$h. Figure 5.7 shows the decay scheme. We will focus in particular on the state in ^{135}Cs at excitation energy $E_x = 249.793$ keV, which decays directly (and only) to the ground state.

Question: If the β decay to the state at $E_x = 249.793$ keV occurs 96% of the time, why does the total decay from the state occur 97% of the time?

Although it may seem like a small difference, this is not a numerical or typographical error. Rather, the 97% decay from the state at $E_x = 249.793$ keV includes the 96% direct feeding from the β decay, as well as β decays to higher-

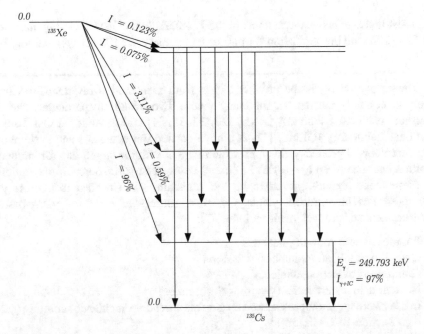

Fig. 5.7 Level diagram showing β^- decay of ^{135}Xe into ^{135}Cs. Excitation and gamma-ray energies are given, along with absolute (i.e., $\gamma + $ IC) transition intensities between the ^{135}Cs levels. Adapted from the ENSDF database

energy states, which then decay to the level at $E_x = 249.793$ keV. Fortunately, the ENSDF evaluation usually provides the total intensity for each decay.

Question: For every 100 beta decays of ^{135}Xe, how many γ rays are produced by the decay of the $E_x = 249.793$ keV level to the ground state?

To answer the question, we need to account for conversion electrons using Eq. (5.8). We can obtain the conversion coefficients from the ICcalc tool (see Sect. 5.7) or use the values supplied in the ENSDF file for the K, L, and M atomic shells for this γ ray (and neglecting smaller conversion coefficients for other shells):

$$\alpha (K) = 0.0623$$

$$\alpha (L) = 0.0091 \qquad (5.9)$$

$$\alpha (M) = 0.00188$$

Of the 100 decays of ^{135}Xe, 97% of them will populate the $E_x = 249.793$ keV level, or a number $N_{tot} = 97$. Then, using Eq. (5.8), we can deduce the corresponding number of γ rays:

$$N_\gamma = \frac{N_{tot}}{1 + \alpha (K) + \alpha (L) + \alpha (M)} \qquad (5.10)$$
$$\approx 90$$

This result also explains the footnote in the ENSDF file that lists the γ-ray intensity as $I_\gamma = 100$ with the stipulation "For absolute intensity per 100 decays, multiply by 0.90."

The extraction of fission product yields from measured γ-ray intensities has been successfully carried out for many years. Two groups in particular, one at Argonne National Laboratory (ANL) [28, 32–38, 61] and the other at Oak Ridge National Laboratory (ORNL) [17–24], have carried out extensive and systematic measurements of product yields for a wide range of actinide targets and for incident neutron energies of up to ~8 MeV. In many cases, the major source of uncertainty in these measurements proved to be the branching ratio needed to convert γ-ray intensities into product yields. Common sources of error for γ-ray-based measurements of product yields include [97]:

- Photopeak area counting statistics
- Overall normalization/number of fissions
- Gamma-ray detection efficiency
- Nuclear data uncertainties (viz., branching ratios)
- Losses from the photopeak caused by gamma cascade coincidence summing (see, e.g., Section 10.7.3 in [53])
- Errors due to peak fitting and peak stripping (peak stripping refers to procedures used to remove background counts)

- Corrections due to gamma absorption
- Room-return corrections
- Outgassing of volatile fission products

More recently, a collaboration between Los Alamos National Laboratory, Lawrence Livermore National Laboratory, and the Triangle Universities Nuclear Laboratory has undertaken a series of measurements of fission product yields using monoenergetic neutrons with incident energies from 0.6 to 14.8 MeV on ^{235,238}U and ^{239}Pu targets [39]. These measurements fill in some of the gaps left by the earlier measurements in the 1970s and 1980s.

5.11 Fragment Angular Momentum Deduced from Measured Gamma Rays

The angular momentum imparted to the fragments during fission has been studied both theoretically [10, 56, 57] and experimentally [8, 59, 76, 80, 82, 96]. In Sect. 4.4, we alluded to the measurement of gamma rays emitted by the fragments as a way of estimating their initial angular momentum. More specifically, the angular momenta of the fragments have previously been determined by measuring the angular distribution of unresolved gammas, the gamma-ray multiplicities, the ratio of isomer yields, and the intensity of ground-state band transitions. For example, Tanikawa et al. [80] used the relative population of isomeric states with very different spins to estimate the fragment angular momenta following the ^{238}U(p, f) reaction, with the help of a statistical model. Wilhelmy et al. [96], on the other hand, measured prompt ground-state band γ rays following the ^{252}Cf (SF) decay to deduce the initial fragment angular momenta (see Exercise 87). Based on measurements with different fissioning systems and for different fission products, the average fragment angular momenta are found to range from a few $\hbar$ to about $12\hbar$ [8, 59, 76, 80, 82, 96]. In general, these measurements show that most of the angular momentum from a high-spin fissioning nucleus is imparted to the relative motion of the fragments, rather than to their internal angular momenta [8, 96]. As far as the partition of angular momentum between heavy and light fragments is concerned, it is found that the heavy fragment carries a larger share of the angular momentum than the light one [59, 80, 96] and that the angular momentum does not depend sensitively on the mass of the fragment (e.g., Bogachev et al. [8] measured a mass dependence that approached the asymptotic $\sim A^{5/6}$ behavior expected from a statistical scission model in the case of fission following the ^{208}Pb$(^{18}$O$, f)$ reaction).

5.12   Questions

43. Name three types of neutron-producing facilities.

44. Why does a white neutron source have to be pulsed in a TOF measurement?

45. What reaction mechanisms are used in neutron detection?

46. What is the shape of the spectrum of evaporated neutron in the fragment's reference frame?

47. What is the shape of the spectrum of evaporated neutron in the laboratory reference frame?

48. What is a major source of uncertainty in PFNS measurements?

49. What is the general trend of fission neutron multiplicity as a function of parent mass number?

50. Why are some beta decays followed by neutron emission? What is a necessary condition for this to occur?

51. What information can gamma rays provide?

52. What states are more likely to be populated in photon-induced reactions?

53. What type of detector would you use to identify discrete gamma rays from fission fragments?

54. Roughly how much energy is released through gamma rays in a fission event?

55. What is a major source of uncertainty in extracting fragment yields from measured gamma-ray intensities?

56. What are two experimental methods that can be used to estimate the initial amount of angular momentum in fission fragments?

5.13   Exercises

Problem 69 What kinetic energy does a neutron need to have for its de Broglie wavelength to equal the radius of a ^{239}Pu nucleus?

Problem 70 The ^{7}Li (p, n) ^{7}Be reaction is sometimes used to generate neutrons. For an incident 4-MeV proton beam, what is the range (minimum and maximum) of neutron energies produced?

Problem 71 The number ν of fission neutrons has an approximately normal (or Gaussian) probability distribution,

$$P(\nu) = \frac{1}{\sqrt{2\pi}\,\sigma} \exp\left[-\frac{(\nu - \bar{\nu})^2}{2\sigma^2}\right] \tag{5.11}$$

where $\bar{\nu}$ is the average number of fission neutrons and $\sigma \approx 1.08$. What is the probability that the thermal neutron-induced fission on a ^{235}U target produces at least one neutron ($\bar{\nu} = 2.43$)?

Problem 72 Consider the compound-nucleus reaction $n + ^{235}\text{U} \rightarrow ^{236}\text{U}^*$ with a 14-MeV incident neutron (the "*" superscript indicates the nucleus is left in an excited state). What is the recoil velocity of the compound nucleus in the laboratory?

Problem 73 The $^3\text{H}(p, n)^3\text{He}$ reaction is sometimes used to generate neutrons. From energy considerations, what is the minimum proton energy needed to produce neutrons?

Problem 74 If the neutron evaporation spectrum in the fragments reference frame has a Weisskopf form, given by Eq. (5.1), what is the average neutron kinetic energy in that frame?

Problem 75 Consider a neutron emission spectrum that is very strongly peaked at a single energy E_0 in the center-of-mass frame, so that it can be represented by a delta function

$$N_n^{\text{com}}(E) = \delta(E - E_0) \tag{5.12}$$

what is the corresponding neutron spectrum in the lab frame?

Problem 76 Referring to Exercise 25 in Chap. 2, assume the excitation energy that will be dissipated by each fragment through neutron emission is proportional to the elongation of that fragment beyond its doubly magic core (i.e., proportional to x and $L - x$ for the light and heavy fragment, respectively). Assume also that the number of neutrons emitted by each fragment is proportional to that excitation energy. The values of the proportionality constants are not important for this exercise, so you can take them to be equal to 1. Plot the neutron multiplicity divided by L as a function of fragment mass, and compare the shape with Fig. 5.4.

Problem 77 Figure 5.8 summarizes relevant information that can be retrieved from the ENSDF database for the β decay of ^{133}Xe to ^{133}Cs.

If, over the course of a measurement, 1000 gamma rays with the characteristic $E_\gamma = 81.0\text{keV}$ energy of the transition in ^{133}Cs from the level at $E_x = 80.9979\text{keV}$ are emitted, how many ^{133}Xe nuclei decayed into ^{133}Cs?

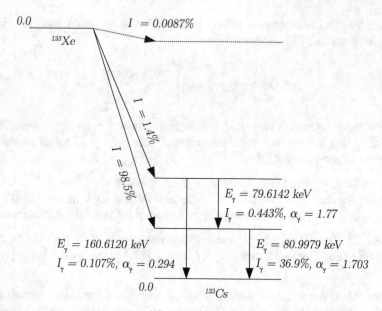

Fig. 5.8 Level diagram for β decay of ^{133}Xe to ^{133}Cs. Data taken from the ENSDF database

Problem 78 Assume that the prompt fission γ-ray spectrum in the fragment reference frame is given as a function of its nuclear temperature T by [52]

$$\Phi\left(E_\gamma\right) = \frac{E_\gamma^2}{2T^3} \exp\left[-\frac{E_\gamma}{T}\right] \tag{5.13}$$

What are the average and most probable γ-ray energies at a given temperature T? If an average prompt γ-ray energy of 3 MeV is measured for this fragment, what is the corresponding temperature? (Assume the spectrum is the same in the lab frame).

Problem 79 Suppose the number of prompt fission γ rays in a fission event (i.e., the multiplicity M_γ) follows the probability distribution [86]

$$P\left(M_\gamma\right) \propto \left(2M_\gamma + 1\right) \exp\left[-\frac{M_\gamma\left(M_\gamma + 1\right)}{B^2}\right] \tag{5.14}$$

where B is a free parameter that can be adjusted to data. Express the average multiplicity, $\bar{M}_\gamma$, in terms of B. (You can treat M_γ as a continuous variable for this problem.)

Problem 80 The quantity B^2 in Problem 79 can be expressed in terms of the moment of inertia $\mathfrak{I}$ of the nucleus and its temperature T,

$$B^2 = \frac{\mathfrak{I}T}{\hbar^2} \tag{5.15}$$

If the moment of inertia remains constant, how does the distribution in Eq. (5.14) vary with increasing temperature (and, therefore, generally speaking, with excitation energy)?

Problem 81 Using the results from Exercises 78 and 79, write an expression for the average total prompt γ-ray energy released per fission event, and state the assumptions you made.

Problem 82 A model calculation [77] gives the angular distribution of γ rays relative to the fission recoil axis emitted by a fragment as

$$W_L\left(\theta\right) = \begin{cases} 1 + \frac{1}{8}\left(\frac{\hbar^2 I}{\mathfrak{I}T}\right)^2 \sin^2\theta & \text{dipole radiation } (L = 1) \\ 1 - \frac{3}{8}\left(\frac{\hbar^2 I}{\mathfrak{I}T}\right)^2 \sin^2\theta & \text{quadrupole radiation } (L = 2) \end{cases} \tag{5.16}$$

where θ is the angle between the γ ray and the fission axis, I is the fragment's spin, $\mathfrak{I}$ is the fragment's moment of inertia, and T is its temperature. Comment on the conditions under which the radiation is approximately isotropic in the fragment's frame of reference.

Problem 83 The ^{235}U (n, f) cross section for thermal neutrons is ≈ 577 b, while the cross section for the absorption of a thermal neutron that does not lead to fission is ≈ 101 b. Thermal fission produces 2.44 neutrons per fission on average. How many fission neutrons are produced per neutron absorbed on average?

Problem 84 Neutrons from a white source (i.e., with a broad energy spectrum) are produced at the LANSCE/WNR facility in Los Alamos using a pulsed spallation source. Assume the pulses are 1.8 μs apart and that the neutrons travel 20 m to reach a sample of target material to be studied. If a 5-MeV neutron from the current pulse and a slower neutron from the previous pulse reach the sample at the same time, what is the energy of the slower neutron from the previous pulse? This effect is called "wrap around" and constitutes an unwanted background in measurements that must be reduced through the use of absorber materials and/or subtracted out.

Problem 85 Table 5.2 gives the probabilities for emitting various numbers ν of neutrons in the spontaneous fission of ^{252}Cf. Calculate the average multiplicity $\bar{\nu}$ and the standard deviation σ_ν of this probability distribution. Plot a Gaussian distribution with this mean $\bar{\nu}$ and standard deviation σ_ν along with the data in Table 5.2: How well are the data reproduced by the Gaussian distribution?

Table 5.2 Neutron
multiplicity probability
distribution, $P(\nu)$, for
^{252}Cf (SF) [9]. Data taken
from the EXFOR database

ν	$P(\nu)$
0	0.002
1	0.026
2	0.127
3	0.273
4	0.304
5	0.185
6	0.066
7	0.015
8	0.002

Problem 86 The following function is a fit to the fission γ-ray spectrum following thermal neutron-induced fission on a ^{235}U target with the discrete γ rays removed [88],

$$
N\left(E_\gamma\right) = \begin{cases} 38.13\left(E_\gamma - 0.085\right)e^{1.648E_\gamma} & 0.085\,\text{MeV} \le E_\gamma < 0.3\,\text{MeV} \\ 26.8e^{-2.30E_\gamma} & 0.3\,\text{MeV} \le E_\gamma < 1.0\,\text{MeV} \\ 8.0e^{-1.10E_\gamma} & 1.0\,\text{MeV} \le E_\gamma < 8.0\,\text{MeV} \end{cases}
$$

(5.17)

According to this formulation of the spectrum, what is the most probable γ-ray energy? What is the average γ-ray energy?

Problem 87 A ^{144}Ba fission fragment is produced in a fission event and decays first by neutron emission to a state with excitation energy $E_x = 6$ MeV and an angular momentum probability distribution [87, 96]

$$
P(J) \propto (2J + 1)\exp\left[-\frac{(J + 1/2)^2}{b^2}\right]
$$

(5.18)

with $b = 7.2$. From there, it decays directly to the members of the ground-state band listed in Table 5.3 via statistical E1 (dipole) transitions. Assume that an E1 transition from an initial spin J_i can populate any state with final spin J_f such that

$$
|J_i - 1| \le J_f \le J_i + 1
$$

(5.19)

and that the probability of decaying to a state is proportional to E_γ^3 (you can take all proportionality constants equal to 1 for this exercise). Starting from the distribution in Eq. (5.18), calculate the relative populations of the ground-state members before they decay.

Problem 88 Referring to Exercise 87, each member of the ground-state band decays to the next one (i.e., $6^+ \to 4^+$, $4^+ \to 2^+$, $2^+ \to 0^+$) by E2 transitions.

J^π	E_x (MeV)
0^+	0.000
2^+	0.199
4^+	0.530
6^+	0.962

Table 5.3 Total angular momentum and parity (J^π) and excitation energy (E_x) of the first few levels in the ground-state band of ^{144}Ba (data taken from the ENSDF database)

Using the relative populations obtained in Exercise 87, calculate the $6^+ \to 4^+$ and $4^+ \to 2^+$ γ-ray intensities relative to the $2^+ \to 0^+$ (remember to account for IC).

References

1. D. Abriola, B. Singh, I. Dillmann, tech. rep. INDC (NDS)–0599 (IAEA, 2011)
2. A.N. Andreyev, K. Nishio, K.-H. Schmidt, Rep. Progr. Phys. **81**, 016301 (2017)
3. G. Audi, F.G. Kondev, M. Wang, W. Huang, S. Naimi, Chin. Phys. C **41**, 030001 (2017)
4. R.E. Azuma, L.C. Carraz, P.G. Hansen, B. Jonson, K.-L. Kratz, S. Mattsson, G. Nyman, H. Ohm, H.L. Ravn, A. Schröder, W. Ziegert, Phys. Rev. Lett. **43**, 1652 (1979)
5. G.C. Baldwin, G.S. Klaiber, Phys. Rev. **71**, 3 (1947)
6. L. Bernstein, D. Brown, A. Hurst, J. Kelly, F. Kondev, E. McCutchan, C. Nesaraja, R. Slaybaugh, A. Sonzogni (2015), arXiv:1511.07772v1
7. D. Blanchet, N. Huot, P. Sireta, H. Serviere, M. Boyard, M. Antony, V. Laval, P. Henrard, Ann. Nucl. Energy **35**, 731 (2008)
8. A. Bogachev et al., Eur. Phys. J. A **34**, 23 (2007)
9. J.W. Boldeman, M.G. Hines, Nucl. Sci. Eng. **91**, 114 (1985)
10. L. Bonneau, P. Quentin, I.N. Mikhailov, Phys. Rev. C **75**, 064313 (2007)
11. H.R. Bowman, S.G. Thompson, J.C.D. Milton, W.J. Swiatecki, Phys. Rev. **126**, 2120 (1962)
12. R. Caballero-Folch et al., Phys. Rev. C **98**, 034310 (2018)
13. N. Carjan, M. Rizea, Phys. Rev. C **82**, 014617 (2010)
14. M. Chadwick et al., Nucl. Data Sheets **112**, 2887 (2011)
15. A. D'Angelo, J.L. Rowlands, Progr. Nucl. Energy **41**, 391 (2002)
16. S. Das, Progr. Nucl. Energy **28**, 209 (1994)
17. J.K. Dickens, Nucl. Sci. Eng. **70**, 177 (1979)
18. J.K. Dickens, J.W. McConnell, Nucl. Sci. Eng. **73**, 42 (1980)
19. J.K. Dickens, J.W. McConnell, Phys. Rev. C **24**, 192 (1981)
20. J.K. Dickens, J.W. McConnell, Phys. Rev. C **23**, 331 (1981)
21. J.K. Dickens, J.W. McConnell, Phys. Rev. C **27**, 253 (1983)
22. J.K. Dickens, J.W. McConnell, Phys. Rev. C **34**, 722 (1986)
23. J.K. Dickens, J.W. McConnell, K.J. Northcutt, Nucl. Sci. Eng. **77**, 146 (1981)
24. J.K. Dickens, J.W. McConnell, K.J. Northcutt, Nucl. Sci. Eng. **80**, 455 (1982)
25. J.M. Eisenberg, W. Greiner, *Nuclear Theory: Microscopic heory of the Nucleus* (North-Holland, London, 1986)
26. N. Feather, tech. rep. BR 335A (U.S. Atomic Energy Commission, 1942)
27. R.B. Firestone, *Table of Isotopes* (Wiley VCH, Weinheim, 1999)
28. K. Flynn, J. Gindler, L. Glendenin, J. Inorg. Nucl. Chem. **39**, 759 (1977)
29. C. Franklyn, C. Hofmeyer, D. Mingay, Phys. Lett. B **78**, 564 (1978)
30. R.W. Fuller, Phys. Rev. **126**, 684 (1962)
31. A. Gagarski, I. Guseva, G. Val'sky, G. Petrov, V. Petrova, T. Zavarukhina, in *Isinn-20* (2012)
32. J.E. Gindler, L.E. Glendenin, D.J. Henderson, J.W. Meadows, Phys. Rev. C **27**, 2058 (1983)
33. J. Gindler, L. Glendenin, D. Henderson, J. Inorg. Nucl. Chem. **43**, 1433 (1981)

34. J. Gindler, L. Glendenin, D. Henderson, J. Inorg. Nucl. Chem. **43**, 1743 (1981)
35. J. Gindler, L. Glendenin, E. Krapp, S. Fernandez, K. Flynn, D. Henderson, J. Inorg. Nucl. Chem. **43**, 445 (1981)
36. J. Gindler, D. Henderson, L. Glendenin, J. Inorg. Nucl. Chem. **43**, 895 (1981)
37. L.E. Glendenin, J.E. Gindler, I. Ahmad, D.J. Henderson, J.W. Meadows, Phys. Rev. C **22**, 152 (1980)
38. L.E. Glendenin, J.E. Gindler, D.J. Henderson, J.W. Meadows, Phys. Rev. C **24**, 2600 (1981)
39. M. E. Gooden, et al., Nucl. Data Sheets **131**, 319 (2016)
40. F. Gönnenwein, *Neutron and Gamma Emission in Fission*, LANL (2014)
41. I. Halpern, in *Proceedings of the Symposium on Physics and Chemistry of Fission*, Vol. II (1965), p. 369
42. J. Hamilton, A. Ramayya, S. Zhu, G. Ter-Akopian, Y. Oganessian, J. Cole, J. Rasmussen, M. Stoyer, Progr. Part. Nucl. Phys. **35**, 635 (1995)
43. D. Hilscher, H. Rossner, Ann. Phys. **17**, 471 (1992)
44. D.C. Hoffman, M.M. Hoffman, Annu. Rev. Nucl. Sci. **24**, 151 (1974)
45. R. Hofstadter, Phys. Rev. **75**, 796 (1949)
46. M.M. Islam, H.-H. Knitter, Nucl. Sci. Eng. **50**, 108 (1973)
47. S.A. Johansson, Nucl. Phys. **60**, 378 (1964)
48. D.T.L. Jones, Rad. Phys. Chem. **61**, 469 (2001)
49. G.R. Keepin, T.F. Wimett, R.K. Zeigler, Phys. Rev. **107**, 1044 (1957)
50. E. Kwan et al., Nucl. Instrum. Methods Phys. Res. Sect. A: Accel. Spectrom. Detect. Assoc. Equip. **688**, 55 (2012)
51. I.-Y. Lee, Nucl. Phys. A **520**, c641 (1990)
52. S. Lemaire, P. Talou, T. Kawano, M.B. Chadwick, D.G. Madland, Phys. Rev. C **73**, 014602 (2006)
53. W.R. Leo, *Techniques for Nuclear and Particle Physics Experiments: A How-to Approach* (Springer, New York, 1992)
54. V.N. Litvinenko, I. Ben-Zvi, D. Kayran, I. Pogorelsky, E. Pozdeyev, T. Roser, V. Yakimenko, IEEE Trans. Plasma Sci. **36**, 1799 (2008)
55. D. Madland, Nucl. Phys. A **772**, 113 (2006)
56. L.G. Moretto, G.F. Peaslee, G.J. Wozniak, Nucl. Phys. A **502**, 453 (1989)
57. L.G. Moretto, R.P. Schmitt, Phys. Rev. C **21**, 204 (1980)
58. J.C. Mouzon, R.D. Park, J.A. Richards, Phys. Rev. **55**, 668 (1939)
59. S. Mukhopadhyay, L.S. Danu, D.C. Biswas, A. Goswami, P.N. Prashanth, L.A. Kinage, A. Chatterjee, R.K. Choudhury, Phys. Rev. C **85**, 064321 (2012)
60. R.L. Murray, K.E. Holbert, in *Nuclear Energy* (Elsevier, Amsterdam, 2015), pp. 331–349
61. S. Nagy, K.F. Flynn, J.E. Gindler, J.W. Meadows, L.E. Glendenin, Phys. Rev. C **17**, 163 (1978)
62. D. Neudecker, T. Taddeucci, R. Haight, H. Lee, M. White, M. Rising, Nucl. Data Sheets **131**, 289 (2016)
63. H. Nifenecker, C. Signarbieux, R. Babinet, and R. Poitou, in *Proceedings of the Symposium on Physics and Chemistry of Fission* (1973)
64. R.W. Peelle, F.C. Maienschein, Phys. Rev. C **3**, 373 (1971)
65. G.A. Petrov et al., in *AIP Conference Proceedings*, **1175**, 289 (2009)
66. P.L. Reeder, R.A. Warner, T.R. Yeh, R.E. Chrien, R.L. Gill, M. Shmid, H.I. Liou, M.L. Stelts, Phys. Rev. Lett. **47**, 483 (1981)
67. R. Reifarth, C. Lederer, F. Käppeler, J. Phys. G: Nucl. Part. Phys. **41**, 053101 (2014)
68. G. Rimpault, D. Bernard, D. Blanchet, C. Vaglio-Gaudard, S. Ravaux, A. Santamarina, Phys. Proc. **31**, 3 (2012)
69. R.B. Roberts, R.C. Meyer, P. Wang, Phys. Rev. **55**, 510 (1939)
70. G. Rudstam, P. Finck, A. Filip, A. D'Angelo, R.D. McKnight, tech. rep. WPEC-6 (NEA, 2002)
71. P. Marmier, E. Sheldon, *Physics of Nuclei and Particles: v. 1* (Academic, London, 1969)
72. J. Simpson, Zeitschrift für Physik A Hadrons Nuclei **358**, 139 (1997)
73. K. Skarsvåg, Nucl. Phys. A **253**, 274 (1975)
74. K. Skarsvåg, Phys. Rev. C **22**, 638 (1980)

75. K. Skarsvåg, I. Singstad, Nucl. Phys. **62**, 103 (1965)
76. A.G. Smith, C.E. Barrett, M.A. Alothman, A. Chatillon, H. Faust, G. Fioni, D. Goutte, H. Goutte, in *AIP Conference Proceedings*, vol. 1175 (2009), p. 193
77. V.M. Strutinsky, Sov. JETP **37**, 861 (1959)
78. C. Sükösd, W. Galster, I. Licot, M. Simonart, Nucl. Instrum. Methods Phys. Res. Sect. A: Accel. Spectrom. Detect. Assoc. Equip. **355**, 552 (1995)
79. P. Talou, T. Kawano, I. Stetcu, J.P. Lestone, E. McKigney, M.B. Chadwick, Phys. Rev. C **94**, 064613 (2016)
80. M. Tanikawa, H. Kudo, H. Sunaoshi, M. Wada, T. Shinozuka, M. Fujioka, Zeitschrift für Physik A Hadrons Nuclei **347**, 53 (1993)
81. A. Tavendale, G. Ewan, Nucl. Instrum. Methods **25**, 185 (1963)
82. G.M. Ter-Akopian et al., Phys. Rev. C **55**, 1146 (1997)
83. G.M. Ter-Akopian, J.H. Hamilton, Y.T. Oganessian, J. Kormicki, G.S. Popeko, A.V. Daniel, A.V. Ramayya, Q. Lu, K. Butler-Moore, W.C. Ma, J.K. Deng, D. Shi, J. Kliman, V. Polhorsky, M. Morhac, W. Greiner, A. Sandelescu, J.D. Cole, R. Aryaeinejad, N.R. Johnson, I.Y. Lee, F.K. McGowan, Phys. Rev. Lett. **73**, 1477 (1994)
84. J. Terrell, in *Proceedings of the Symposium on Physics and Chemistry of Fission*, vol. II (1965), p. 3
85. J. Terrell, Phys. Rev. **113**, 527 (1959)
86. J.L. Ullmann et al., Phys. Rev. C **87**, 044607 (2013)
87. R. Vandenbosch, J.R. Huizenga, *Nuclear Fission* (Academic, London, 1973)
88. J. Verbeke, C. Hagmann, D.M. Wright, tech. rep. UCRL-AR-228518 (LLNL, 2007)
89. V.V. Verbinski, H. Weber, R.E. Sund, Phys. Rev. C **7**, 1173 (1973)
90. A.S. Vorobyev, O.A. Shcherbakov, Y.S. Pleva, A.M.Gagarski, G.V. Val'ski, G.A. Petrov, V.I. Petrova, T.A. Zavarukhina, in *ISINN-17* (2009)
91. A. Vorobyev, O. Shcherbakov, A. Gagarski, G. Val'ski, G. Petrov, *EPJ Web of Conferences*, **8**, 03004 (2014)
92. A. Vorobyev, O. Shcherbakov, Y. Pleva, A. Gagarski, G. Val'ski, G. Petrov, V. Petrova, T. Zavarukhina, Nucl. Instrum. Methods Phys. Res. Sect. A Accel. Spectrom. Detect. Assoc. Equip. **598**, 795 (2009)
93. C. Wagemans, *The Nuclear Fission Process* (CRC Press, Boca Raton, 1991)
94. B.E. Watt, Phys. Rev. **87**, 1037 (1952)
95. V. Weisskopf, Phys. Rev. **52**, 295 (1937)
96. J.B. Wilhelmy, E. Cheifetz, R.C. Jared, S.G. Thompson, H.R. Bowman, J.O. Rasmussen, Phys. Rev. C **5**, 2041 (1972)
97. W. Younes, J.J. Ressler, J.A. Becker, tech. rep. LLNL-TR-648488 (LLNL, 2014)

Chapter 6
Fission Models Revisited: Structure

Abstract This chapter deals with the nuclear structure aspects of fission at an advanced level. The discussion starts with the isotropic harmonic oscillator potential and describes in detail the Strutinsky shell correction method and its application. Pairing, which plays an important role in fission and more generally in many nuclear physics phenomena, is discussed in detail. Next, we consider the concept of level densities which are used in both nuclear structure and nuclear reaction models and are a critical component of fission cross-section calculations. Therefore, the chapter includes a detailed discussion of the calculation of level densities in nuclei. In the advanced topics section of the chapter, we build up to the Hartree-Fock approximation, the foundation of modern (so-called microscopic) theories of fission. Starting with solutions of the Schrödinger equation for many particles moving in a common potential, we arrive at the concept of a "mean field," where the particles themselves generate that common potential in a self-consistent way.

We introduced several basic models of nuclear fission in Chap. 3. In this chapter we dive deeper into the formalism behind these models. This closer look will serve to justify some of the results in Chap. 3 and also to lay the groundwork for more advanced approaches in Sect. 6.5. In this chapter, we focus on the nuclear structure aspects of fission, leaving the discussion of fission reactions and dynamics for Chap. 7.

6.1 Single-Particle Models

For a more fundamental understanding of the fission process, we turn to models that deal with the individual protons and neutrons of the fissioning nucleus. For this, we begin with the time-independent Schrödinger equation for a particle of mass m moving in a potential $V(\mathbf{r})$ [32, 47],

© The Author(s), under exclusive license to Springer Nature Switzerland AG 2021 115
W. Younes, W. D. Loveland, *An Introduction to Nuclear Fission*, Graduate Texts
in Physics, https://doi.org/10.1007/978-3-030-84592-6_6

$$-\frac{\hbar^2}{2m}\nabla^2\Psi\,(\mathbf{r}) + V\,(\mathbf{r})\,\Psi\,(\mathbf{r}) = E\Psi\,(\mathbf{r}) \tag{6.1}$$

The solution of this second-order differential equation yields the single-particle energy levels E and the corresponding wave functions $\Psi\,(\mathbf{r})$. The symbol ∇^2 for the *Laplace operator* denotes a second-order derivative with respect to position and takes a different form depending on the choice of coordinate system (e.g., spherical, cylindrical, Cartesian [25]). The potential $V\,(\mathbf{r})$ is a crucial ingredient; in the present context, it represents the average influence felt by a typical nucleon and caused by its combined interactions with all the other nucleons (in Sect. 6.5, we will use the term "mean field" in this context). This is an *independent particle model*, where the nucleons are affected by the potential but otherwise move independently of one another (other than the fact that they obey the Pauli exclusion principle, which prevents like nucleons from occupying the same quantum state). In essence, this chapter will deal with three basic questions:

1. How do we come up with a potential $V\,(\mathbf{r})$ for Eq. (6.1)?
2. How do we solve Eq. (6.1) for that potential?
3. How do we use the solution of Eq. (6.1) to better understand and predict fission observables?

To motivate the discussion, we will tackle the third question first, and we will use a toy model of the nucleus to illustrate the concepts and methods. The first question will be addressed in Sect. 6.5.1, and we leave the more technical answer to Question 2 for Sect. 6.5.2. We will assume the potential felt by a typical nucleon has the form of an isotropic harmonic oscillator potential. We choose this form because it is well studied and the Schrödinger equation can be solved exactly for this potential [32, 47].

For the isotropic harmonic oscillator potential

$$V\,(\mathbf{r}) = \frac{1}{2}m\omega_0^2 r^2 \tag{6.2}$$

written in spherical coordinates, the single-particle levels are given by (see, e.g., Section 3.2.1 in [46]),

$$E_{n\ell} = \left(2n + \ell + \frac{3}{2}\right) = \left(N + \frac{3}{2}\right)\hbar\omega_0 \tag{6.3}$$

where

$$N = 2n + \ell \tag{6.4}$$

is the *major shell number* and where $n = 0, 1, 2, \ldots$ is the principal quantum number, which gives the number of nodes of the wave function. The orbital angular momentum number ℓ is a positive integer. For historical reasons originating in atomic spectroscopy, the value of ℓ is sometimes replaced by a letter with s, p, d, and f, standing for $\ell = 0, 1, 2, 3$, respectively, and alphabetically thereafter (i.e., g for $\ell = 4$, h for $\ell = 5$, etc.). The wave functions have the following form in spherical coordinates

$$\psi_{n\ell m_\ell m_s}(\mathbf{r}) = g_{n\ell}(r) Y_{\ell m_\ell}(\theta, \varphi) \chi_{m_s} \tag{6.5}$$

where the radial part is

$$g_{n\ell}(r) = \sqrt{\frac{2n!}{b^3 \Gamma(n + \ell + 3/2)}} \left(\frac{r}{b}\right)^\ell e^{-r^2/2b^2} L_n^{(\ell+1/2)}\left(\frac{r^2}{b^2}\right), \quad b = \sqrt{\frac{\hbar}{m\omega_0}} \tag{6.6}$$

expressed using the gamma function, $\Gamma(x)$, and the generalized Laguerre polynomial $L_n^{(\ell+1/2)}(x)$ [1]. The $Y_{\ell m_\ell}(\theta, \varphi)$ functions are spherical harmonics [12], and the χ_{m_s} represent the intrinsic spin of the particle with $m_s = \pm 1/2$.

At this point, it is useful to make a distinction between a quantum state and a quantum level in discussing the implications of Eq. (6.3). Equation (6.3) gives the energy of the levels of the isotropic harmonic oscillator. Each level corresponds to many states that are degenerate in energy (i.e., have the same energy), with each state uniquely labeled by the quantum numbers (n, ℓ, m_ℓ, m_s). The degeneracy of a level is the number of states with the energy of that level. For the isotropic harmonic oscillator, the degeneracy of a level for a given value of N in Eq. (6.4), including a factor of 2 for the two possible values of m_s, is given by

$$\Omega_N = (N + 1)(N + 2) \tag{6.7}$$

Equation (6.3) describes a spectrum of evenly spaced levels, separated by an energy gap

$$\left(N + 1 + \frac{3}{2}\right)\hbar\omega_0 - \left(N + \frac{3}{2}\right)\hbar\omega_0 = \hbar\omega_0 \tag{6.8}$$

This spectrum of levels can also be thought of as groups of states bunched together and separated by gaps of energy $\hbar\omega_0$, where each individual state can hold at most one nucleon of a given type (neutron or proton). Now imagine building up a nucleus in this independent particle model by filling those states with neutrons (and a separate set of states with protons), starting with the lowest energy states and going up in energy. Table 6.1 shows the result of this filling operation. Thus, for example, the 9th through the 20th particles occupy the $N = 2$ level, which can hold $\Omega_2 = 12$ particles and has a single-particle energy of $7/2\hbar\omega_0$.

Table 6.1 Properties of the
isotropic harmonic oscillator
levels. The last column shows
the range of particle number
of a given type (proton or
neutron) that each level
(labeled by the value of N
defined by Eq. (6.4)) can
contain

N	Ω_N	$E_{n\ell}$	Nucleons
0	2	$\frac{3}{2}\hbar\omega_0$	$1\dots2$
1	6	$\frac{5}{2}\hbar\omega_0$	$3\dots8$
2	12	$\frac{7}{2}\hbar\omega_0$	$9\dots20$
3	20	$\frac{9}{2}\hbar\omega_0$	$21\dots40$
4	30	$\frac{11}{2}\hbar\omega_0$	$41\dots70$
5	42	$\frac{13}{2}\hbar\omega_0$	$71\dots112$
6	56	$\frac{15}{2}\hbar\omega_0$	$113\dots169$

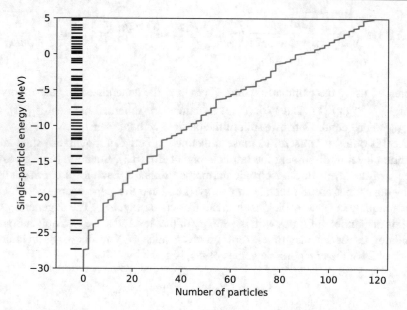

Fig. 6.1 Single-particle energy as a function of the number of particles for protons in ^{240}Pu. Data taken from the RIPL database

6.2   The Strutinsky Shell Correction Method Revisited

Figure 6.1 shows the single-particle energy of the level occupied by the last added particle, as a function of the number of particles added for the ^{240}Pu nucleus. As particles are added, they fill a level until its degeneracy is exhausted and the next particle starts filling the next level up, causing a jump in energy in the plot. The result is the staircase function plotted in Fig. 6.1 for protons and Fig. 6.2 for neutrons.

The first step in the Strutinsky procedure is to find a smooth function that approximates the staircase shape in Figs. 6.1 and 6.2. For this, we will use the prescription in [6]. The idea is to expand the staircase function in an infinite series where the terms become increasingly oscillatory. Thus, keeping the first few terms in the series yields a smooth function while the remaining higher-order terms generate the irregular part of the plot in Figs. 6.1 and 6.2. The mathematical

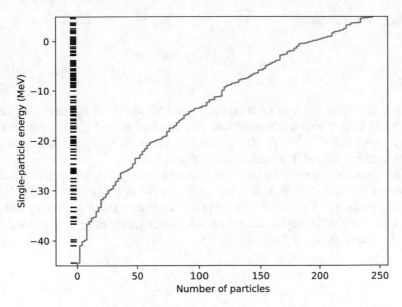

Fig. 6.2 Single-particle energy as a function of the number of particles for neutrons in ^{240}Pu. Data taken from the RIPL database

separation into smooth and oscillatory behaviors is at the heart of the Strutinsky procedure, with the smooth trend identified with the classical liquid-drop model and the oscillatory behavior identified with a quantum-mechanical correction for shell effects. Hermite polynomials $H_m(x)$ [1, 19] have precisely the desired property whereby they become increasingly oscillatory with increasing polynomial order. So as not to interrupt the flow of the discussion, we simply quote the result here and leave its derivation as an exercise for the student (see Problems 90–94). We express the result as the averaged (i.e., "smoothed") number of particles as a function of the energy of the last filled single-particle level (the plots in Fig. 6.1 and 6.2 are the inverse of this function, which amounts to exchanging the x and y axes in the figures) [6]

$$\bar{n}\,(\varepsilon) = \sum_{n=1}^{\infty} \left\{ \frac{1}{2}\,[1 + \mathrm{erf}\,(u_n)] - e^{-u_n^2} \sum_{m=1}^{p} c_m H_{m-1}\,(u_n) \right\} \qquad (6.9)$$

where

$$u_n \equiv \frac{\varepsilon - \varepsilon_n}{\gamma} \qquad (6.10)$$

is an energy dependent quantity defined in terms of the single-particle energies ε_n and a positive scaling factor γ. The scaling factor γ and the integer number p in Eq. (6.9) are chosen to give the desired amount of smoothing. The c_m coefficients in Eq. (6.9) are given by

$$c_m = \begin{cases} \dfrac{(-1)^{m/2}}{\sqrt{\pi}\,2^m\,(m/2)!} & m \text{ even} \\ 0 & m \text{ odd} \end{cases} \tag{6.11}$$

and erf (x) denotes the error function [1]. Figure 6.3 shows the staircase function of Fig. 6.1 overlayed with a smoothed function obtained using Eq. (6.9) with $p = 6$ and $\gamma = 41 \times 240^{-1/3}$ MeV. The procedure is applied in Fig. 6.4 to the neutron states from Fig. 6.2, with the same values of p and γ.

The Hermite polynomial expansion technique above can be applied to other quantities of interest as well. It is especially useful to apply it to the density of single-particle states, $g(\varepsilon)$, which counts the number of states per energy interval (see Exercise 92). This state density can be expressed formally in terms of Dirac delta functions [6]

$$g(\varepsilon) = \sum_i \delta(\varepsilon - \varepsilon_i) \tag{6.12}$$

where the summation extends over all single-particle states with energies ε_i. Note that, because of the degeneracy of the levels, the individual terms in the sum can appear more than once. Thus, in our isotropic oscillator toy model, we have (see Table 6.1)

$$g(\varepsilon) = 2\delta\left(\varepsilon - \frac{3}{2}\hbar\omega_0\right) + 6\delta\left(\varepsilon - \frac{5}{2}\hbar\omega_0\right) + \ldots \tag{6.13}$$

In order to perform the expansion and extract the smooth component of $g(\varepsilon)$, it is useful to reformulate it to include the scaling factor γ,

$$g(\varepsilon) = \frac{1}{\gamma} \sum_i \delta\left(\frac{\varepsilon - \varepsilon_i}{\gamma}\right) \tag{6.14}$$

Note that the integral of Eq. (6.14) over any energy interval yields the same result as Eq. (6.12) and therefore the two expressions are functionally equivalent. We find the smooth component (see Exercise 92)

$$\bar{g}(\varepsilon) = \frac{1}{\gamma} \sum_{n=1}^{\infty} e^{-u_n^2} \sum_{m=0}^{p} c_m H_m(u_n) \tag{6.15}$$

with u_n given by Eq. (6.10) and c_m by Eq. (6.11). Once $\bar{g}(\varepsilon)$ has been calculated, we can use it to estimate any number of average quantities by using it as a weight in an integral. For example, the average number of particles in Eq. (6.9) can also be

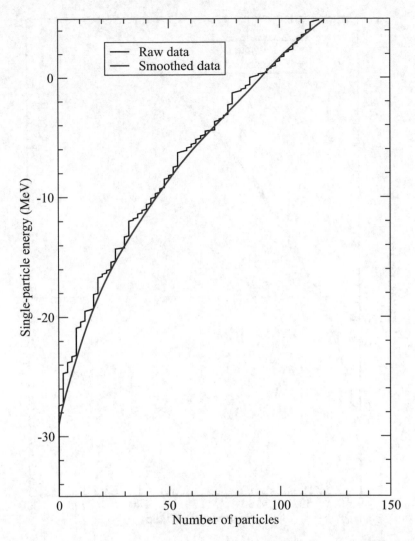

Fig. 6.3 Strutinsky smoothing procedure applied to Eq. (6.9) for the proton states in ^{240}Pu

written as (see Exercise 93)

$$\bar{n}\left(\varepsilon\right) = \int_{-\infty}^{\varepsilon} d\varepsilon' \, \bar{g}\left(\varepsilon'\right) \tag{6.16}$$

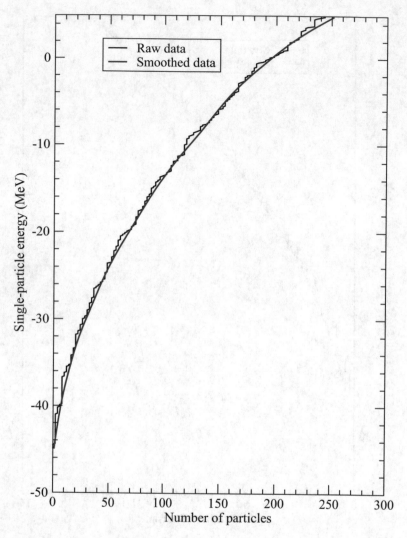

Fig. 6.4 Strutinsky smoothing procedure applied to Eq. (6.9) for the neutron states in ^{240}Pu

For a nucleus with a given number $\mathcal{N}$ of particles (which can be either protons or neutrons), we can find the average energy $\bar{\lambda}$ of the last filled level through

$$\mathcal{N} = \bar{n}\left(\bar{\lambda}\right)$$
$$= \int_{-\infty}^{\bar{\lambda}} d\varepsilon\, \bar{g}\left(\varepsilon\right) \tag{6.17}$$

We can use a very similar form to obtain the exact energy of the last filled level (assuming all levels are filled to capacity)

$$
\begin{aligned}
\mathcal{N} &= n\,(\lambda) \\
&= \int_{-\infty}^{\lambda} d\varepsilon\, g\,(\varepsilon)
\end{aligned}
\tag{6.18}
$$

with $g\,(\varepsilon)$ given by Eq. (6.12). The quantity λ is also called the Fermi energy. It is straightforward to check that, by property of Dirac delta functions, Eq. (6.18) reduces to the self-evident form

$$
\mathcal{N} = \sum_{i=1}^{\mathcal{N}} 1
\tag{6.19}
$$

We now have all the necessary ingredients to calculate the Strutinsky shell correction to the LDM energy.

In the spirit of Eq. (6.18), we define the exact shell model energy [41]

$$
\begin{aligned}
E_{\text{sh}} &\equiv \int_{-\infty}^{\lambda} d\varepsilon\, \varepsilon g\,(\varepsilon) \\
&= \sum_{\varepsilon_i \le \lambda} \varepsilon_i
\end{aligned}
\tag{6.20}
$$

where the summation extends over those single-particle states below the Fermi energy λ. Similarly, we define the averaged shell model energy

$$
\bar{E}_{\text{sh}} \equiv \int_{-\infty}^{\bar{\lambda}} d\varepsilon\, \varepsilon \bar{g}\,(\varepsilon)
\tag{6.21}
$$

with $\bar{g}\,(\varepsilon)$ given by Eq. (6.15). The shell energy defined in Eq. (6.20) is simply the sum of the energies of the occupied single-particle states. This quantity approximates the ground-state binding energy of the nucleus, but it is not exactly equal to it in general (as we will see in Sect. 6.5.3). As we did for $\bar{n}\,(\varepsilon)$ in Eq. (6.9), we can work out an explicit formula for $\bar{E}_{\text{sh}}$ [6] (see also Exercise 94),

$$
\begin{aligned}
\bar{E}_{\text{sh}} = \sum_{n=1}^{\infty} \Bigg\{ & \frac{1}{2}\varepsilon_n \left[1 + \text{erf}\,(\bar{u}_n) \right] - \gamma\, \frac{e^{-\bar{u}_n^2}}{2\sqrt{\pi}} \\
& - e^{-u_n^2} \sum_{m=1}^{p} c_m \left[\frac{1}{2}\gamma\, H_m\,(\bar{u}_n) \right. \\
& \left. \left. + \varepsilon_n H_{m-1}\,(\bar{u}_n) + m\gamma\, H_{m-2}\,(\bar{u}_n) \right] \right\}
\end{aligned}
\tag{6.22}
$$

where

$$\bar{u}_n \equiv \frac{\bar{\lambda} - \varepsilon_n}{\gamma} \tag{6.23}$$

If E_{sh} gives the exact energy of the nucleus and $\bar{E}_{sh}$ a "smoothed" version of that energy which corresponds to the LDM prediction, then the difference

$$\Delta E_{sh} \equiv E_{sh} - \bar{E}_{sh} \tag{6.24}$$

represents a correction that can be added to the LDM energy to account for shell effects.

We illustrate this discussion of the Strutinsky method using our toy model nucleus. For this, we assume $\mathcal{N} = 20$ particles which fill the $N = 0, 1, 2$ levels (see Table 6.1), and we take $\hbar\omega_0 = 1$ MeV. Using Eq. (6.20), we calculate the exact shell energy

$$\begin{aligned} E_{sh} &= 2 \times \frac{3}{2}\hbar\omega_0 + 6 \times \frac{5}{2}\hbar\omega_0 + 12 \times \frac{7}{2}\hbar\omega_0 \\ &= 60\hbar\omega_0 \\ &= 60\,\text{MeV} \end{aligned} \tag{6.25}$$

Next, we use $\bar{n}(\varepsilon)$ from Eq. (6.9) with $p = 4$ and $\gamma = 1$ MeV and vary $\bar{\lambda}$ in Eq. (6.17) until $\bar{n}(\bar{\lambda})$ matches the number of particles, $\mathcal{N} = 20$. We find $\bar{\lambda} = 3.924$ MeV. Using this value of $\bar{\lambda}$ in Eq. (6.22), we find the average shell model energy $\bar{E}_{sh} = 58.973$ MeV, and from there the Strutinsky shell correction given by Eq. (6.24) is $\Delta E_{sh} = (60 - 58.973)$ MeV $= 1.027$ MeV.

6.3   Pairing

As previously noted, the shell model energy in Eq. (6.20) is not quite the same thing as the ground-state energy of the nucleus. One reason for this discrepancy is that it does not account for the observed tendency for like nucleons in the ground state of the nucleus to pair off by aligning their angular momenta in opposite directions. Thus, two paired protons or two paired neutrons will occupy *time-reversed states*, such that one state has an angular momentum projection K, and the other $-K$ on the symmetry axis of the nucleus, as depicted in Fig. 6.5. One of the more striking consequences of this pairing is that all nuclei with even numbers of protons and neutrons have zero angular momentum in their ground state, without exception. Another observed manifestation of pairing is that even-even nuclei near closed shells

Fig. 6.5 Schematic
representation of nucleons in
time-reversed orbitals

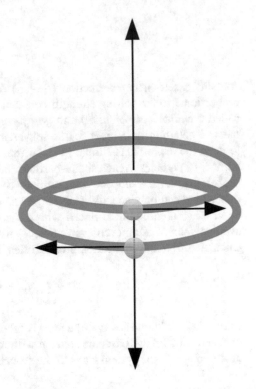

typically display an energy gap of 1–2 MeV between their ground state and first
excited state.

We will eschew the more rigorous derivation of the pairing formalism which
can be found elsewhere [8, 20, 31, 39, 41, 42] and focus instead on the concepts
and formulas that can be readily applied. We will also restrict ourselves to the
simplest form of the theory developed by Bardeen, Cooper, and Schrieffer (BCS) for
superconductors [3] using a "pure pairing force" [41] of strength G. BCS theory can
be applied once sets of proton and neutron single-particle states have been obtained.
For a nucleus with axial symmetry, the single-particle states with energies ε_ν are
doubly degenerate, so that a state (with projection quantum number K_ν) and its
time-reversed partner (with projection quantum number $-K_\nu$) have the same energy
ε_ν. BCS theory then assigns a probability v_ν^2 that the time-reversed states at energy
ε_ν are occupied, and a probability u_ν^2 that they are unoccupied (with $u_\nu^2 + v_\nu^2 = 1$),
given by

$$u_\nu^2 = \frac{1}{2}\left(1 + \frac{\varepsilon_\nu' - \lambda}{\sqrt{\left(\varepsilon_\nu' - \lambda\right)^2 + \Delta^2}}\right)$$

$$v_\nu^2 = \frac{1}{2}\left(1 - \frac{\varepsilon_\nu' - \lambda}{\sqrt{\left(\varepsilon_\nu' - \lambda\right)^2 + \Delta^2}}\right)$$

(6.26)

where

$$\varepsilon_\nu' \equiv \varepsilon_\nu - G v_\nu^2 \tag{6.27}$$

contains a self-energy correction $(-G v_\nu^2)$ to the single-particle energy which is proportional to the pairing strength constant G and where the *Fermi level* λ (also called the *chemical potential*) and *gap parameter* Δ are unknown quantities that must be determined. In practice, the self-energy correction, $-G v_\nu^2$, is often ignored to keep the expressions simple [41], so that $\varepsilon_\nu' = \varepsilon_\nu$ in Eq. (6.26). Note that the number of particles in each single-particle state has been replaced with a population probability. As a result, the total number of particles cannot be known precisely. Instead we can only specify its average value, with the understanding that we are describing an ensemble of nuclei with a distribution of particle numbers with this average value. For an even-even nucleus, the average number of particles can be calculated from the occupation probabilities in Eq. (6.26) as

$$2 \sum_{\nu > 0} v_\nu^2 = N \tag{6.28}$$

where "$\nu > 0$" indicates that the sum is taken over only one member of the time-reversed pairs (in this case, the states ν with $K_\nu > 0$). In addition, the minimization of the energy of the system leads to the so-called gap equation,

$$\sum_{\nu > 0} \frac{1}{\sqrt{\left(\varepsilon_\nu' - \lambda\right)^2 + \Delta^2}} = \frac{2}{G} \tag{6.29}$$

where G is the strength of the pairing interaction and ε_ν' is given by Eq. (6.27). Thus, given a set of single-particle states with energies ε_ν, a number of particles N occupying a subset of those states, and a pairing strength G, Eqs. (6.28) and (6.29) can be solved simultaneously to yield values for the unknown parameters λ and Δ. The numerical solution of this system of two equations can be obtained using an iterative Newton-Raphson method, and a code to do this can be found, for example, in Section 9.8 of [23].

The BCS formalism turns the problem of describing particles interacting through a residual pairing interaction into that of a system of independent *quasiparticles* with energies

$$E_\nu = \sqrt{\left(\varepsilon_\nu' - \lambda\right)^2 + \Delta^2} \tag{6.30}$$

thereby restoring the simplicity of the independent particle model. This simplification comes at the cost of sacrificing exact particle properties, such as the numbers of protons and neutrons, which can only be constrained to have some given average value.

The ground-state energy of the nucleus in the BCS formalism is given by [13, 23]

$$E_{GS} = 2 \sum_{\nu>0} v_\nu^2 \varepsilon_\nu' - \frac{\Delta^2}{G} \qquad (6.31)$$

In order not only to apply the Strutinsky procedure to the pairing part of the energy of the nucleus but also to satisfy ourselves of internal consistency of the BCS formalism, we next look at the limit of Eq. (6.31) where pairing is turned off. The limit without pairing is obtained by letting $\Delta \to 0$ in Eqs. (6.31) and (6.26), which leads to

$$E_{GS}^{(0)} = \sum_{\nu>0} 2 \left(v_\nu^{(0)}\right)^2 \varepsilon_\nu' \qquad (6.32)$$

where the occupations $\left(v_\nu^{(0)}\right)^2$ in the $\Delta \to 0$ limit are 1 for $\varepsilon_\nu' < \lambda$ and 0 for $\varepsilon_\nu' > \lambda$. We neglect the effect of $\Delta \to 0$ on the corrected single-particle energies in Eq. (6.27) and keep the same ε_ν' values in both Eqs. (6.31) and (6.32). In effect, $E_{GS}^{(0)}$ is the shell energy we calculated in Eq. (6.20). The *pairing correlation* energy is then given by [13]

$$E_{\text{pc}} \equiv E_{GS} - E_{GS}^{(0)}$$
$$= \sum_{\nu>0} 2 \left[v_\nu^2 - \left(v_\nu^{(0)}\right)^2 \right] \tilde{\varepsilon}_\nu - \frac{\Delta^2}{G} \qquad (6.33)$$

The pairing correlation energy can also be thought of as the correction that must be added to the shell energy to obtain a ground-state energy that takes into account the effect of the residual pairing interaction.

Just as with shell corrections, we can also define a correction to the LDM energies due to pairing effects. The pairing correction is given by the difference between the pairing correlation energy in Eq. (6.33) and a pairing correlation energy calculated with a smoothed density of states,

$$\Delta E_{\text{P}} = E_{\text{pc}} - \bar{E}_{\text{pc}} \qquad (6.34)$$

To calculate the smoothed energy $\bar{E}_{\text{pc}}$, we assume a uniform density of single-particle states in some range around the Fermi energies, e.g., from $\lambda - \delta\lambda$ to $\lambda + \delta\lambda$ for some large enough $\delta\lambda > 0$. For convenience, and without loss of generality, we will set $\lambda = 0$. We will also ignore the self-energy term in Eq. (6.27). The gap equation, Eq. (6.29), then becomes

$$\frac{1}{2} g \int_{-\delta\lambda}^{\delta\lambda} \frac{d\varepsilon}{\sqrt{\varepsilon^2 + \bar{\Delta}^2}} = \frac{2}{G} \qquad (6.35)$$

where g is the average density of states (a number, which does not depend on the energy ε), $\bar{\Delta}$ is a gap parameter, and where the factor of $1/2$ in front of the integral comes from the fact that summation in Eq. (6.29) is only over the states $\nu > 0$. The integral can be evaluated explicitly, and we obtain

$$g \sinh^{-1}\left(\frac{\delta\lambda}{\bar{\Delta}}\right) = \frac{2}{G} \tag{6.36}$$

This equation can be solved for $\bar{\Delta}$,

$$\bar{\Delta} = \frac{\delta\lambda}{\sinh\left(\frac{2}{gG}\right)} \tag{6.37}$$

Next, we follow the same procedure for the pairing correlation energy in Eq. (6.33). We have

$$\begin{aligned}
\bar{E}_{\mathrm{pc}} &= \frac{1}{2}g\left[\int_{-\delta\lambda}^{\delta\lambda} d\varepsilon\left(1 - \frac{\varepsilon}{\sqrt{\varepsilon^2 + \bar{\Delta}^2}}\right)\varepsilon - 2\int_{-\delta\lambda}^{0} d\varepsilon\,\varepsilon\right] - \frac{\bar{\Delta}^2}{G} \\
&= \frac{g}{2}\left[\delta\lambda^2 - \delta\lambda\sqrt{\delta\lambda^2 + \bar{\Delta}^2} + \bar{\Delta}^2\sinh^{-1}\left(\frac{\delta\lambda}{\bar{\Delta}}\right)\right] - \frac{\bar{\Delta}^2}{G}
\end{aligned} \tag{6.38}$$

Using Eq. (6.36) and assuming $\delta\lambda \gg \bar{\Delta}$ leaves a much simpler expression (see, e.g., Section 3.5 in [8])

$$\bar{E}_{\mathrm{pc}} \approx -\frac{g}{4}\bar{\Delta}^2 \tag{6.39}$$

which can then be used in Eq. (6.34) to calculate the pairing correction.

6.4 Level Densities

Level densities play an important role in the description of the fission process. The decay probability of a compound nucleus into the various exit channels (e.g., fission, neutron, gamma) is proportional to the available *phase space* in that channel, which in turn is given by the density of levels.

6.4.1 The Many-Body State Density: Energy Dependence

We have so far only considered the ground state of the nucleus and relied on a simple picture to calculate its energy where single-particle states are akin to shelves and the

Fig. 6.6 Ground-state and
particle-hole excitations in an
independent particle model

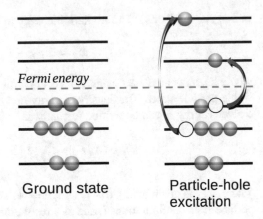

Fermi energy

Ground state   Particle-hole
excitation

ground state is constructed by placing the nucleons on those shelves starting with the
lowest and moving up, until all nucleons have been placed. We can also construct
rudimentary excitations of the nucleus within this same simple picture by moving
individual nucleons from a lower state to a higher one (assuming the higher one is
not fully occupied), leaving holes behind in the lower states. In this way, many-body
states can be constructed by arranging nucleons (and the holes they leave behind)
among the single-particle states. Ignoring pairing and any other residual interaction
between the nucleons for now, each particle promoted to a state of energy ε_p and
leaving a hole in its initial state of energy ε_h contributes an energy

$$\Delta E_{ph} = \varepsilon_p - \varepsilon_h \tag{6.40}$$

to the energy of the many-body state. Figure 6.6 depicts the ground-state configuration and some particle-hole excited states that can be constructed in this type of
basic independent particle model.

The number of many-body states that can be generated in this way increases
exponentially as a function of excitation energy. When the number of levels becomes
too great, it is more convenient to work with a state density, rather than the individual
states. Note that here we are speaking of a many-body state density, in contrast to the
single-particle densities discussed in Sect. 6.2. For n particles arranged in a generic
configuration of particles and holes, labeled "i," we calculate the shell energy as

$$E_{n,i} = \sum_\nu n_\nu^{(i)} \varepsilon_\nu \tag{6.41}$$

where the index ν runs over all single-particle states (occupied or not), ε_ν is the
energy of the state, and $n_\nu^{(i)}$ is the occupation number of the state (in the limit of no

pairing, $n_v^{(i)}$ is 0 or 1). The excitation energy in this case is

$$E_x = E_{n,i} - E_{n,0} \tag{6.42}$$

where $E_{n,0}$ is calculated using Eq. (6.41) with the nucleons occupying the lowest single-particle states. The density of many-body states for a system of A particles at a given energy E can be written formally as

$$\rho(A, E) = \sum_i \delta(E - E_{A,i}) \tag{6.43}$$

where $E_{A,i}$ is given by Eq. (6.41) with $n = A$. The summation runs over all possible particle-hole configurations i, and as a result, the number of terms in the sum grows very rapidly with E when counting states in some energy interval about E.

There is a very clever procedure [2, 5, 51][1] to evaluate the right-hand side of Eq. (6.43) which consists of (1) taking the Laplace transform of $\rho(A, E)$ which gives the so-called grand-canonical partition function of the system,

$$Z(\alpha, \beta) \equiv \int_0^\infty dA \int_0^\infty dE \, \rho(A, E) \, e^{\alpha A - \beta E} \tag{6.44}$$

(2) factorizing the partition into the product of a relatively small number of terms, and (3) taking the inverse Laplace transform,

$$\rho(A, E) = \left(\frac{1}{2\pi i}\right)^2 \int_{-i\infty}^{i\infty} d\alpha \int_{-i\infty}^{i\infty} d\beta \, Z(\alpha, \beta) \, e^{-\alpha A + \beta E} \tag{6.45}$$

of the factorized partition function to recover the state density $\rho(A, E)$ [5, 27, 49]. The inverse Laplace transform in the last step can be carried out, for example, using the saddle-point approximation method [33, 35]. We leave the details of this approach to the cited references and give instead the final results (see also Exercises 95–98). The partition function is given by

$$Z(\alpha, \beta) = \left[1 + e^{\alpha - \beta \varepsilon_1}\right] \times \left[1 + e^{\alpha - \beta \varepsilon_2}\right] \times \ldots \times \left[1 + e^{\alpha - \beta \varepsilon_s}\right]$$

$$= \prod_{v=1}^{s} \left[1 + e^{\alpha - \beta \varepsilon_v}\right] \tag{6.46}$$

where α and β are parameters to be determined and s is the number of single-particle states, where each state can hold at most one particle (if a level has a degeneracy $\Omega > 1$, it can be thought of as Ω states of the same energy each with degeneracy 1

---

[1] In more technical terms, this is a thermodynamic approach in the grand-canonical ensemble to the state density problem.

for the purpose of evaluating the right-hand side of Eq. (6.46)). The state density is
then given by

$$\rho\left(A, E\right) \approx \frac{e^{S_0}}{2\pi\sqrt{D_0}} \tag{6.47}$$

where we have defined the *entropy*

$$S_0 \equiv \ln Z\left(\alpha_0, \beta_0\right) - \alpha_0 A + \beta_0 E \tag{6.48}$$

evaluated at the point $(\alpha, \beta) = (\alpha_0, \beta_0)$, which is determined by solving the system
of equations

$$\frac{\partial}{\partial\beta} \ln Z\left(\alpha, \beta\right) + E = 0$$
$$\frac{\partial}{\partial\alpha} \ln Z\left(\alpha, \beta\right) - A = 0 \tag{6.49}$$

and D_0 is the determinant of the 2×2 matrix of 2nd order derivatives (also known
as a *Hessian*),

$$D_0 = \det \begin{pmatrix} \frac{\partial^2}{\partial\alpha^2} \ln Z\left(\alpha, \beta\right) & \frac{\partial^2}{\partial\alpha\partial\beta} \ln Z\left(\alpha, \beta\right) \\ \frac{\partial^2}{\partial\beta\partial\alpha} \ln Z\left(\alpha, \beta\right) & \frac{\partial^2}{\partial\beta^2} \ln Z\left(\alpha, \beta\right) \end{pmatrix} \tag{6.50}$$

evaluated at $(\alpha, \beta) = (\alpha_0, \beta_0)$.

6.4.2 The Many-Body State Density: Angular Momentum and Parity Dependence

The state density calculated above is only one component of the full level density
which must also depend on the angular momentum and parity. The level density, as
a function of energy E, total angular momentum J, and parity Π is usually written

$$\rho\left(E, J, \Pi\right) = \rho\left(E\right) F\left(E, J\right) P\left(E, \Pi\right) \tag{6.51}$$

where $\rho\left(E\right)$ is the energy dependent part given by Eq. (6.47) (we omit the particle
number argument A to simplify the notation), the function $F\left(E, J\right)$ contains the
dependence on angular momentum, and $P\left(E, \Pi\right)$ is the probability of a state at
energy E having parity Π.

The angular momentum dependence of the level density can be calculated using a statistical argument due to Bethe [4] which gives

$$F(E, J) = \frac{2J + 1}{2\sqrt{2\pi}\sigma^3} \exp\left[-\frac{(J + 1/2)^2}{2\sigma^2}\right] \tag{6.52}$$

where σ^2 is known as the *spin cutoff* parameter and usually depends on energy $\sigma^2 = \sigma^2(E)$. The dependence on parity in Eq. (6.51) can be modeled [48], but often one makes the simplifying assumption that, at a sufficiently high excitation energy E, there are as many positive-parity as negative-parity states, in which case (but see also [2])

$$P(E, \Pi) = \frac{1}{2} \tag{6.53}$$

6.4.3   The Many-Body State Density: Collective Enhancements

The states counted by the state density in Eq. (6.47) consist of multi-particle multi-hole excitations. There are, however, residual interactions that can couple these *intrinsic* states, giving rise to *collective* (or coherent) vibrations and rotations of the nucleus. This coupling in turn will tend to bring down in energy the collective states, thereby affecting the level density calculation. A common prescription to try to account for the effect of the collective levels consists in adding a collective enhancement factor $K(E)$ to the level density formula,

$$\rho(E, J, \Pi) = \rho(E) F(E, J) P(E, \Pi) K(E) \tag{6.54}$$

This enhancement factor can be written as the contribution from rotational and vibrational levels,

$$K(E) = K_{rot}(E) K_{vib}(E) \tag{6.55}$$

There are various prescriptions for estimating these enhancement factors [9, 27], but we will not go any further into the details of this formalism here.

6.4.4   Additional Improvements

Finally, we briefly discuss two additional improvements that can be made to the level density formulation described thus far: accounting for the effects of pairing and taking into account both types of nucleons (protons and neutrons). Including pairing into the formalism is not much more complicated than what has already been shown in Sect. 6.4.1; a partition function similar to the one in Eq. (6.46) is obtained in terms of quasiparticle energies using the same procedure outlined in the section. The inclusion of both types of particles amounts to replacing the entropy S_0 in Eq. (6.47) by the sum of the proton and neutron entropies, each given by Eq. (6.48), and extending the denominator term D_0 into the determinant of a 3×3 matrix whose elements contain both proton and neutron contributions [2, 5, 51].

There are other, even more sophisticated methods to calculate level densities: the combinatorial method [24, 48], the moment method [34], and the shell model Monte Carlo method [26] to name a few. We will not discuss these methods here, but an instructive overview can be found in [2].

6.5 Advanced Topics

6.5.1 More Realistic Single-Particle Models

Although the harmonic oscillator potential offers the convenience of analytic solutions, these solutions do not agree well with experimental data. In particular, one of the crucial contributions made by Maria Goeppert Mayer was to show the importance of the spin-orbit interaction to reproduce realistic shell gaps among the single-particle states [15–17]. These shell gaps explain much of the stability and spectroscopy of nuclei with magic numbers of protons and neutrons (e.g., 2, 8, 20, 28, 50, 82, and 126). In addition, because of the short range of the nuclear force (~ 1 fm) compared to the nuclear radius ($r_0 A^{1/3} \sim 7.5$ fm for a typical actinide), we expect the nuclear potential to be relatively constant inside the nucleus and to fall off rapidly beyond its surface. A harmonic oscillator potential does not fit this expected behavior. By contrast, a Woods-Saxon potential [50] has a more realistic shape. In spherical coordinates (r, θ, φ), it has the form

$$V(r) = -\frac{V_0}{1 + \exp\left(\frac{r - R_0}{a}\right)} \tag{6.56}$$

with $V_0, a > 0$, and $R_0 = r_0 A^{1/3}$. The Woods-Saxon potential in Eq. (6.56) is valid for spherical nuclei but can be readily generalized to deformed systems by replacing the parameter R_0 with a function of spherical-coordinate angles θ and φ (i.e., $R_0 \rightarrow R(\theta, \varphi)$).

Alternately, the axial harmonic oscillator potential can be modified to mimic the effects of a Woods-Saxon shape and to include a spin-orbit coupling contribution, as was done by Sven G. Nilsson in 1955 [38]. The Nilsson potential has the following form in Cartesian coordinates [38, 41]

$$V(x, y, z) = \frac{1}{2}m\omega_\perp^2 \left(x^2 + y^2\right) + \frac{1}{2}m\omega_z^2 z^2 + C\boldsymbol{\ell} \cdot \mathbf{s} + D\left(\boldsymbol{\ell}^2 - \left\langle \boldsymbol{\ell}^2 \right\rangle\right) \tag{6.57}$$

where ℓ is the orbital angular momentum of the nucleon and $\mathbf{s}$ its intrinsic spin. The quantity $\langle \ell^2 \rangle = N(N+3)/2$ is an average taken over the shell N [41]. The values of the C and D parameters are determined by experimental data. The parameters C, D, $\omega_\perp$, and ω_z in Eq. (6.57) take on different values as a function of the deformation of the nucleus.

The potentials discussed above can be used to calculate single-particle states for simple shapes of the nucleus. In the case of fission, however, the nucleus explores more exotic shapes eventually leading to its division into two (or more) nuclei. Therefore, and especially in the later stages of fission, the potentials discussed so far may not be appropriate.

One way of generating a potential $V(\mathbf{r})$ that closely follows the shape of the nucleus is to convolve a "generating potential" $g(\mathbf{r}')$, which is constant inside the sharply defined surface of the nucleus and zero outside, with a folding function $f(\mathbf{r}, \mathbf{r}') = f(|\mathbf{r} - \mathbf{r}'|)$ [22]

$$V(\mathbf{r}) = \int d^3 r' \, g(\mathbf{r}') \, f(\mathbf{r}, \mathbf{r}') \tag{6.58}$$

In effect the right-hand side of Eq. (6.58) is proportional to the integral of the folding function taken over the volume of the nucleus. A realistic description of the nucleus should allow for a diffuse nuclear surface with the nucleon density falling off quickly (but not sharply) with the distance from the surface. This type of nucleon distribution, constant inside and falling off outside, is known as a leptodermous distribution [22, 36, 37]. One way of allowing for such a diffuse surface is to use a short-ranged folding function in Eq. (6.58). For example, Bolsterli et al. [6, 22] used a Yukawa function

$$f(r) \equiv \frac{1}{4\pi a^3} \frac{e^{-r/a}}{r/a} \tag{6.59}$$

with parameter $a > 0$. The folding procedure then yields the nuclear potential

$$V_1(\mathbf{r}) = -\frac{V_0}{4\pi a^3} \int_{\text{vol}} d^3 r' \frac{e^{-|\mathbf{r} - \mathbf{r}'|/a}}{|\mathbf{r} - \mathbf{r}'|/a} \tag{6.60}$$

where the three-dimensional integral is carried out over the volume of the nucleus. The choice of a Yukawa function to perform the folding operation is not unique. Other common choices of short-ranged folding functions include the Yukawa plus an exponential and the Gaussian distribution [22].

For a general shape of the nucleus, it may not be possible to obtain an analytical form of the folding integrals. In the very special case of a spherical shape though, the integral in Eq. (6.60) can be evaluated explicitly giving [6]

$$V_1(r) = \begin{cases} -V_0 \left[1 - \left(1 + \frac{R_0}{a} \right) e^{-R_0/a} \frac{\sinh(r/a)}{r/a} \right] & r \leq R_0 \\ -V_0 \left[\frac{R_0}{a} \cosh \frac{R_0}{a} - \sinh \frac{R_0}{a} \right] \frac{e^{-r/a}}{r/a} & r > R_0 \end{cases} \tag{6.61}$$

while the integral of the Coulomb potential gives the well-known form for a spherical uniform charge distribution [6],

$$V_C\left(r\right) = \begin{cases} \frac{1}{2}\frac{Ze^2}{R_0}\left[3 - \left(\frac{r}{R_0}\right)^2\right] & r \leq R_0 \\ \frac{Ze^2}{r} & r > R_0 \end{cases}$$

(6.62)

with $e^2 \approx 1.44\,\text{MeV} \cdot \text{fm}$.

6.5.2 Matrix Formulation of the Schrödinger Equation

The appeal of the harmonic oscillator potential discussed in Chap. 6 is that exact solutions can be written down explicitly (see Sect. 6.1). For the Nilsson and folding potentials (Sect. 6.5.1), this will not generally be the case, and a numerical solution must be sought instead. One way to proceed is to solve the Schrödinger equation by using a set of basis functions. For a deformed axially symmetric nucleus, the spherical solutions given by Eq. (6.5) are not very convenient, while the deformed harmonic oscillator functions $\psi_{n_r,\Lambda,n_z,\sigma}\left(\rho, \varphi, z\right)$, with

$$\psi_{n_r,\Lambda,n_z}\left(\rho, \varphi, z\right) = \Phi_{n_r,|\Lambda|}\left(\rho\right)\frac{e^{i\Lambda\varphi}}{\sqrt{2\pi}}\Phi_{n_z}\left(z\right)\chi_\sigma$$

(6.63)

where

$$\Phi_{n_r,|\Lambda|}\left(\rho\right) = \frac{1}{b_\perp}\sqrt{\frac{2n_r!}{\left(n_r + |\Lambda|\right)!}}\left(\frac{\rho^2}{b_\perp^2}\right)^{|\Lambda|/2}\exp\left(-\frac{\rho^2}{2b_\perp^2}\right)L_{n_r}^{|\Lambda|}\left(\frac{\rho^2}{b_\perp^2}\right)$$

(6.64)

and

$$\Phi_{n_z}\left(z\right) \equiv \frac{1}{\sqrt{b_z\sqrt{\pi}2^{n_z}n_z!}}\exp\left(-\frac{z^2}{2b_z^2}\right)H_{n_\xi}\left(\frac{z}{b_z}\right)$$

(6.65)

form a more useful set of orthonormal basis states [47].

The solutions of the Schrödinger equation can then be expanded on these basis states,

$$\Psi\left(\rho, \theta, \varphi\right) = \sum_{n_r,\Lambda,n_z,\sigma} c_{n_r,\Lambda,n_z,\sigma}\psi_{n_r,\Lambda,n_z,\sigma}\left(\rho, \varphi, z\right)$$

(6.66)

where the coefficients $c_{n_r,\Lambda,n_z,\sigma}$ are unknown quantities to be determined. To make the notation more compact, we label the quantum numbers $(n_r, \Lambda, n_z, \sigma)$ by a single index (e.g., $i \equiv \left(n_r^{(i)}, \Lambda^{(i)}, n_z^{(i)}, \sigma^{(i)} \right)$). The Schrödinger equation (Eq. (6.1)) can be written in the form

$$H\psi = E\psi \tag{6.67}$$

where

$$H \equiv -\frac{\hbar^2}{2m} \nabla^2 + V\,(\mathbf{r}) \tag{6.68}$$

is the *Hamiltonian* of the system. Equation (6.67) can also be expressed in a very useful matrix form,

$$\begin{pmatrix} H_{1,1} & H_{1,2} & \cdots \\ H_{2,1} & H_{2,2} & \cdots \\ \vdots & \vdots & \ddots \end{pmatrix} \begin{pmatrix} c_1 \\ c_2 \\ \vdots \end{pmatrix} = \varepsilon \begin{pmatrix} c_1 \\ c_2 \\ \vdots \end{pmatrix} \tag{6.69}$$

where the c_i are the coefficients in Eq. (6.66) and where the elements of the *Hamiltonian matrix* on the left-hand side consist of a kinetic and potential term:

$$H_{i,j} = T_{i,j} + V_{i,j} \tag{6.70}$$

The kinetic term is given by the integral of the kinetic energy operator between the basis functions,

$$T_{i,j} = \int_0^\infty \rho \, d\rho \int_0^{2\pi} d\varphi \int_{-\infty}^\infty dz \, \psi^\dagger_{n_r^{(i)}, \Lambda^{(i)}, n_z^{(i)}, \sigma^{(i)}} (\rho, \varphi, z)$$
$$\times \left(-\frac{\hbar^2}{2m} \nabla^2 \right) \psi_{n_r^{(j)}, \Lambda^{(j)}, n_z^{(j)}, \sigma^{(j)}} (\rho, \varphi, z) \tag{6.71}$$

where the adjoint operation (denoted by "†") indicates that we must take the complex conjugate of the spatial part of the wave function Ψ and the transpose of the spin part of the wave function, which is represented by a vector:

$$\begin{pmatrix} 1 \\ 0 \end{pmatrix} \text{ for spin up,} \quad \begin{pmatrix} 0 \\ 1 \end{pmatrix} \text{ for spin down} \tag{6.72}$$

Then, for the spin-up state, for example,

$$\begin{pmatrix} 1 \\ 0 \end{pmatrix}^\dagger = (1\ 0) \tag{6.73}$$

The matrix elements of the potential are likewise given by an integral

$$
\begin{aligned}
V_{i,j} = \int_0^\infty \rho d\rho \int_0^{2\pi} d\varphi \int_{-\infty}^\infty dz\, \psi_{n_r^{(i)},\Lambda^{(i)},n_z^{(i)},\sigma^{(i)}}^\dagger (\rho,\varphi,z) \\
\times V(\rho,\varphi,z)\, \psi_{n_r^{(j)},\Lambda^{(j)},n_z^{(j)},\sigma^{(j)}} (\rho,\varphi,z)
\end{aligned}
\tag{6.74}
$$

which, in general, must be evaluated numerically (see, e.g., [10]). In practice, only a finite number of harmonic oscillator states $\psi_{n_r,\Lambda,n_z,\sigma}(\rho,\varphi,z)$ can be included in the basis because of computational limitations, and a truncation scheme is usually adopted to limit the size of the basis. The matrix form of the Schrödinger equation, Eq. (6.69), is a well-known eigenvalue problem which can be solved by standard numerical techniques (see, e.g., [40]). The eigenvalues ε give the single-particle energies of the nucleus, and the eigenvectors are their corresponding wave functions expanded on the harmonic-oscillator basis via Eq. (6.66). In an independent particle approach, the ground state of the nucleus is then constructed by filling the lowest eigenvalue states with the A nucleons.

To illustrate the formulation of the Schrödinger equation in matrix form, consider the one-dimensional case

$$
-\frac{\hbar^2}{2m}\frac{d^2\psi(x)}{dx^2} + V(x)\psi(x) = \varepsilon\psi(x)
\tag{6.75}
$$

and suppose $\psi(x)$ is expanded as a linear combination of two basis functions $\Phi_1(x)$ and $\Phi_2(x)$:

$$
\psi(x) = c_1\Phi_1(x) + c_2\Phi_2(x)
\tag{6.76}
$$

We choose the basis functions so that they are orthonormalized, that is, they satisfy

$$
\int_{-\infty}^\infty dx\, \Phi_i^*(x)\,\Phi_j(x) = \begin{cases} 1 & i = j \\ 0 & i \neq j \end{cases}
\tag{6.77}
$$

Inserting Eq. (6.76) into Eq. (6.75), we obtain after grouping terms

$$
\begin{aligned}
\left[-\frac{\hbar^2}{2m}\frac{d^2\Phi_1(x)}{dx^2} + V(x)\Phi_1(x) \right] c_1 \\
+ \left[-\frac{\hbar^2}{2m}\frac{d^2\Phi_2(x)}{dx^2} + V(x)\Phi_2(x) \right] c_2 = \varepsilon\left[c_1\Phi_1(x) + c_2\Phi_2(x)\right]
\end{aligned}
\tag{6.78}
$$

Now, we multiply both sides of Eq. (6.78) by $\Phi_1^*(x)$ and integrate over all x to obtain

$$(T_{11} + V_{11}) c_1 + (T_{12} + V_{12}) c_2 = \varepsilon c_1 \tag{6.79}$$

where we have used Eq. (6.77) in simplifying the right-hand side and where we have defined

$$T_{ij} \equiv -\frac{\hbar^2}{2m} \int_{-\infty}^{\infty} dx\, \Phi_i^*(x)\, \frac{d^2 \Phi_j(x)}{dx^2} \tag{6.80}$$

and

$$V_{ij} \equiv \int_{-\infty}^{\infty} dx\, \Phi_i^*(x)\, V(x)\, \Phi_j(x) \tag{6.81}$$

Similarly, if we multiply Eq. (6.78) by $\Phi_2^*(x)$ and integrate over all x, we obtain

$$(T_{21} + V_{21}) c_1 + (T_{22} + V_{22}) c_2 = \varepsilon c_2 \tag{6.82}$$

Finally, Eqs. (6.79) and (6.82) can be written in compact matrix form

$$\begin{pmatrix} T_{11} + V_{11} & T_{12} + V_{12} \\ T_{21} + V_{21} & T_{22} + V_{22} \end{pmatrix} \begin{pmatrix} c_1 \\ c_2 \end{pmatrix} = \varepsilon \begin{pmatrix} c_1 \\ c_2 \end{pmatrix} \tag{6.83}$$

The same procedure can be applied to the 3D Schrödinger equation and for any number of terms in the expansion of the wave function; the dimension of the Hamiltonian matrix is then given by the number of terms in the expansion.

6.5.3   The Hartree-Fock Approximation

In our previous discussion of single-particle models, we used the time-independent Schrödinger equation (TISE) (Eq. (6.1)) to calculate single-particle states corresponding to a given potential. We discussed several types of potentials that can be used for this, including the more sophisticated folding potential method that follows the shape of the nucleus. Now, we will take a more fundamental approach, starting from protons and neutrons and some kind of interaction between them [29, 43]. In this way, the spatial distribution of the nucleons will determine the shape of the nucleus, instead of the other way around.

In the Hartree-Fock approximation [20, 21, 28, 41], each nucleon moves under the influence of a potential that represents the averaged effect of the remaining nucleons, referred to as a *mean field*. Thus, the TISE for the ith state of the system in the Hartree-Fock approximation is

$$-\frac{\hbar^2}{2m}\nabla^2\psi_i\left(\mathbf{r}\right) + V\left(\mathbf{r}\right)\psi_i\left(\mathbf{r}\right) = \varepsilon_i\psi_i\left(\mathbf{r}\right) \tag{6.84}$$

where $V\left(\mathbf{r}\right)$ is the potential generated by the combined effect of the nucleons. This type of potential is also known as a *mean field*.

We can define a spatial density distribution of the nucleons by

$$\rho\left(\mathbf{r}\right) \equiv \sum_{i\in\text{occ}} \psi_i^*\left(\mathbf{r}\right)\psi_i\left(\mathbf{r}\right) \tag{6.85}$$

where the notation "$i \in$ occ" indicates that the sum extends only over those states occupied by nucleons. If we are given an interaction $V\left(\mathbf{r}, \mathbf{r}'\right)$ between two nucleons at $\mathbf{r}$ and $\mathbf{r}'$, then it seems natural to define the potential $V\left(\mathbf{r}\right)$ in the same spirit as the folding potential by

$$V\left(\mathbf{r}\right) \equiv \int d^3r'\, \rho\left(\mathbf{r}'\right) V\left(\mathbf{r}, \mathbf{r}'\right)$$
$$= \sum_{j\in\text{occ}} \int d^3r'\, \psi_j^*\left(\mathbf{r}'\right) V\left(\mathbf{r}, \mathbf{r}'\right) \psi_j\left(\mathbf{r}'\right) \tag{6.86}$$

This is almost the potential needed in Eq. (6.84) save for one important detail: the Pauli exclusion principle. To fix this shortcoming, we start by writing the term $V\left(\mathbf{r}\right)\psi_i\left(\mathbf{r}\right)$ in Eq. (6.84) explicitly using Eq. (6.86)

$$V\left(\mathbf{r}\right)\psi_i\left(\mathbf{r}\right) = \sum_{j\in\text{occ}} \int d^3r'\, \psi_j^*\left(\mathbf{r}'\right) V\left(\mathbf{r}, \mathbf{r}'\right) \psi_j\left(\mathbf{r}'\right) \psi_i\left(\mathbf{r}\right) \tag{6.87}$$

The product $\psi_j\left(\mathbf{r}'\right)\psi_i\left(\mathbf{r}\right)$ on the right-hand side of Eq. (6.87) is the root of the problem because it does not explicitly respect antisymmetrization rules under exchange of the indices i and j (i.e., this term should change sign when i and j are exchanged). This product must therefore be replaced by $\psi_j\left(\mathbf{r}'\right)\psi_i\left(\mathbf{r}\right) - \psi_i\left(\mathbf{r}'\right)\psi_j\left(\mathbf{r}\right)$, and Eq. (6.87) then takes the properly antisymmetrized form[2]

$$V_i\left(\mathbf{r}\right)\psi_i\left(\mathbf{r}\right) = \sum_{j\in\text{occ}} \int d^3r'\, \psi_j^*\left(\mathbf{r}'\right) V\left(\mathbf{r}, \mathbf{r}'\right) \psi_j\left(\mathbf{r}'\right) \psi_i\left(\mathbf{r}\right)$$
$$- \sum_{j\in\text{occ}} \int d^3r'\, \psi_j^*\left(\mathbf{r}'\right) V\left(\mathbf{r}, \mathbf{r}'\right) \psi_i\left(\mathbf{r}'\right) \psi_j\left(\mathbf{r}\right) \tag{6.88}$$

---

[2] A more natural derivation of the Hartree-Fock equation, with antisymmetrization built in from the get-go, requires the use of second quantization, which is beyond the scope of this textbook (but see, e.g., [20, 41]).

To make the equations more compact and readable, we can define

$$\sum_{j \in \mathrm{occ}} \int d^3 r' \, \psi_j^* \left(\mathbf{r}'\right) V \left(\mathbf{r}, \mathbf{r}'\right) \psi_j \left(\mathbf{r}'\right) \psi_i \left(\mathbf{r}\right) \equiv \Gamma_D \left(\mathbf{r}\right) \psi_i \left(\mathbf{r}\right) \tag{6.89}$$

and

$$-\sum_{j \in \mathrm{occ}} \int d^3 r' \, \psi_j^* \left(\mathbf{r}'\right) V \left(\mathbf{r}, \mathbf{r}'\right) \psi_i \left(\mathbf{r}'\right) \psi_j \left(\mathbf{r}\right) \equiv \int d^3 r' \, \Gamma_E \left(\mathbf{r}, \mathbf{r}'\right) \psi_i \left(\mathbf{r}'\right)$$

$$\tag{6.90}$$

where $\Gamma_D \left(\mathbf{r}\right)$ and $\Gamma_E \left(\mathbf{r}, \mathbf{r}'\right)$ are the *direct* and *exchange* potentials. The Hartree-Fock Eq. (6.84) can now be rewritten as [41]

$$-\frac{\hbar^2}{2m} \nabla^2 \psi_i \left(\mathbf{r}\right) + \Gamma_D \left(\mathbf{r}\right) \psi_i \left(\mathbf{r}\right) + \int d^3 r' \, \Gamma_E \left(\mathbf{r}, \mathbf{r}'\right) \psi_i \left(\mathbf{r}'\right) = \varepsilon_i \psi_i \left(\mathbf{r}\right)$$

$$\tag{6.91}$$

and it must be solved for all states i to obtain the single-particle wave functions $\psi_i \left(\mathbf{r}\right)$ and energies ε_i. Note that Eq. (6.91) can no longer be written as a simple eigenvalue problem, as in Eq. (6.69), because the potentials $\Gamma_D \left(\mathbf{r}\right)$ and $\Gamma_E \left(\mathbf{r}, \mathbf{r}'\right)$ depend themselves on the wave functions $\psi_i \left(\mathbf{r}\right)$ that we are solving for. Instead, Eq. (6.91) must be solved through an iterative algorithm. First, the same trick that was used to generate Eq. (6.69) can be applied here by expanding the $\psi_i \left(\mathbf{r}\right)$ on a basis (e.g., of deformed harmonic oscillator states), and the problem is then reduced to that of finding the coefficients of that expansion. The iterative algorithm proceeds as follows:

1. Start with a guess for the expansion coefficients.
2. Solve Eq. (6.91) by casting it in matrix form (as in Eq. (6.69)) and treating it as a standard matrix eigenvalue problem.
3. Check to see whether the $\psi_i \left(\mathbf{r}\right)$ have changed significantly from the previous iteration:

   a. If they have not changed, then stop: the algorithm has converged.
   b. If they have changed, then proceed to the next step.

4. Calculate new potentials $\Gamma_D \left(\mathbf{r}\right)$ and $\Gamma_E \left(\mathbf{r}, \mathbf{r}'\right)$ using the new $\psi_i \left(\mathbf{r}\right)$ solutions.
5. Go back to step 2.

The interaction $V \left(\mathbf{r}, \mathbf{r}'\right)$ used in Eq. (6.91) is not the fundamental nucleon-nucleon interaction, but rather an *effective interaction* that accounts for the restrictions imposed on the motion of the nucleons by the surrounding nuclear medium, through the Pauli exclusion principle. These effective interactions contain terms for the nuclear, spin-orbit, and Coulomb (in the case of protons) potentials.[3] There are two often-used forms for the nuclear term: delta functions [44, 45] and Gaussian functions [11, 14].

Once the converged ψ_i (**r**) solutions have been obtained, the many-body wave function for the ground state of the nucleus can be constructed as the (antisymmetrized) Slater determinant of the A occupied ψ_i (**r**) states:

$$\Psi\left(\mathbf{r}_1, \mathbf{r}_2, \ldots, \mathbf{r}_A\right) = \frac{1}{\sqrt{A!}} \det \begin{pmatrix} \psi_1\left(\mathbf{r}_1\right) & \psi_1\left(\mathbf{r}_2\right) & \cdots & \psi_1\left(\mathbf{r}_A\right) \\ \psi_2\left(\mathbf{r}_1\right) & \psi_2\left(\mathbf{r}_2\right) & \cdots & \psi_2\left(\mathbf{r}_A\right) \\ \vdots & \vdots & \ddots & \vdots \\ \psi_A\left(\mathbf{r}_1\right) & \psi_A\left(\mathbf{r}_2\right) & \cdots & \psi_A\left(\mathbf{r}_A\right) \end{pmatrix} \tag{6.92}$$

The binding energy of the nucleus is no longer simply the shell model energy of Eq. (6.20) because the single-particle energies ε_i now contain an implicit contribution from the mean-field potential, which must be subtracted out in order to not count the same interactions multiple times. The energy is instead given by

$$E = \sum_{i \in \text{occ}} \varepsilon_i - \frac{1}{2} \sum_{i, j \in \text{occ}} V_{ij} \tag{6.93}$$

where

$$V_{ij} \equiv \int d^3r \int d^3r' \, \psi_i^*\left(\mathbf{r}\right) \psi_j^*\left(\mathbf{r}'\right) V\left(\mathbf{r}, \mathbf{r}'\right) \psi_i\left(\mathbf{r}\right) \psi_j\left(\mathbf{r}'\right)$$
$$- \int d^3r \int d^3r' \, \psi_i^*\left(\mathbf{r}\right) \psi_j^*\left(\mathbf{r}'\right) V\left(\mathbf{r}, \mathbf{r}'\right) \psi_j\left(\mathbf{r}\right) \psi_i\left(\mathbf{r}'\right) \tag{6.94}$$

The Hartree-Fock approximation can be extended to include pairing, either by applying the BCS model directly to the solutions of Eq. (6.91) or by treating the mean filed and pairing potential on the same footing in the Hartree-Fock-Bogoliubov (HFB) approximation [20, 41]. The HFB can then be used to construct self-consistent configurations of the parent nucleus that are relevant to fission, under the assumption of a solution consisting of independent quasi-particles, by constraining collective parameters of the nucleus (e.g., quadrupole and octupole moments) [20, 41]. This type of approach falls under the larger category of density functional theory (DFT) [7]. Figure 6.7 shows nucleon densities for various configurations of ^{240}Pu along the most likely path to scission and the corresponding HFB energies calculated using a Gaussian form for the effective nuclear interaction [14, 52].

---

[3] Most effective interactions used at present also typically include a term that depends nonlinearly on the nucleon density ρ (**r**). These terms modify somewhat the form of Eq. (6.91), and we will not discuss them here (see [20, 41] for more details).

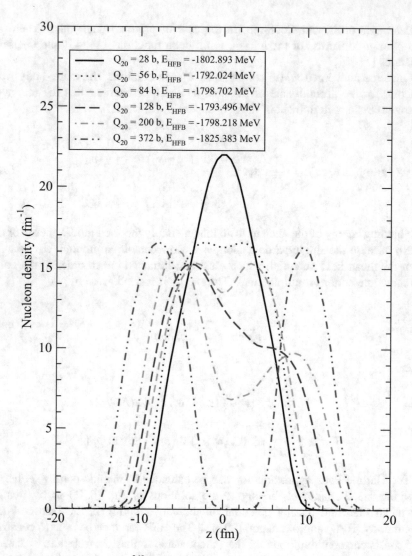

Fig. 6.7 Nucleon densities for ^{240}Pu as a function of quadrupole moment of the nucleus. The corresponding HFB energies are given in the legend

The quadrupole (Q_{20}) and octupole (Q_{30}) moment constraints are commonly used in HFB calculations for fission and can be defined in terms of the nucleon density $\rho(\mathbf{r})$ [18, 52]. In cylindrical coordinates (r, φ, z), and assuming axial symmetry (so that $\rho(\mathbf{r}) = \rho(r, z)$ does not depend on the azimuthal angle φ), the quadrupole moment is[4]

---

[4] Some authors use a definition that is $1/2$ the one in Eq. (6.95).

$$Q_{20} \equiv 2\pi \int_0^\infty r\, dr \int_{-\infty}^\infty dz\, \rho\,(r,z) \left[2\,(z - \bar{z})^2 - r^2 \right] \tag{6.95}$$

where $\bar{z}$ is the center of mass position, which for a nucleus with A nucleons is given by

$$\bar{z} = \frac{1}{A} \times 2\pi \int_0^\infty r\, dr \int_{-\infty}^\infty dz\, \rho\,(r,z)\, z \tag{6.96}$$

and where the number of particles can also be expressed in terms of the density:

$$A = 2\pi \int_0^\infty r\, dr \int_{-\infty}^\infty dz\, \rho\,(r,z) \tag{6.97}$$

Normally in HFB calculations of fission, the $\bar{z}$ value for the parent nucleus is held fixed at $\bar{z} = 0$. The octupole moment is given by

$$Q_{30} \equiv 2\pi \int_0^\infty r\, dr \int_{-\infty}^\infty dz\, \rho\,(r,z) \left[(z - \bar{z})^3 - \frac{3}{2}\,(z - \bar{z})\, r^2 \right] \tag{6.98}$$

The quadrupole moment in Eq. (6.95) is one of the five components of the $Q_{2\mu}$ moment, where $\mu = -2, -1, 0, 1, 2$. The Q_{22} and $Q_{2,-2}$ moments in particular can be used to characterize shapes that deviate from axial symmetry, which may be of interest near the first fission barrier.

There are various ways to extract fission observables using the microscopic method. For example, the set of HFB energies as a function of collective parameters can be used within a scission point model [30] or to construct a PES as the basic framework for the methods discussed in Sect. 3.3. They can also be used in dynamical descriptions of fission (Sect. 7.4.2). A more advanced survey of the application of microscopic approaches to the fission problem can be found in the review by Schunck and Robledo [43].

6.6 Questions

57. What are some methods that can be used to calculate single-particle levels for a nucleus?

58. How many harmonic-oscillator major shells are needed to hold the 92 protons and 144 neutrons of ^{236}U?

59. Why does the filling of single-particle levels lead to jumps in the energy of the nucleus?

60. How can Hermite polynomials be used to smooth the "staircase" shape of the single-particle energy of the nucleus?

61. How is the exact shell model energy of a nucleus calculated?

62. What is the purpose of the Strutinsky procedure?

63. What is the occupation probability of the Fermi level in BCS theory?

64. If two states have the same quasiparticle energy, do they also have the same single-particle energy? Why or why not?

65. In the uniform state density model, does a higher single-particle state density near the Fermi level lead to a larger or smaller pairing gap?

66. Which is larger at a given energy: state density or level density?

67. What is the relationship between state density and partition function?

68. Justify the name "spin cutoff" for the parameter σ^2 in the angular momentum dependence of the level density.

69. What are some advantages of the folding potential method?

6.7 Exercises

Problem 89 Assuming it is a uniformly charged sphere, calculate the Coulomb energy of ^{236}U, and compare it to its binding energy.

Problem 90 Strutinsky1. The Hermite polynomials $H_n(x)$ [1, 19] satisfy the orthonormality condition

$$\int_{-\infty}^{\infty} dx\, e^{-x^2} H_m(x) H_n(x) = \sqrt{\pi} 2^n n! \delta_{mn} \tag{6.99}$$

where δ_{mn} is the Kronecker delta symbol. A delta function can be expanded in terms of the $H_n(x)$ as

$$\delta(x) = e^{-x^2} \sum_{n=0}^{\infty} c_n H_n(x) \tag{6.100}$$

Evaluate the coefficients c_n of this expansion.

Problem 91 Strutinsky2. Show that the delta functions

$$\delta(x) \tag{6.101}$$

and

$$\frac{1}{\gamma} \delta \left(\frac{x}{\gamma} \right) \tag{6.102}$$

where $\gamma \geq 0$ integrate to 1 in any range that includes $x = 0$ and can therefore be used interchangeably.

Problem 92 Strutinsky3. Consider the function

$$f_N (x) = e^{-x^2} \sum_{n=0}^{N} c_n H_n (x) \tag{6.103}$$

where the c_n coefficients are those obtained in Exercise 90. Plot $f_N (x)$ for $N = 0, 4,$ and 8. Comment on the limit where $N \to \infty$. Use this result, along with Exercises 90 and 91, to explain Eq. (6.15) for $\bar{g} (\varepsilon)$.

Problem 93 Strutinsky4. Derive Eq. (6.9) for $\bar{n} (\varepsilon)$ using Eq. (6.16) and Exercises 90, 91, and 92. Hint: use recurrence relations for the Hermite polynomials and their derivatives to show that, for $m \geq 1$,

$$\int_{-\infty}^{x} dt\, e^{-t^2} H_m (t) = -e^{-x^2} H_{m-1} (x) \tag{6.104}$$

and use the definition of the error function for $m = 0$.

Problem 94 Strutinsky5. Derive Eq. (6.22) for $\bar{E}_{sh}$ using Eq. (6.21) and Exercises 90, 91, and 92.

Problem 95 LevelDensity1. Consider two particles distributed among the four levels in Fig. 6.8 with energies $\varepsilon_1 = 0$ MeV, $\varepsilon_2 = 0.5$ MeV, $\varepsilon_3 = 1.0$ MeV, and $\varepsilon_4 = 1.5$ MeV. Assume an independent particle model and a degeneracy $\Omega = 2$ for each level.

1. What is the ground-state energy E_0 of this system?
2. What is the excitation energy of the highest excited state of this system?

Fig. 6.8 Single-particle levels

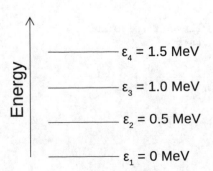

Problem 96 LevelDensity2. Write an expression for the entropy S_0 of the system in Exercise 95, given by Eq. (6.48), for arbitrary α_0 and β_0 values in terms of the single-particle energies. Note: do not forget about the degeneracy of the states.

Problem 97 LevelDensity3. Assuming a temperature of $T = 0.5$ MeV (and an inverse temperature $\beta = 1/T$) in Problems 95 and 96:

1. Verify that $\alpha = 0.106$ satisfies (see Eq. (6.49))

$$\frac{\partial}{\partial \alpha} \ln Z (\alpha, \beta) - A = 0 \qquad (6.105)$$

2. What is the corresponding energy E (see Eq. (6.49))?

Problem 98 LevelDensity4. For the values of α and β in Exercise 97:

1. What is the value of the denominator term D_0 in Eq. (6.50)?
2. What is the value of the state density in the approximation of Eq. (6.47)?

Problem 99 Consider four independent identical particles occupying the three levels in Fig. 6.9 with energies $\varepsilon_1 = 1$ MeV, $\varepsilon_2 = 2$ MeV, and $\varepsilon_3 = 3$ MeV and degeneracies $\Omega_1 = 4$, $\Omega_2 = 2$, and $\Omega_3 = 2$.

1. List the excitation energies of all possible distinct multi-particle multi-hole configurations that can be constructed for this system.
2. Plot the state density for this system from excitation energy 0.5 MeV to 6.5 MeV in 2-MeV steps.

Problem 100 Consider a nucleus near scission that will break at some point $z = z_0$ along the neck. The nascent fragments can be defined as the parts of the nucleon density $\rho (\mathbf{r})$ on either side of a plane perpendicular to the z axis and cutting the neck at $z = z_0$. Show that the quadrupole moment in Eq. (6.95) can be decomposed into a sum the quadrupole moments of the nascent fragments (with the Q_{20} moment of each fragment measured relative to its own center of mass, $\bar{z}_L$ and $\bar{z}_R$) and a term proportional to the square of the separation distance $d = \bar{z}_R - \bar{z}_L$ between the centers of mass of the fragments. Hint: you may find it useful to prove the identity

$$A\bar{z} = A_L\bar{z}_L + A_R\bar{z}_R \qquad (6.106)$$

Fig. 6.9 Single-particle levels

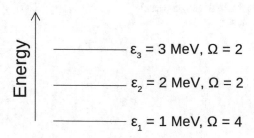

where A_L is the integral of $\rho(\mathbf{r})$ to the left of the plane through $z = z_0$ and A_R the integral to the right of it.

Problem 101 Plot the particle density given by

$$\rho(r, \theta, \phi) = \frac{1}{4\pi}\left[a_0^2 g_{0,0}^2(r) + a_1^2 g_{1,0}^2(r)\right] \tag{6.107}$$

in spherical coordinates, where $a_0 = 2\sqrt{3}$, $a_1 = 2\sqrt{2}$, and the $g_{n,\ell}(r)$ radial functions are given by Eq. (6.6) with $b = 2$ fm. How many particles does this density represent?

Problem 102 Consider the set of single-particle levels with degeneracy $\Omega = 2$ (i.e., each level can hold at most one time-reversed pair of particles) and evenly spaced energies

$$\varepsilon_i = 0.5i \text{ MeV} \tag{6.108}$$

with $i = 1, 2, \ldots, 10$. Given the Fermi energy $\lambda = 2.75$ MeV and the pairing gap parameter $\Delta = 0.5$ MeV, calculate the average number of particles. Ignore the self-energy correction.

Problem 103 For the system described in Exercise 102 (and ignoring the self-energy correction):

1. Calculate the strength G of the pairing interaction.
2. Calculate the ground-state energy.
3. Calculate the pairing correlation energy.

Problem 104 Consider a single level with energy ε, degeneracy Ω, and containing A particles (with $A \leq \Omega$).

1. Write the grand canonical partition function $Z(\alpha, \beta)$ for this system.
2. Show that Eq. (6.49) gives an energy $E = A\varepsilon$.
3. Show that $D_0 = 0$ in Eq. (6.50). What does that imply in terms of the level density formula in Eq. (6.47)?

Problem 105 Consider two particles occupying a state in one dimension. The state is described by a wave function $\psi(x)$. The interaction between the particles is given by a delta function,

$$V(x, x') = V_0 \delta(x - x') \tag{6.109}$$

1. Write the spatial density $\rho(x)$ for this system.
2. Write the direct and exchange potentials $\Gamma_D(x)$ and $\Gamma_E(x, x')$, respectively, for the system.

References

1. M. Abramowitz, I. Stegun, *Handbook of Mathematical Functions: With Formulas, Graphs, and Mathematical Tables* (Dover Publications, 1965)
2. Y. Alhassid, in *6th International Workshop on Compound-Nuclear Reactions and Related Topics CNR*18* (2021), pp. 97–112
3. J. Bardeen, L.N. Cooper, J.R. Schrieffer, Phys. Rev. **108**, 1175 (1957)
4. H.A. Bethe, Rev. Mod. Phys. **9**, 69 (1937)
5. A. Bohr, B.R. Mottelson, *Nuclear Structure vol. I: Single-Particle Motion* (World Scientific, 1998)
6. M. Bolsterli, E.O. Fiset, J.R. Nix, J.L. Norton, Phys. Rev. C **5**, 1050 (1972)
7. D.M. Brink, Nucl. Phys. News **12**, 27 (2002)
8. D.M. Brink, R.A. Broglia, *Nuclear Superfluidity: Pairing in Finite Systems* (Cambridge University Press, 2005)
9. R. Capote et al., Nucl. Data Sheets **110**, 3107 (2009)
10. J. Damgaard, H. Pauli, V. Pashkevich, V. Strutinsky, Nucl. Phys. A **135**, 432 (1969)
11. J. Dechargé, D. Gogny, Phys. Rev. C **21**, 1568 (1980)
12. A.R. Edmonds, *Angular Momentum in Quantum Mechanics* (Princeton University Press, 1996)
13. J.M. Eisenberg, W. Greiner, *Nuclear Theory: Microscopic Theory of the Nucleus* (North-Holland, 1986)
14. D. Gogny, in *Proceedings of the International Conference on Self-Consistent Fields* (1975), p. 333
15. M. Goppert-Mayer, Phys. Rev. **74**, 235 (1948)
16. M. Goppert-Mayer, Phys. Rev. **78**, 16 (1950)
17. M. Goppert-Mayer, Phys. Rev. **78**, 22 (1950)
18. H. Goutte, J.F. Berger, P. Casoli, D. Gogny, Phys. Rev. C **71**, 024316 (2005)
19. I.S. Grashteyn, I.M. Ryzhik, *Table of Integrals, Series and Products, Corrected and Enlarged Edition* (Academic Press, 1980)
20. W. Greiner, J.A. Maruhn, *Nuclear Models* (Springer, 1996)
21. M. Harvey, A. Jensen, Nucl. Phys. A **164**, 641 (1971)
22. R.W. Hasse, W.D. Myers, *Geometrical Relationships of Macroscopic Nuclear Physics* (Springer, 1988)
23. K.L.G. Heyde, *The Nuclear Shell Model* (Springer, 1990)
24. S. Hilaire, M. Girod, S. Goriely, A.J. Koning, Phys. Rev. C **86**, 064317 (2012)
25. J.D. Jackson, *Classical Electrodynamics* (Wiley, 1998)
26. S. Koonin, D. Dean, K. Langanke, Physics Reports **278**, 1 (1997)
27. H.J. Krappe, K. Pomorski, *Theory of Nuclear Fission* (Springer-Verlag GmbH, 2012)
28. S.J. Krieger, Phys. Rev. C **1**, 76 (1970)
29. D. Lacroix, e-prints (2010). arXiv:1001.5001v1
30. J.-F. Lemaître, S. Panebianco, J.-L. Sida, S. Hilaire, S. Heinrich, Phys. Rev. C **92**, 034617 (2015)
31. J.A. Maruhn, P.-G. Reinhard, E. Suraud, *Simple Models of Many-Fermion Systems* (Springer, 2010)
32. A. Messiah, *Quantum Mechanics* (Dover Publications, Mineola, NY, 1999)
33. A.B. Migdal, *Qualitative Methods in Quantum Theory* (Advanced Book Program, Perseus Pub, Cambridge, MA, 2000)
34. K. Mon, J. French, Ann. Phys. **95**, 90 (1975)
35. P. Morse, *Methods of Theoretical Physics* (McGraw-Hill, New York, 1953)
36. W.D. Myers, W.J. Swiatecki, Ann. Phys. **55**, 395 (1969)
37. W.D. Myers, W.J. Swiatecki, Ann. Phys. **84**, 186 (1974)
38. S.G. Nilsson, Mat. Fys. Medd. Dan. Vid. Selsk. **29** (1955)
39. S.G. Nilsson, I. Ragnarsson, *Shapes and Shells: Nuclear Structure* (Cambridge University Press, 1995)

40. W. Press, *Numerical Recipes in C: The Art of Scientific Computing* (Cambridge University Press, Cambridge Cambridgeshire New York, 1992)
41. P. Ring, P. Schuck, *The Nuclear Many-Body Problem* (Springer, 1980)
42. D.J. Rowe, J.L. Wood, *Fundamentals of Nuclear Models* (World Scientific, 2010)
43. N. Schunck, L.M. Robledo, Rep. Progress Phys. **79**, 116301 (2016)
44. T.H.R. Skyrme, Philosophical Magazine **1**, 1043 (1956)
45. T. Skyrme, Nucl. Phys. **9**, 615 (1958)
46. J. Suhonen, *From Nucleons to Nucleus* (Springer, Berlin, Heidelberg, 2007)
47. C. Tannoudji, *Quantum Mechanics* (Wiley, New York, 1977)
48. H. Uhrenholt, S. Åberg, A. Dobrowolski, T. Døssing, T. Ichikawa, P. Möller, Nucl. Phys. A **913**, 127 (2013)
49. D. Vautherin, in *Advances in Nuclear Physics* (Kluwer Academic Publishers, 2002), pp. 123–172
50. R.D. Woods, D.S. Saxon, Phys. Rev. **95**, 577 (1954)
51. W. Younes, tech. rep. UCRL-TR-205363 (LLNL, 2004)
52. W. Younes, *A Microscopic Theory of Fission Dynamics Based on the Generator Coordinate Method* (Springer, Cham, Switzerland, 2019)

Chapter 7
Fission Models Revisited: Reactions and Dynamics

Abstract Chapter 7 is an advanced discussion of models of fission reactions and dynamics. The chapter starts with a general discussion of scattering theory, resonances, and widths to arrive at the formal definition of a cross section and to set the stage for Hauser-Feshbach theory, the framework widely used for the calculation of fission (and other) cross sections. Hauser-Feshbach theory is treated in detail with an extensive discussion of transition state theory and its application to fission. The discussion of the Hauser-Feshbach formalism highlights the importance of two crucial ingredients in the calculations: level densities, already discussed in Chap. 6, and transition strengths introduced in this chapter. Quasifission is discussed along with direct-reaction-induced fission. A brief survey of advanced topics gives a discussion of semi-classical and quantum mechanical methods used in the description of fission dynamics and which represent the state of the art in current research to describe the evolution of the fissioning nucleus to scission and beyond.

We expand on the discussions of fission reactions and fission dynamics in Chap. 3. We discuss standard methods to calculate fission cross sections, as well as more advanced time-dependent approaches.

7.1 Scattering Theory

7.1.1 Introduction

In previous chapters we discussed the tools needed to calculate the energy of the nucleus in its various configurations along the fission path. In the present section, we turn our attention to the formation of the parent nucleus and to the probability that this nucleus will fission. In particular, we will be concerned with processes which, unlike spontaneous fission, involve fission induced by a particle incident on a target. For example, in a fission experiment where a beam of low-energy

© The Author(s), under exclusive license to Springer Nature Switzerland AG 2021   151
W. Younes, W. D. Loveland, *An Introduction to Nuclear Fission*, Graduate Texts in Physics, https://doi.org/10.1007/978-3-030-84592-6_7

neutrons impinges on an actinide target, the most likely outcomes are typically fission, neutron emission (followed or not by fission), and gamma emission. We are interested in quantifying the likelihood of these various outcomes.

The formal treatment of this dynamic problem begins with the time-dependent Schrödinger equation (TDSE)

$$-\frac{\hbar^2}{2m}\nabla^2\Psi\left(\mathbf{r},t\right) + V\left(r\right)\Psi\left(\mathbf{r},t\right) = \hbar i \frac{\partial}{\partial t}\Psi\left(\mathbf{r},t\right) \tag{7.1}$$

where the solution $\Psi\left(\mathbf{r},t\right)$ now contains an explicit dependence on time t. By assuming a continuously incident plane-wave beam,

$$\Psi\left(\mathbf{r},t\right) \propto e^{i\left(\mathbf{k}\cdot\mathbf{r}-\omega t\right)} \tag{7.2}$$

we can bring the problem back to a time-independent Schrödinger equation (see Section 1.A.2 in [41]),

$$-\frac{\hbar^2}{2m}\nabla^2\Psi\left(\mathbf{r}\right) + V\left(r\right)\Psi\left(\mathbf{r}\right) = \frac{\hbar^2 k^2}{2m}\Psi\left(\mathbf{r}\right) \tag{7.3}$$

where $k = |\mathbf{k}|$ is the wave number and the solution $\Psi\left(\mathbf{r}\right)$ can be complex-valued. Far from the target (where $V\left(r\right) \approx 0$), we expect the wave function to consist of the incident plane wave $e^{i\mathbf{k}\cdot\mathbf{r}}$, and a scattered spherical wave, leading to the asymptotic form in spherical coordinates (r,θ,ϕ)

$$\Psi_{\text{asy}}\left(\mathbf{r}\right) = e^{i\mathbf{k}\cdot\mathbf{r}} + f\left(\theta,\phi\right)\frac{e^{ikr}}{r} \tag{7.4}$$

where $f\left(\theta,\phi\right)$ is the *scattering amplitude*. The scattered wave is

$$\Psi_{\text{sca}}\left(\mathbf{r}\right) = f\left(\theta,\phi\right)\frac{e^{ikr}}{r} \tag{7.5}$$

Even though we dropped the explicit time dependence in going from Eqs. (7.1) to (7.3), there is still an implicit "flow" associated with the wave function which we can quantify by calculating its *probability current*

$$\mathbf{j} = \frac{\hbar i}{2m}\left[\nabla\Psi^*\left(\mathbf{r}\right)\Psi\left(\mathbf{r}\right) - \Psi^*\left(\mathbf{r}\right)\nabla\Psi\left(\mathbf{r}\right)\right] \tag{7.6}$$

For example, we can calculate the current corresponding to the incident plane wave $e^{i\mathbf{k}\cdot\mathbf{r}}$ using Eq. (7.6),

$$\mathbf{j}_{\text{inc}} = \frac{\hbar}{m}\mathbf{k} \tag{7.7}$$

Another useful quantity is the *flux* of particles through a given area. The flux through a surface S is formally defined as the surface integral

$$\Phi \equiv \int_S \mathbf{j} \cdot d\mathbf{A} \tag{7.8}$$

where $\mathbf{j}$ is the current given by Eq. (7.6) and $d\mathbf{A}$ is surface element vector perpendicular to that surface [41, 42, 46, 58]. We can now define the *absorption cross section* as the net loss of flux divided by the magnitude of the incident current (see, e.g., Section VIII.2 in [10]),

$$\sigma_{abs} \equiv -\frac{1}{|\mathbf{j}_{inc}|} \int_S \mathbf{j}_{asy} \cdot d\mathbf{A} \tag{7.9}$$

where $\mathbf{j}_{asy}$ is obtained by plugging Eq. (7.4) into Eq. (7.6). The overall minus sign in Eq. (7.9) ensures that flux into the target region is counted as positive. We can calculate the (elastic) scattering cross section in a similar manner with the definition

$$\sigma_{sca} \equiv \frac{1}{|\mathbf{j}_{inc}|} \int_S \mathbf{j}_{sca} \cdot d\mathbf{A} \tag{7.10}$$

with an overall positive sign to count the outgoing flux as positive. Here, $\mathbf{j}_{sca}$ represents the scattered current, and is obtained by plugging Eq. (7.5) into Eq. (7.6). The total cross section is the sum of σ_{sca} and σ_{abs},

$$\sigma_{tot} \equiv \sigma_{sca} + \sigma_{abs} \tag{7.11}$$

The formal definitions in Eqs. (7.9) and (7.10) are useful in theoretical developments but less so for the actual measurement of cross sections of interest. In an experiment, the particles (e.g., gamma rays, neutrons, fission fragments) resulting from a beam of particles incident on some target material will be detected by an instrument subtending a solid angle Ω. For the sake of argument, we will assume a perfect detector so that every particle reaching the detector is counted (otherwise, the counts must be corrected for efficiency and deadtime [39, 42]). For a flux Φ_{inc} of particles incident on a target of surface area A and *areal density* a (i.e., a is the number of target nuclei per unit area), and an average rate $dR/d\Omega$ of particles[1] recorded by the detector per unit time and per unit solid angle, we can define the resulting *differential cross section* as [42]

$$\frac{d\sigma}{d\Omega} \equiv \frac{1}{a A \Phi_{inc}} \frac{dR}{d\Omega} \tag{7.12}$$

---

[1] Note that the beam and detected particles do not have to be the same, unless the reaction is elastic.

Integrating over all solid angles, we obtain the cross section

$$\sigma = \int d\Omega \, \frac{d\sigma}{d\Omega} \tag{7.13}$$

7.1.2 Resonances and Widths

The formal definitions in Sect. 7.1.1 do not give us an easy way to predict a cross section from a model. In theory, we could solve the TISE in Eq. (7.3) with some potential to deduce, wave functions, currents, fluxes, and ultimately cross sections. In practice, this procedure is not always straightforward to carry out. In this section, we will connect the cross section to level densities so that we can arrive at a more practical formulation in Sect. 7.2.

The parent nucleus just before scission can be seen as occupying a many-body state with a finite lifetime inside the fission barrier(s). Quantum mechanically, a state with a finite lifetime τ will not have a sharply defined energy. Instead, this *resonance* state can be found at an energy E_r with a probability distribution $P(E)$ proportional to a Lorentzian form with a width $\Gamma = \hbar/\tau$ (see Exercises 107–108),

$$P(E) \propto \frac{1}{(E - E_r)^2 + \frac{1}{4}\Gamma^2} \tag{7.14}$$

We can also look at this situation as a barrier transmission problem [25], characterized by a transmission coefficient T, that can in principle be obtained by solving the corresponding Schrödinger equation. Provided that the average width Γ of neighboring levels is small compared to their average energy separation D, it is possible to derive the important relationship (see, e.g., Chapter VIII in [10])

$$T = 2\pi \frac{\Gamma}{D} \tag{7.15}$$

This remarkable equation connects the Schrödinger equation (through the transmission coefficient), the width (which is related to the decay rate), and the level density (which is $1/D$).

Having made these preliminary remarks on resonant states, we now consider a reaction between a projectile a and a target A leading to the formation of a resonant compound nuclear state C, before decaying into products b and B. Schematically, we write the process as $a + A \rightarrow C \rightarrow b + B$. If the incident energy is very low, the only decay *channel* open for the compound-nucleus reaction $a + A \rightarrow C$ will be the inverse process, $C \rightarrow a + A$. However, at higher incident energies, other channels may become accessible. To each channel α, representing a possible outcome of the reaction, we can associate a *partial width* Γ_α. The sum of all possible partial widths for a given *entrance channel* $a + A$ gives the total width,

$$\Gamma = \sum_\alpha \Gamma_\alpha \qquad (7.16)$$

The probability P_α of decay from a resonant state into a channel α is given by the ratio

$$P_\alpha = \frac{\Gamma_\alpha}{\Gamma} \qquad (7.17)$$

For a particle with orbital angular momentum ℓ, the cross section for transmission through a barrier is related to the transmission coefficient T_ℓ via [58]

$$\sigma_\ell = \frac{\pi}{k^2} (2\ell + 1) T_\ell \qquad (7.18)$$

To obtain the cross section σ_{CN} for the formation of a *compound nucleus* through the absorption into a single resonance, we multiply the average cross section in Eq. (7.18) by the Lorentzian probability distribution in Eq. (7.14). After proper normalization of the probability distribution, we find

$$\sigma_{CN} = \sigma_\ell \frac{D\Gamma}{2\pi} \frac{1}{(E - E_r)^2 + \left(\frac{\Gamma}{2}\right)^2} \qquad (7.19)$$

Combining Eqs. (7.15), (7.17)–(7.19) leads to

$$\sigma_{CN} = \frac{\pi}{k^2} (2\ell + 1) \frac{\Gamma_\alpha \Gamma}{(E - E_r)^2 + \left(\frac{\Gamma}{2}\right)^2} \qquad (7.20)$$

where α here designates the entrance channel, with orbital angular momentum ℓ. If we now consider a generic compound-nuclear reaction $a + A \rightarrow C \rightarrow b + B$, we can use the *Bohr independence hypothesis*—the assumption that the reaction proceeds in two essentially independent steps:[2] formation, followed by decay—and Eq. (7.20) to write the reaction cross section:

$$\sigma_{\alpha\beta} = \sigma_{CN} \times \frac{\Gamma_\beta}{\Gamma}$$

$$= \frac{\pi}{k^2} (2\ell + 1) \frac{\Gamma_\alpha \Gamma_\beta}{(E - E_r)^2 + \left(\frac{\Gamma}{2}\right)^2} \qquad (7.21)$$

---

[2] With the caveat that energy, angular momentum, and parity must be conserved throughout the process.

where we have labeled the entrance channel $a + A$ as α and the exit channel $b + B$ as β. This is the Breit-Wigner resonant cross-section formula that Bohr and Wheeler used in their seminal 1939 paper [12]. Since there are only two open channels for the decay—α if the reaction is elastic and $\beta \neq \alpha$ if it is non-elastic—the total width is

$$\Gamma = \Gamma_\alpha + \Gamma_\beta \tag{7.22}$$

Thus, by studying resonances, we have made the connection between the lifetime of the compound nuclear state, its width, the density of states near the resonance, and the decay probability. In the next section, we include angular momentum and parity and formulate the Hauser-Feshbach theory of reaction cross sections.

7.2 Hauser-Feshbach Theory

7.2.1 General Formalism

Hauser-Feshbach (HF) theory [19, 30] is a statistical theory of compound-nuclear reactions. In the HF approach, we will be interested in energy-averaged cross sections; therefore we will generalize Eq. (7.21) to the case of multiple resonances and take its average over an energy range containing many resonances with angular momentum J_C. We will skip over the derivation to present the resulting reaction cross section for entrance channel α and exit channel β (but see Exercises 112–117)[3]

$$\langle \sigma_{\alpha\beta}(E) \rangle = \frac{\pi}{k^2} \sum_{J_C} \frac{2J_C + 1}{(2j_a + 1)(2J_A + 1)} W_{\alpha\beta} \frac{T_\alpha(E) T_\beta(E)}{\sum_\gamma T_\gamma(E)} \tag{7.23}$$

In Eq. (7.23) k is the wave number in the entrance channel; j_a and J_A are the total angular momenta of the projectile and target, respectively; and Eq. (7.15) has been used to express the result in terms of transmission coefficients $T_\alpha(E)$, $T_\beta(E)$, and $T_\gamma(E)$ rather than widths. The coefficient $W_{\alpha\beta}$ is a *width fluctuation correction factor* [48], which accounts for the fact that the average of a ratio of widths is not in general equal to the ratio of their averages. The transmission-coefficient sum $\sum_\gamma T_\gamma(E)$ in the denominator extends over all open exit channels.

The transmission coefficients in Eq. (7.23) contain the important physics of the problem. They can be written as a sum over all possible final states of individual transmission coefficients. At sufficiently high excitation energies, where it becomes

---

[3] We ignore some of the details for the sake of simplicity, namely, parity and the angular momenta of the individual channels α, β, and γ. A more complete version of Eq. (7.23) would include cumbersome summations over those quantum numbers as well [19].

too cumbersome to enumerate individual levels, this sum becomes an integral of individual transmission coefficients, multiplied by the level density in the exit channel. For low-energy fission, the transmission coefficients that are usually needed are those for neutron emission, gamma decay, and fission. We will focus our discussion to the transmission coefficient for fission, which we present in greater detail next.

7.2.2 Transition State Theory

We mentioned in Sect. 3.6 the existence of saddles on the PES which present bottlenecks in the evolution of the parent nucleus toward scission. In a very simplified mathematical model, the potential at a saddle can be written as[4]

$$V(x, y) = V_0 - \frac{1}{2}M_x\omega_x x^2 + \frac{1}{2}M_y\omega_y y^2 \tag{7.24}$$

with x representing the collective coordinate along which the nucleus moves toward scission (e.g., x could be the deformation of the nucleus) and M_x represents the *inertial mass* of the nucleus associated with the collective motion in the x coordinate. The coordinate y is used to describe motion in a transverse direction with an associated inertial mass M_y. The solution of the Schrödinger equation with the potential in Eq. (7.24) can be decomposed into a transverse wave associated with the discrete energies of a harmonic oscillator, $\hbar\omega_y (n + 1/2)$ with $n = 0, 1, 2, \ldots$, and a wave in the direction of motion [17]. The discrete states associated with the transverse motion can be viewed as the transition states.

The transmission coefficient for an inverted harmonic oscillator potential, such as the one in the x coordinate of Eq. (7.24), is well known and given by Hill and Wheeler [33]

$$T(E_x) = \frac{1}{1 + \exp\left[-\frac{2\pi}{\hbar\omega}(E_x - E_b)\right]} \tag{7.25}$$

where E_x is the excitation energy of the compound nucleus and E_b is the barrier height. Equation (7.25) holds both above and below the barrier top: Below the barrier, it can be derived using the WKB approximation, while above the barrier, a more sophisticated treatment in the complex plane is required [8, 14, 24, 31, 38].

The transition state model [11, 22, 69] posits a set of states on top of the fission barrier(s) that regulate the passage of the system across the barrier(s). The effect of these transition states is calculated by assuming that each state acts as a separate

---

[4] In realistic calculations, the situation is usually more complicated with M_x and M_y themselves functions of x and y and the presence of a cross term in xy with its own inertial mass.

barrier in the fission process. By adding up the contributions from the individual transition states, we arrive at an effective transmission coefficient for fission from a state (E_x, J, π) given by the integral of the Hill-Wheeler transmission coefficients in Eq. (7.25) over the density of transition states,

$$
T_f (E_x, J, \pi) = \int_0^\infty d\varepsilon \, \rho \, (\varepsilon, J, \pi) \, \frac{1}{1 + \exp\left[-\frac{2\pi}{\hbar\omega} (E_x - E_b - \varepsilon)\right]} \tag{7.26}
$$

where $\rho \, (\varepsilon, J, \pi)$ is the density of transition states at an energy ε measured from the top of the barrier. Although the quantity given by Eq. (7.26) is referred to as a transmission coefficient, it is essentially the sum of many individual transmission coefficients given by Eq. (7.25) and can therefore exceed 1; it is perhaps more appropriately referred to as the *effective number of open channels*. Near the top of the barrier, we expect to find a discrete spectrum of transition states, and the continuous density function in Eq. (7.26) can be replaced with a discrete sum,

$$
T_f (E_x, J, \pi) = \sum_i \frac{\delta_{J, J_i} \delta_{\pi, \pi_i}}{1 + \exp\left[-\frac{2\pi}{\hbar\omega} (E_x - E_b - \varepsilon_i)\right]} \tag{7.27}
$$

The Kronecker delta symbols δ_{J, J_i} and δ_{π, π_i} in Eq. (7.27) ensure that the sum extends only over those transition states with angular momentum J and parity π.

The discrete transition states consist of rotational bands built on top of vibrational band heads that are of the same type as those observed in the low-energy spectrum of well-deformed nuclei (e.g., ground state, γ vibration, etc.) [9, 28, 44]. We show typical values of some properties of the discrete transition states for the inner and outer barriers in Tables 7.1 and 7.2 such as their type, the K^π quantum numbers, their excitation energy above the top of the barrier ε, the moment of inertia $\mathcal{J}$ for the rotational band, and the first few rotational states built on top of the band heads.[5] As the energy above the barriers increases, the discrete transition states merge into a quasi-continuum of states and give rise to the level density in Eq. (7.26). In actinide nuclei, the inner barrier typically corresponds to a breaking of axial symmetry of the nuclear shape (i.e., leading to triaxial shapes, or shapes with three symmetry axes), whereas the outer barrier corresponds to a restoration of axial symmetry but a breaking of reflection symmetry (i.e., shapes that begin to resemble light and heavy pre-fragments connected by a neck).

In addition to the transition states on top of the barriers, there also states in the first and second wells (Fig. 7.1). The states in the first well are the "normal" states of the nucleus that can be observed, for example, by γ-ray spectroscopy, and in the context of fission, they are referred to as *class I* states. The states in the second well

---

[5] For alternate values of the transition state energies and moments of inertia, see, for example, the user's manual for the TALYS reaction code (available at https://tendl.web.psi.ch/tendl_2019/talys.html).

Table 7.1 Properties of typical transition states on top of the *inner* barrier for even-even nuclei (adapted from [44])

Type	K^π	ε (keV)	$\hbar^2/2\mathcal{J}$ (keV)	Rotational states
Ground	0^+	0	3.5	$2^+, 4^+, \ldots$
γ vibration	2^+	200		$3^+, 4^+, \ldots$
γ vibration	0^+	400–500		$2^+, 4^+, \ldots$
γ vibration	4^+	400–500		$5^+, 6^+, \ldots$
Mass asymmetry vibration	0^-	700		$1^-, 3^-, \ldots$
Bending vibration	1^-	800		$2^-, 3^-, \ldots$

Table 7.2 Properties of typical transition states on top of the *outer* barrier for even-even nuclei (adapted from [44])

Type	K^π	ε (keV)	$\hbar^2/2\mathcal{J}$ (keV)	Rotational states
Ground	0^+	0	2.5	$2^+, 4^+, \ldots$
Mass asymmetry vibration	0^-	100		$1^-, 3^-, \ldots$
γ vibration	2^+	800		$3^+, 4^+, \ldots$
γ vibration	0^+	1500		$2^+, 4^+, \ldots$
γ vibration	4^+	1500		$5^+, 6^+, \ldots$
Bending vibration	1^-	800		$2^-, 3^-, \ldots$

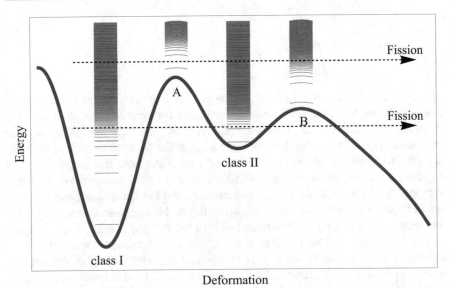

Fig. 7.1 Double-humped fission barrier with class I, class II, and transition states. The inner and outer barriers are labeled "A" and "B," respectively

are called *class II* states, and they contribute to resonances in fission cross sections, especially when the excitation energy of the nucleus is below the fission barriers. The class II states are also responsible for fission isomers. More importantly for cross-section calculations, the class I and class II states can couple in which case fission must be described as a two-step process: first taking the system from the class I states across barrier A to the class II states, followed by a second step from the class II states across barrier B and then on to scission. In particular, well above the top of the barriers, we expect the coupling to be weak, and the effect of the two barriers can reproduced by using a single effective transmission coefficient calculated from the transmission coefficients of the individual barriers,

$$T_f \equiv \frac{T_A T_B}{T_A + T_B} \tag{7.28}$$

For stronger coupling, different formulas can be derived which we will not discuss here for the sake of simplicity (but see, e.g., Lynn and Back [45]).

Since the fission channel usually competes with neutron and gamma-ray emission, we briefly discuss the transmission coefficients for those channels as well, and which are needed in Eq. (7.23). The transmission coefficient for the neutron channel is given by the formula [72]

$$
\sum_{\beta} T_\beta = \sum_{\ell=0}^{\infty} \sum_{j=\left|\ell-\frac{1}{2}\right|}^{\ell+\frac{1}{2}} \sum_{I=|J_C-j|}^{J_C+j} \sum_{\pi_I} \delta_{\pi_C(-1)^\ell, \pi_I}
$$
$$
\times \int_0^{E_x - S_n} d\varepsilon \, T_{\ell j} \left(E_x - S_n - \varepsilon\right) \rho\left(\varepsilon, I, \pi_I\right) \tag{7.29}
$$

On the left-hand side of Eq. (7.29), the sum over β represents the sum over all levels in the nucleus reached by neutron emission (or neutron capture in the case of an entrance channel). On the right-hand side, we sum over the orbital angular momentum (ℓ) and total angular momentum (j) of the neutron, as well as the total angular momentum I of the residual nucleus resulting from the coupling of the angular momentum J_C of the compound state and j of the neutron. We also sum over the parities π_I (± 1) of the residual nucleus, and the Kronecker delta function enforces parity conservation (with π_C the parity of the compound state and $(-1)^\ell$ the parity corresponding to the neutron). Next, the integral is carried out over the excitation energy ε of the residual nucleus, up to the maximum energy $E_x - S_n$ where E_x is the excitation energy of the compound state and S_n is the neutron separation energy in the compound system. The $T_{\ell j}(E)$ are neutron transmission coefficients (calculated, e.g., using an optical model; see, e.g., [58]), and $E_x - S_n - \varepsilon$ represents the kinetic energy of the neutron. The level density $\rho(\varepsilon, I, \pi_I)$ in the residual nucleus is a continuous function at higher excitation energies and a sum of Dirac delta functions at lower energies where discrete levels can be identified (in which case the integral in Eq. (7.29) reverts to a discrete sum).

The transmission coefficient for gamma-ray emission is given by Young et al. [72]

$$\sum_\beta T_\beta = \sum_{X=E,M} \sum_{L=0}^{\infty} \sum_{I=|J_C-L|}^{J_C+L} \sum_{\pi_I} \delta_{\pi_C \pi_{XL}, \pi_I}$$

$$\times \int_0^{E_x} d\varepsilon_\gamma \, f_{XL}\left(\varepsilon_\gamma\right) \rho \left(E_x - \varepsilon_\gamma, I, \pi_I\right) \tag{7.30}$$

where the summations on the right-hand side are carried out over the character ($X = E$ for electric and M for magnetic) and multipole (L) of the radiation, as well as the angular momentum (I) and parity (π_I) of the residual nucleus. As in Eq. (7.29), a Kronecker delta function enforces parity conservation with the parity of the gamma radiation given by

$$\pi_{XL} = \begin{cases} (-1)^L & X = E \\ (-1)^{L+1} & X = M \end{cases} \tag{7.31}$$

The integral in Eq. (7.30) is carried out over the gamma-ray energy (ε_γ); $f_{XL}\left(\varepsilon_\gamma\right)$ is a gamma-ray strength function, which often has an empirical form (e.g., a Lorentzian distribution or a more sophisticated expression) that is fit to data; and $\rho\left(E_x - \varepsilon_\gamma, I, \pi_I\right)$ is the level density in the residual nucleus (which also becomes discrete at lower excitation energies).

7.3 Other Types of Reactions Leading to Fission

The discussion so far in this chapter has focused exclusively on fission induced through compound reactions, since this is one of the principal mechanisms through which the fission process has been traditionally studied. However, fission also plays an important role in other reaction mechanisms that proceed (initially at least) through a non-equilibrated step. Spallation was already mentioned in Chap. 2 as a mechanism distinct from fission but which, after an intranuclear cascade, can be followed by fission. In this section, we briefly consider two other processes and their corresponding cross sections: *quasifission* and direct reactions followed by fission.

7.3.1 Quasifission Reactions

As we saw in Chap. 1, the eventual discovery of fission was set in motion in the early 1930s by Enrico Fermi's unsuccessful attempts to form transuranic elements through neutron-induced reactions. More modern approaches have met with greater success by relying on heavier projectiles and targets to form superheavy elements (SHE) and extend the periodic table far beyond uranium. For example, element 116 (livermorium, or Lv) was recently synthesized using a beam of ^{48}Ca ions incident on a ^{248}Cm target [52–54]. In such reactions, a process known as quasifission can inhibit the formation of the desired SHE. Quasifission occurs when the projectile + target system fails to fully equilibrate before re-separating [1, 3, 32, 34, 36, 43, 65]. Compared to fission from a compound nucleus, quasifission produces wider mass distributions and more forward-peaked fission-fragment angular distributions [2, 34]. The formation of a SHE can be described as a three-step process: (1) the capture of the projectile onto the target with cross section σ_{cap}, (2) the equilibration into a compound system with probability P_{CN} (this is the step where competition with the quasifission process takes place), and (3) the formation of the SHE as an evaporation residue following neutron emission and surviving the competition with "normal" fission with probability W_{sur}. The cross section for forming a SHE is then given by Loveland [43]

$$\sigma_{SHE} = \sum_{J=0}^{J_{max}} \sigma_{cap}\,(E_{cm}, J)\, P_{CN}\,(E_x, J)\, W_{sur}\,(E_x, J) \tag{7.32}$$

where we have explicitly written the dependence on the center-of-mass energy E_{cm}, the excitation energy E_x, and total angular momentum J of the compound nucleus. As can be seen in Eq. (7.32), the three steps of the SHE formation cannot be treated independently since they all depend on the same angular momentum J which must be summed over. In practice, the allowed values of J are largely constrained by the $W_{sur}\,(E_x, J)$ factor. To model the quasifission reaction, Swiatecki [63] introduced the notion of an "extra push" needed for the projectile to be captured by the target and an "extra extra push" needed for them to form a compound nucleus. This model gives threshold energies for capture and compound-nucleus formation to occur.

One of the current challenges in estimating SHE formation cross section lies in modeling the quasifission process accurately enough to generate reliable calculations of the $P_{CN}\,(E_x, J)$ term. Different models can give estimates of P_{CN} that differ by orders of magnitude [43]. At the same time, reliable calculations of Eq. (7.32) are crucial to the planning of future experimental searches for SHEs. The study of quasifission therefore remains an active research area, with ongoing programs, for example, by the Dubna group [37, 40] and the group at the Australian National University [35].

7.3.2 Direct-Reaction-Induced Fission

Some neutron-induced fission cross sections are impossible to measure in the laboratory. For example, the ^{233}Th (n, f) reaction cannot be studied directly because the target, ^{233}Th, has a half-life of only 22.3 min. As a result, it is impossible to gather enough ^{233}Th to manufacture a reasonable target for an experiment or to find a neutron beam with a sufficiently high flux to compensate for the limited amount of target material. Although the ^{233}Th (n, f) cross section cannot be directly measured, an indirect measurement is possible, for example, through the ^{232}Th (t, pf) reaction.[6] The ^{232}Th target is stable, and the ^{232}Th (t, pf) proceeds in two steps: first the (t, p) reaction transfers two neutrons to the target forming a composite ^{234}Th nucleus, the same one formed by the $n +^{233}$ Th reaction, and then after the system equilibrates, fission follows. This technique of substituting one reaction for another is known as the surrogate method [21] and has been used to extract cross sections for a variety of reactions that cannot be directly measured. The reaction that serves as the surrogate, the (t, p) reaction in our example, proceeds through a direct transfer of nucleons to the target rather than a compound-nucleus process because of the Coulomb barrier which inhibits the statistical evaporation of the proton. The direct transfer prepares the desired nucleus with an initial excitation energy E_x which can be known precisely if the energy and angle of the outgoing particle (the proton in our case) is measured, and with a distribution of total angular momentum and parity J^π which can be calculated from reaction theory. After this direct step, the nucleus can equilibrate into a compound state which may subsequently fission.

In a typical application of the surrogate method, the fission probability following the surrogate reaction is measured as a function of the excitation energy E_x of the fissioning nucleus. For example, in the case of the surrogate ^{232}Th (t, pf), the fission probability $P_{(t,pf)}$ is obtained at each energy as the ratio of the number of coincident events where both proton and fission were detected to the total number of protons detected (whether accompanied by fission or not). This fission probability can be decomposed into the sum of contributions from different J^π values [71]

$$P_{(t,pf)}(E_x) = \sum_{J^\pi} P_{(t,p)}\left(E_x, J^\pi\right) P_f\left(E_x, J^\pi\right) \tag{7.33}$$

where $P_{(t,p)}(E_x, J^\pi)$ is the probability of populating a nuclear state with excitation energy E_x and total spin/parity J^π and $P_f(E_x, J^\pi)$ is the probability of fission from the compound state (E_x, J^π). The probabilities $P_{(t,p)}(E_x, J^\pi)$ and $P_f(E_x, J^\pi)$ can be calculated using reaction theory, and in particular the fission model parameters in $P_f(E_x, J^\pi)$, such as barrier heights, are adjusted to best match the measured $P_{(t,pf)}(E_x)$ values. Next, the cross section for the reaction of interest, ^{233}Th (n, f) in our example, can also be decomposed in a similar fashion to Eq. (7.33) [71],

$$\sigma_{(n,f)}(E_n) = \sum_{J^\pi} \sigma_{CN}\left(E_n, J^\pi\right) P_f\left(E_x, J^\pi\right) \tag{7.34}$$

---

[6] It is worth noting that tritium beams are difficult to come by nowadays, but many (t, pf) reaction studies were performed in the 1960s [21].

where $E_n = E_x - S_n$ is the incident neutron energy and $\sigma_{CN}(E_n, J^\pi)$ is the compound-nucleus formation cross section. The fission probabilities $P_f(E_x, J^\pi)$ are the same as those obtained from Eq. (7.33) by tuning the fission model to reproduce the measured surrogate fission probabilities. The surrogate method relies on the Bohr hypothesis to separate the formation and decay (fission in this case) processes to essentially substitute one formation process (the surrogate reaction's) for another. The usefulness of the surrogate method is due to the fact that the formation process, i.e., $P_{(t,p)}(E_x, J^\pi)$ and $\sigma_{CN}(E_n, J^\pi)$, can often be modeled more reliably than the fission process, i.e., $P_f(E_x, J^\pi)$, which is then constrained by the surrogate fission probability measurement. For a comprehensive review of the surrogate technique and its applications, see [21].

Fission induced by charged particle beams also has applications beyond the surrogate method. For example, in recent work by Swinton-Bland et al. [64], protons incident on various Pb targets were used to induce fission of Bi isotopes at energies close to the fission barrier. In the paper, the study of near-barrier fission in these non-actinide nuclei was used to gain a better understanding of the competition between symmetric and asymmetric mass division.

7.4 Brief Survey of Advanced Topics

7.4.1 Semi-classical Methods

In Chaps. 3 and 6, we discussed various models that can be used to generate a potential energy surface (PES) of the nucleus as a function of shape parameters. The PES is an important ingredient in many modern fission calculations. For example, the PES can be used to locate fission barriers [49] and estimate fission rates based on their properties [4]. More importantly, the PES is the starting point for many dynamical (i.e., time dependent) calculations of the fission process.

One approach to modeling fission dynamics consists in performing a random walk across the PES. For example, Ward et al. [68] used a Metropolis random walk algorithm, and transition rates between points on the PES calculated using level densities obtained by a combinatorial technique, to compute fission-fragment charge distributions. Their results, obtained for a wide range of actinide nuclei, were found to be in good agreement with the experimental data.

Another approach to fission dynamics, the Langevin equation, treats the process in analogy to the evolution of a Brownian particle coupled to a heat bath [23]. The Brownian particle in this case represents the collective properties of the nucleus, while the heat bath represents the single-particle degrees of freedom that the collective motion can couple to. The Langevin approach has been successfully applied to a variety of fission problems. In work by Mazurek et al. [47], the Langevin calculations were coupled at each time step to a statistical particle emission model and used to study particles emitted before scission. The Langevin approach has also been applied to the computation of a wide range of fission properties including fragment mass, charge, and angular distributions, as well as probabilities and cross sections for the induced fission reaction [20, 55, 60, 66].

The Langevin approach has also been applied to the study of quasifission reactions and the formation of superheavy elements. Zagrebaev and Greiner, for example [73], have calculated potential energy surfaces for nucleus-nucleus collisions using a two-core model which depicts the system as two nuclear cores, each with given numbers of protons and neutrons and a number of shared nucleons that move within the volume occupied by the cores. The Langevin equations for the degrees of freedom of the system (distance between nuclear centers, deformations and spatial orientations of the cores, and the mass asymmetry of the system) then give the time evolution of the reaction as trajectories in the space described by these degrees of freedom.

7.4.2   Quantum-Mechanical Methods

Some of the most advanced approaches to fission dynamics are built up from a fully microscopic framework, starting from protons, neutrons, and some interaction between them. The time-dependent Hartree-Fock (TDHF) method is a powerful extension of the microscopic Hartree-Fock procedure discussed in Sect. 6.5.3 that can be used to describe the evolution of the parent nucleus to scission [51, 61]. One limitation of the TDHF approach is that it assumes that the nucleus can be well represented by a single Slater-determinant wave function at all times. In actuality, a full description of the nucleus requires a linear superposition of Slater determinants. However, two significant benefits of the TDHF method are the following: (1) it does not require that collective parameter constraints (such as the quadrupole moment of the nucleus) be imposed to guide its evolution, and (2) elementary single-particle degrees of freedom are automatically included [57]. In other words, the TDHF does not require the calculation of a PES. The TDHF approach has evolved significantly from its first applications to fission [51] to more recent calculations that include the effects of pairing [59, 61]. The time-dependent superfluid local density approximation (TDSLDA) [15, 16, 62] is a fully microscopic alternative to the TDHF method that includes pairing, treats all degrees of freedom on an equal footing, and respects all the symmetries of the nucleus. This approach was recently applied to the study of fission dynamics in ^{240}Pu [16].

As mentioned above, one of the main limitations of the TDHF approach (and of the static Hartree-Fock approximation) is its restriction to a single Slater determinant wave function. There are generally two options to overcome this limitation, and both amount to constructing the wave function that describes the parent nucleus as a linear superposition of microscopic states (instead of a single microscopic state, such as a Slater determinant). The two approaches differ by whether the linear superposition is taken between states labeled by continuous parameters, as in the case of the generator coordinate method (GCM) [27, 29, 33, 70], or between microscopic states from a discrete basis, as in the case of the configuration interaction method [7, 18]. The GCM builds the nuclear wave function out of HFB solutions (see p. 141) constrained by various parameters to generate the PES, such as the quadrupole and octupole moments discussed in Sect. 6.5.3. In practice, because the constraint parameters can take on a continuous set of values, the linear superposition of the HFB solutions is given by an integral over the constraint parameters, which are referred to as the generator coordinates in this context. The wave function of the nucleus Ψ is then expressed as

$$\Psi = \int dq \, f(q) \, \Phi(q) \tag{7.35}$$

where q stands for one (or more) generator coordinates, $\Phi(q)$ is the HFB wave function solution for the value(s) q of its constraint parameter(s), and $f(q)$ is a weight function to be determined. The unknown weight function $f(q)$ is determined by a variational principle applied to the total energy of the system [27, 57]. This variational procedure leads to the so-called Hill-Wheeler (HW) equation [33], which is integro-differential, non-local in the generator coordinates, and nonlinear. In practice, the full HW equation is very difficult to solve for more than one generator coordinate [13]. A standard simplification, known as the Gaussian overlap approximation (GOA), reduces the HW equation to a Schrödinger-like equation with respect to the generator coordinates. All the components of this Schrödinger-like equation (i.e., the collective mass and the potential) are calculated from the underlying microscopic HFB wave functions. It is in this sense that the GCM builds collective motion, such as fission, from microscopic degrees of freedom. The GCM wave function in Eq. (7.35) is readily extended to a time-dependent form by replacing the weight function $f(q)$ by a time-dependent one, $f(q, t)$, and applying a variational procedure as in the static case. The variational procedure then leads to a time-dependent Schrödinger-like equation that can be solved by standard techniques to predict many fission-fragment properties, including mass distributions, TKE, TXE, etc. [5, 26, 56, 70]. We have limited our discussion to generator coordinates of a collective nature, such as the quadrupole and octupole moments of the nucleus, but it is possible to extend the GCM to include single-particle degrees of freedom [50]. Recent work has even started to allow GCM calculations without the simplifications imposed by the GOA [6]. The standing challenge for these GCM approaches today is to include all relevant collective and single-particle degrees of freedom in a computationally tractable way [6, 67].

7.5 Questions

70. What is the probability current for a wave function $\psi(x) = \psi^*(x)$?

71. Why does the Hauser-Feshbach cross-section formula require a width fluctuation correction factor while the Breit-Wigner formula does not?

72. Does a higher level density in the exit channel imply a higher or lower cross section?

73. What is the effective fission transmission coefficient for two weakly coupled barriers when the fission transmission coefficient for one barrier is much larger than the other?

74. How does quasifission differ from the usual fusion-fission process discussed throughout this book?

75. Name the principle used in the surrogate method to decompose the fission probability into products of formation and decay contributions.

76. What is one advantage and one limitation of the standard TDHF approach to fission dynamics?

7.6 Exercises

Problem 106 A ^{235}U sample with areal density of $0.5\,\mathrm{g/cm^2}$ is bombarded by a beam of 10^5 neutrons/s over its entire surface area. A detector subtending 10^{-1} Sr sees a rate of 30 fragments/hr. Assume 100% intrinsic efficiency of the detector, no deadtime, that each detection corresponds to one fission event, and that the fragments are emitted isotropically. What is the measured fission cross section? (Molecular weight of ^{235}U $= 235.04393$ g/mol).

Problem 107 Lorentzian1. Consider a nuclear state described by the wave function

$$\Psi(t) = \Psi(0)\, e^{-i E_r t/\hbar} e^{-t/(2\tau)} \tag{7.36}$$

Show that the probability of finding the nucleus in this state follows an exponential decay law.

Problem 108 Lorentzian2. Find the energy dependence of the wave function in Exercise 107 by taking its Fourier transform. As a reminder, the Fourier transform of a function $f(t)$ is given by

$$F(\omega) = \frac{1}{\sqrt{2\pi}} \int_{-\infty}^{\infty} dt\, e^{i\omega t} f(t) \tag{7.37}$$

where the energy is $E = \hbar\omega$. What is the probability of finding the nucleus in this state at a given energy E? (Hint: note that $\Psi(t) = 0$ for $t < 0$).

Problem 109 Width1. This next set of exercises is adapted from a traditional model of decay as repeated "attacks" of a trapped system on a barrier (see, e.g.,

[10, 46]). Show that if the wave functions $\psi_n(x)$ with energies E_n are solutions of the TISE, then

$$\psi(x, t) = \sum_n a_n \psi_n(x) e^{-iE_n t/\hbar} \tag{7.38}$$

where the a_n are coefficients, is a solution of the corresponding TDSE.

Problem 110 Width2. Consider a set of equidistant states with energies

$$E_n = \bar{E} + nD \tag{7.39}$$

where $\bar{E}$ is a reference energy, D is the energy spacing, and $n = 0, 1, 2, \ldots$. Using the wave function $\psi(x, t)$ in Exercise 109, show that the corresponding probability repeats at regular intervals and calculate this time interval Δt.

Problem 111 Width3. Consider a nucleus trapped inside a fission barrier and "attacking" this barrier at regular intervals.

1. Using the result from Exercise 110, write down the average rate R_b of attacks on the barrier.
2. On any given attack, the probability of crossing the barrier is given by a transmission coefficient T. Express the lifetime τ_f of the fission process in terms of R_b and T.
3. Rewrite your result for τ_f in terms of the fission width Γ_f, and compare with Eq. (7.15).

Problem 112 HauserFeshbach1. The purpose of this set of exercises is to obtain the Hauser-Feshbach formula, Eq. (7.23). Consider the reaction

$$a + A \rightarrow C \tag{7.40}$$

for a projectile a incident on a target A forming a compound nucleus C. Take the beam direction as the z axis. How many different orientations are there for the total angular momentum J_C of the compound state (determined by its projection quantum number m_C on the z axis)?

Problem 113 HauserFeshbach2. Referring to the reaction in Exercise 112, the total angular momenta of the projectile and target are labeled j_a and J_A, respectively, and the orbital angular momentum transferred in the reaction is ℓ.

1. If all orientations of the individual vectors are allowed, according to quantum mechanics, how many ways are there of forming the vector sum

$$\mathbf{J} = \mathbf{j}_a + \mathbf{J}_A + \boldsymbol{\ell} \tag{7.41}$$

2. Using the results from Exercise 112 and the previous question in this exercise, what is the probability that $J = J_C$, if all orientations are random?

Problem 114 HauserFeshbach3. The result from Exercise 113 is called the statistical spin factor $g(J_C)$ [58]. This factor gives the probability that all angular momenta in the entrance channel are properly oriented to form a compound state with total angular momentum J_C. Multiply Eq. (7.21) by $g(J_C)$ to obtain $\sigma_{\alpha\beta}(E, J_C)$.

Problem 115 HauserFeshbach4. In the Hauser-Feshbach formalism, we are interested in energy averages of the cross section. The energy average over some interval ΔE of the Lorentzian form in Eq. (7.21) and in the result to Exercise 114 is given by

$$\frac{1}{\Delta E} \int_{E-\Delta E/2}^{E+\Delta E/2} dE' \frac{\Gamma_\alpha \Gamma_\beta}{(E' - E_r)^2 + \left(\frac{\Gamma}{2}\right)^2} \tag{7.42}$$

Find an approximate formula for this energy average by taking the limits of the integral to infinity, i.e., by letting

$$\int_{E-\Delta E/2}^{E+\Delta E/2} \rightarrow \int_{-\infty}^{\infty} \tag{7.43}$$

Use this to write the energy average $\langle \sigma_{\alpha\beta}(E, J_C) \rangle$ of $\sigma_{\alpha\beta}(E, J_C)$ from Exercise 114.

Problem 116 HauserFeshbach5. The result from Exercise 115 can be generalized to the case of many nearby resonances, all having total angular momentum J_C by letting

$$\Gamma_\alpha \rightarrow \Gamma_\alpha^{(n)}$$

$$\Gamma_\beta \rightarrow \Gamma_\beta^{(n)} \tag{7.44}$$

$$\Gamma \rightarrow \Gamma^{(n)} = \Gamma_\alpha^{(n)} + \Gamma_\beta^{(n)}$$

and summing over the n. Write down this more general form of $\langle \sigma_{\alpha\beta}(E, J_C) \rangle$.

Problem 117 HauserFeshbach6. In the expression for $\langle \sigma_{\alpha\beta}(E, J_C) \rangle$ in Exercise 116, there will be a term of the form

$$\frac{1}{\Delta E} \sum_n \frac{A_n B_n}{C_n} \tag{7.45}$$

By replacing A_n, B_n, and C_n with continuous functions of energy $A(E)$, $B(E)$, and $C(E)$, the sum can be approximated by an integral weighted by the level density

$\rho(E) = 1/D$, where D is the average level spacing,

$$\frac{1}{\Delta E} \sum_n \frac{A_n B_n}{C_n} \approx \frac{1}{\Delta E} \int_{E-\Delta E/2}^{E+\Delta E/2} dE' \frac{1}{D} \frac{A(E') B(E')}{C(E')}$$

$$= \frac{1}{D} \left\langle \frac{A(E) B(E)}{C(E)} \right\rangle \tag{7.46}$$

It is then convenient to introduce the width fluctuation factor W defined by

$$\left\langle \frac{A(E) B(E)}{C(E)} \right\rangle \equiv W \frac{\langle A(E) \rangle \langle A(E) \rangle}{\langle A(E) \rangle} \tag{7.47}$$

Use this factor W, along with Eq. (7.15), to rewrite $\langle \sigma_{\alpha\beta}(E, J_C) \rangle$ and compare with Eq. (7.23).

Problem 118 Consider the transition states shown in Table 7.3, over a barrier of height $E_b = 6$ MeV and curvature $\hbar\omega = 1$ MeV. Plot the Hill-Wheeler transmission coefficient for $J^\pi = 2^+$ as a function of excitation energy from 5 to 7.5 MeV.

At what excitation energy does the Hill-Wheeler transmission coefficient reach 1/2 of its maximum value?

Problem 119 Below the barriers, the fission transmission coefficient is affected by class II states (see Fig. 7.1). Assuming complete absorption into the 2nd well and equidistant class II states, Lynn and Back [45] derived the following relation for the energy-averaged fission probability:

$$\bar{P}_f = \left[1 + \left(\frac{T'}{\bar{T}_f} \right)^2 + 2 \left(\frac{T'}{\bar{T}_f} \right) \coth \left(\frac{T_A + T_B}{2} \right) \right]^{-1/2} \tag{7.48}$$

where T' is the transmission coefficient for all channels other than fission (e.g., neutrons and gammas) and $\bar{T}_f$ is given by Eq. (7.28). What is $\bar{P}_f$ in the limit of large T_A and T_B (i.e., near the top of the barrier and above)?

Table 7.3 Example of a selected subset of transition states over the outer fission barrier. The total angular momentum and parity (J^π) are given in the first column, and the corresponding excitation energies (E_x) measured from the top of the barrier are listed in the second column

J^π	E_x (MeV)
0^+	0.0, 0.5, 0.6, 0.9, 1.2
1^-	0.01, 0.25, 0.55, 0.76
2^+	0.03, 0.27, 0.3, 0.45, 0.5, 0.53, 0.63, 0.72, 0.9, 0.93, 1.23
4^+	0.1, 0.34, 0.37, 0.52, 0.57, 0.6, 0.7, 0.74, 0.79, 0.9, 0.97, 1.0, 1.3

References

1. B.B. Back, Phys. Rev. C **31**, 2104 (1985)
2. B.B. Back, et al., Phys. Rev. C **53**, 1734 (1996)
3. B. Back, Rev. Mod. Phys. **86**, 317 (2014)
4. X. Bao, H. Zhang, G. Royer, J. Li, Nucl. Phys. A **906**, 1 (2013)
5. J. Berger, M. Girod, D. Gogny, Comput. Phys. Commun. **63**, 365 (1991)
6. R. Bernard, H. Goutte, D. Gogny, W. Younes, Phys. Rev. C **84**, 044308 (2011)
7. G.F. Bertsch, W. Younes, L.M. Robledo, Phys. Rev. C **100**, 024607 (2019)
8. E.R. Bittner, *Quantum Mechanics: Lecture Notes for Chemistry 6312 Quantum Chemistry*. University of Houston, Department of Chemistry (2003)
9. S. Bjørnholm, J.E. Lynn, Rev. Mod. Phys. **52**, 725 (1980)
10. J.M. Blatt, V.F. Weisskopf, *Theoretical Nuclear Physics* (Dover Publications, Mineola, 1991)
11. A. Bohr, *Proceedings of the International Conference on the Peaceful Uses of Atomic Energy*, vol. 2 (United Nations, New York, 1956), pp. 151–154
12. N. Bohr, J.A. Wheeler, Phys. Rev. **56**, 426 (1939)
13. P. Bonche, J. Dobaczewski, H. Flocard, P.-H. Heenen, J. Meyer, Nucl. Phys. A **510**, 466 (1990)
14. D.M. Brink, *Lecture Notes in Physics*, vol. 581 (Springer, Berlin, 2001)
15. A. Bulgac, Ann. Rev. Nuclear Particle Sci. **63**, 97 (2013)
16. A. Bulgac, P. Magierski, K.J. Roche, I. Stetcu, Phys. Rev. Lett. **116**, 122504 (2016)
17. M. Büttiker, Phys. Rev. B **41**, 7906 (1990)
18. E. Caurier, G. Martínez-Pinedo, F. Nowacki, A. Poves, A.P. Zuker, Rev. Modern Phys. **77**, 427 (2005)
19. F.S. Dietrich, Technical Report UCRL-TR-201718 (LLNL, 2004)
20. D.O. Eremenko, V.A. Drozdov, O.V. Fotina, S.Y. Platonov, O.A. Yuminov, Phys. Rev. C **94**, 014602 (2016)
21. J.E. Escher, J.T. Burke, F.S. Dietrich, N.D. Scielzo, I.J. Thompson, W. Younes, Rev. Mod. Phys. **84**, 353 (2012)
22. H. Eyring, J. Chem. Phys. **3**, 107 (1935)
23. P. Fröbrich, I. Gontchar, Phys. Rep. **292**, 131 (1998)
24. N. Fröman, Ö. Dammert, Nucl. Phys. **A147**, 627 (1970)
25. R. Gilmore, *Elementary Quantum Mechanics in One Dimension* (Johns Hopkins University Press, Baltimore, 2004)
26. H. Goutte, J.F. Berger, P. Casoli, D. Gogny, Phys. Rev. C **71**, 024316 (2005)
27. W. Greiner, J.A. Maruhn, *Nuclear Models* (Springer, Berlin, 1996)
28. J.J. Griffin, *Proceedings of the Symposium on Physics and Chemistry of Fission*, vol. I (1965)
29. J.J. Griffin, J.A. Wheeler, Phys. Rev. **108**, 311 (1957)
30. W. Hauser, H. Feshbach, Phys. Rev. **87**, 366 (1952)
31. D.M. Heim, W.P. Schleich, P.M. Alsing, J.P. Dahl, S. Varro, Phys. Lett. A **377**, 1822 (2013)
32. B. Heusch, C. Volant, H. Freiesleben, R.P. Chestnut, K.D. Hildenbrand, F. Puehlhofer, W.F.W. Schneider, B. Kohlmeyer, W. Pfeffer, Zeit. Phys. A **288**, 391 (1978)
33. D.L. Hill, J.A. Wheeler, Phys. Rev. C **89**, 1102 (1953)
34. D.J. Hinde, M. Dasgupta, A. Mukherjee, Phys. Rev. Lett. **89**, 282701 (2002)
35. D. Hinde, M. Dasgupta, D. Jeung, G. Mohanto, E. Prasad, C. Simenel, E. Williams, I. Carter, K. Cook, S. Kalkal, D. Rafferty, E. Simpson, H. David, C. Düllmann, J. Khuyagbaatar, in *EPJ Web of Conferences*, ed. by E. Simpson, C. Simenel, K. Cook, I. Carter, **163**, 00023 (2017)
36. M.G. Itkis, et al., Nucl. Phys. **A787**, 150 (2007)
37. Y.M. Itkis, A. Karpov, G.N. Knyazheva, E.M. Kozulin, N.I. Kozulina, K.V. Novikov, K.B. Gikal, I.N. Diatlov, I.V. Pchelintsev, I.V. Vorobiov, A.N. Pan, P.P. Singh, Bull. Russ. Acad. Sci. Phys. **84**, 938 (2020)
38. E.C. Kemble, Phys. Rev. **48**, 549 (1935)
39. G.F. Knoll, *Radiation Detection and Measurement* (Wiley, Hoboken, 2010)

40. E.M. Kozulin, G.N. Knyazheva, T.K. Ghosh, A. Sen, I.M. Itkis, M.G. Itkis, K.V. Novikov, I.N. Diatlov, I.V. Pchelintsev, C. Bhattacharya, S. Bhattacharya, K. Banerjee, E.O. Saveleva, I.V. Vorobiev, Phys. Rev. C **99**, 014616 (2019)
41. R.H. Landau, *Quantum Mechanics II: A Second Course in Quantum Theory* (Wiley, Hoboken, 1990)
42. W.R. Leo, *Techniques for Nuclear and Particle Physics Experiments: A How-to Approach* (Springer, Berlin, 1992)
43. W. Loveland, J. Phys. Conf. Ser. **420**, 012004 (2013)
44. J.E. Lynn, *Fission Cross Section Theory*. Lecture at the FIESTA School, LANL, LANL (2014)
45. J.E. Lynn, B.B. Back, J. Phys. A **7**, 395 (1974)
46. P. Marmier, *Physics of Nuclei and Particles* (Academic, New York, 1969)
47. K. Mazurek, C. Schmitt, P.N. Nadtochy, A.V. Cheredov, Phys. Rev. C **94**, 064602 (2016)
48. P. Moldauer, Phys. Rev. **123**, 968 (1961)
49. P. Möller, A.J. Sierk, A. Iwamoto, Phys. Rev. Lett. **92**, 072501 (2004)
50. H. Müther, K. Goeke, K. Allaart, A. Faessler, Phys. Rev. C **15**, 1467 (1977)
51. J.W. Negele, S.E. Koonin, P. Möller, J.R. Nix, A.J. Sierk, Phys. Rev. C **17**, 1098 (1978)
52. Y.T. Oganessian, et al., Phys. Rev. C **69**, 054607 (2004)
53. Y.T. Oganessian, et al., Phys. Rev. C **74**, 044602 (2006)
54. Y.T. Oganessian, A. Sobiczewski, G.M. Ter-Akopian, Phys. Scr. **92**, 023003 (2017)
55. J. Randrup, P. Möller, A.J. Sierk, Phys. Rev. C **84**, 034613 (2011)
56. D. Regnier, N. Dubray, N. Schunck, M. Verrière, Phys. Rev. C **93**, 054611 (2016)
57. P. Ring, P. Schuck, *The Nuclear Many-Body Problem* (Springer, Berlin, 1980)
58. G.R. Satchler, *Introduction to Nuclear Reactions* (Macmillan, Basingstoke, 1990)
59. G. Scamps, C. Simenel, D. Lacroix, Phys. Rev. C **92**, 011602 (2015)
60. A.J. Sierk, Phys. Rev. C **96**, 034603 (2017)
61. C. Simenel, A.S. Umar, Phys. Rev. C **89**, 031601 (2014)
62. I. Stetcu, A. Bulgac, P. Magierski, K.J. Roche, Phys. Rev. C **84**, 051309 (2011)
63. W.J. Swiatecki, Phys. Scr. **24**, 113 (1981)
64. B.M.A. Swinton-Bland, M.A. Stoyer, A.C. Berriman, D.J. Hinde, C. Simenel, J. Buete, T. Tanaka, K. Banerjee, L.T. Bezzina, I.P. Carter, K.J. Cook, M. Dasgupta, D.Y. Jeung, C. Sengupta, E.C. Simpson, K. Vo-Phuoc, Phys. Rev. C **102**, 054611 (2020)
65. J. Töke, R. Bock, G. Dai, A. Gobbi, S. Gralla, K. Hildenbrand, J. Kuzminski, W. Müller, A. Olmi, H. Stelzer, S.B.B.B. Back, Nucl. Phys. **A440**, 327 (1985)
66. E. Vardaci, et al., Phys. Rev. C **92**, 034610 (2015)
67. M. Verrière, N. Dubray, N. Schunck, D. Regnier, P. Dossantos-Uzarralde, *EPJ Web Conferences*, **146**, 04034 (2017)
68. D.E. Ward, B.G. Carlsson, T. Døssing, P. Möller, J. Randrup, S. Åberg, Phys. Rev. C **95**, 024618 (2017)
69. E. Wigner, Trans. Faraday Soc. **34**, 29 (1938)
70. W. Younes, *A Microscopic Theory of Fission Dynamics Based on the Generator Coordinate Method* (Springer, Cham, 2019)
71. W. Younes, H.C. Britt, Phys. Rev. C **67**, 024610 (2003)
72. P.G. Young, E.D. Arthur, M.B. Chadwick, Technical Report LA-12343-MS (LANL, 1992)
73. V. Zagrebaev, W. Greiner, J. Phys. G Nuclear Particle Phys. **31**, 825 (2005)

Chapter 8
Solutions to Selected Problems

8.1 Solutions for Chap. 1

1 $Q_\alpha = 4.130\,\text{MeV}$ and $Q_f = 192.261\,\text{MeV}$ using QCalc, and therefore $Q_f \gg Q_\alpha$.

2 $E_C = 237.9\,\text{MeV}$, high by $\approx 23.7\%$.

3 $E_x = 4.8\,\text{MeV}$, i.e., below the critical energy.

4 Looks roughly parabolic, with $B_A \approx 1.61\,\text{MeV/c}^2$.

5 $\tau_{\text{SF}} \approx 3.59 \times 10^{16}\,\text{y}$.

6 $d \approx 27.3\,\text{keV}$.

7

$$\frac{v}{c} = \frac{pc}{\sqrt{(pc)^2 + (mc^2)^2}} \approx 0.9987 \tag{8.1}$$

8 $\rho \approx 0.14\,\text{nucleons/fm}^3$.

9 Highest $Z = 92$, so we do not reach new elements this way.

10 $d_\alpha \approx 0.00196\,\text{cm}$ and $d_{\text{Ba}} \approx 0.00110\,\text{cm}$; therefore, an absorber thick enough to stop α particles will also likely stop the Ba fragment.

8.2   Solutions for Chap. 2

11 $R_{SF} \approx 1.29$ events/s.

12 $T_{1/2} = 28.7$ s, which is off by a factor of $\approx 77,530$.

13 ^{135}I, ^{131}Xe, ^{134}Xe, ^{136}Xe, ^{133}Cs.

14 Largest deviation is for ^{143}Nd produced in the fission of ^{239}U. The deviation in that case is $\approx 9\%$.

15 $B_{rel}\left(^{209}\text{Pb}\right)/B_{rel}\left(^{240}\text{Pu}\right) \approx 2.74$.

16 Q (symm) ≈ 193.2 MeV and $Q\left(^{134}\text{Te}/^{102}\text{Zr}\right) \approx 196.6$ MeV.

17 Four neutrons, with the last neutron emitted leaving ^{232}U with an excitation energy of 1.67 MeV.

18 ^{91}Kr/^{142}Ba $+ 3n$, ^{100}Nb/^{133}Sb $+ 3n$, ^{95}Sr/^{139}Xe $+ 2n$, ^{99}Zr/^{135}Te $+ 2n$.

19 $R_{SF} \approx 176.58 \, \text{y}^{-1}$.

20 Denoting by $f_{238}(t)$ the fraction of ^{238}U at time t, we can show that

$$f_{238}(0) = \frac{\exp\left(\frac{t}{\tau_{238}} - \frac{t}{\tau_{235}}\right) f_{238}(t)}{1 + \left[\exp\left(\frac{t}{\tau_{238}} - \frac{t}{\tau_{235}}\right) - 1\right] f_{238}(t)} \tag{8.2}$$

where τ_{235} and τ_{238} are the lifetimes of ^{235}U and ^{238}U, respectively, and therefore $f_{238}(0) \approx 76.6\%$ and $f_{235}(0) \approx 23.4\%$.

21 $P \approx 18.3$ W.

22 $Q\left(^{237}\text{Np} \rightarrow \, ^{79}\text{Ga} + \, ^{79}\text{Ga} + \, ^{79}\text{Ga}\right) \approx 232.4$ MeV.

23 Typical household could be powered for ≈ 2.3 y.

24 For the β^- decay chain from ^{102}Zr to ^{102}Ru, we have $Q = 4.7170 + 7.2620 + 1.0070 + 4.5340 = 17.52$ MeV, and for the chain from ^{134}Te to ^{134}Xe, we have $Q = 1.5100 + 4.0820 = 5.592$ MeV.

25

$$A_L = A_L^{(0)} + \left(A - A_L^{(0)} - A_H^{(0)}\right)\frac{x}{L}$$
$$A_H = A_H^{(0)} + \left(A - A_L^{(0)} - A_H^{(0)}\right)\frac{L - x}{L} \tag{8.3}$$

Table 8.1 Total kinetic energies (TKE) calculated using the Viola empirical relation

Nucleus	TKE (MeV)
^{235}U	170.380
^{238}U	169.692
^{239}Pu	176.593
^{252}Cf	188.087

The resulting mass distribution for $A = 236$ has the form

$$Y(A) \propto \begin{cases} 1 & 78 \leq A \leq 104 \text{ or } 132 \leq A \leq 158 \\ 0 & \text{otherwise} \end{cases} \tag{8.4}$$

26 A_H remains fairly constant as A_L varies, as expected.

27 The calculated TKE are summarized in Table 8.1:

28

$$E_C = \frac{(eZ)^2}{4 \times 2^{2/3} r_0 A^{1/3}} \tag{8.5}$$

which explains the $Z^2/A^{1/3}$ dependence in the Viola formula.

8.3 Solutions for Chap. 3

29 Since volume is conserved, the volume at scission is the same as when the nucleus is in a spherical shape, $V = \frac{4}{3}\pi r_0^3 A \approx 1737.2 \, \text{fm}^3$.

30 $E_S = 633.0 \, \text{MeV}$.

31 $E_C = 266.1 \, \text{MeV}$.

32 $E_{\text{def}} = -627.8 \, \text{MeV}$.

33

1. $a = R_0 \left(1 - \varepsilon^2\right)^{1/6}$ and $c = R_0 \left(1 - \varepsilon^2\right)^{-1/3}$.
2. $\alpha = \frac{8}{5}\pi\sigma (1 - x)$
3. $\alpha \left(^{102}\text{Zr}\right) \approx 3.55 \, \text{MeV/fm}^2$ and $\alpha \left(^{134}\text{Te}\right) \approx 3.09 \, \text{MeV/fm}^2$.

34 The energies E_n for a number of nucleons n are given by

$$E_n = \begin{cases} 0.025 - 0.05n + 0.025n^2 & n = 1, 3, 5, \ldots, 19 \\ -0.05n + 0.025n^2 & n = 2, 4, 6, \ldots, 20 \end{cases} \tag{8.6}$$

35 The energies E_n for a number of nucleons n are given approximately by

$$E_n = -0.0842889 + 0.100008n - 0.0272505n^2$$
$$+ 0.00447501n^3 - 0.000111875n^4 \tag{8.7}$$

for $n = 1, 3, 5, \ldots, 19$, and by

$$E_n = -0.153594 + 0.122386n - 0.0302714n^2$$
$$+ 0.00458689n^3 - 0.000111875n^4 \tag{8.8}$$

for $n = 2, 4, 6, \ldots, 20$.

36 The top of the barrier is located at $x = 0$, where $V(x, y)$ reaches a maximum for a fixed y. For fixed x, $V(x, y)$ is minimized when $y = 0$ and the fission path is therefore along the x axis (i.e., with $y = 0$ and x free to vary).

37 The most probable division corresponds to $A_H = 140$.

38

- $E_f < B_n - 0.5\,\text{MeV} \Rightarrow$ relatively low fission barrier $\Rightarrow$ neutron emission is counterbalanced by fission.
- $E_f > B_n - 0.5\,\text{MeV} \Rightarrow$ relatively high fission barrier $\Rightarrow$ neutron emission wins out initially, but eventually fission starts to pick up.

39 The LDM calculation gives $S_n^{(\text{LDM})} \approx 16.217\,\text{MeV}$ which is larger than the experimental value $S_n \approx 6.545\,\text{MeV}$ by a factor of ≈ 2.5.

40 We find a difference in binding energies between the cube and sphere of $\text{BE}_{\text{cube}} - \text{BE}_{\text{sphere}} = -1889.23 - (-2760.9) = 871.67\,\text{MeV}$. Therefore the cubic shape is highly energetically unfavorable compared to the spherical one.

41 The neck is thinnest at $z = 9.55\,\text{fm}$.

42 $E_L^{\text{lab}} = M_H Q / (M_L + M_H)$, $E_H^{\text{lab}} = M_L Q / (M_L + M_H)$, and $\theta_L^{\text{lab}} = \theta_H^{\text{lab}}$.

43 $\Gamma = 0.66\,\text{eV}$.

44 $\sigma = 681\,\text{b} = 68100\,\text{fm}^2$, while $A = \pi R^2 = 172.3\,\text{fm}^2$.

45 The energy released by fission is $Q\left(^{239}\text{U} \rightarrow\, ^{119}\text{Pd} +^{120}\text{Pd}\right) = 179.7\,\text{MeV}$, while the energy that must be supplied to completely break up the nucleus is $92m_p c^2 + 147m_n c^2 - M\left(^{239}\text{U}\right) c^2 = 2161.06\,\text{MeV}$.

47 $E_{\text{thr}} = 5.320\,\text{MeV}$.

48 $E_n = 6 - 4.80638 = 1.194\,\text{MeV}$.

49 $v_{\text{max}}^{\text{lab}} \approx 0.0269\,\text{cm/ns}$.

50 There are two possible solutions for the Q value:

- $Q = -8.20643 \, \text{MeV} \Rightarrow \text{TXE} = 2.798 \, \text{MeV} - Q = 11.004 \, \text{MeV}.$
- $Q = -8.49803 \, \text{MeV} \Rightarrow \text{TXE} = 2.798 \, \text{MeV} - Q = 11.296 \, \text{MeV}.$

51 $v \approx 0.013 \, \text{cm/ns}.$

52

1. There are two possible solutions:

   (a) Using the "+" sign in Eq. (3.27):$E_L^{\text{lab}} = 125.122 \, \text{MeV}$ and $E_H^{\text{lab}} = 92.170 \, \text{MeV}.$

   (b) Using the "−" sign in Eq. (3.27): $E_L^{\text{lab}} = 121.549 \, \text{MeV}$ and $E_H^{\text{lab}} = 95.743 \, \text{MeV}.$

2. The two possibilities are:

   (a) For the "+" case: $\theta_H^{\text{lab}} = 61.7°.$

   (b) For the "−" case: $\theta_H^{\text{lab}} = 58.4°.$

8.4   Solutions for Chap. 4

53 $K_1 = M_2 \times \text{TKE}/(M_1 + M_2)$ and $K_2 = M_1 \times \text{TKE}/(M_1 + M_2).$

54

$$
\frac{dN}{dt} \equiv \frac{dN_A}{dt} + \frac{dN_B}{dt}
$$

$$
= -\frac{\lambda_A \lambda_B}{\lambda_B - \lambda_A} N_A(0) \left[e^{-\lambda_A t} - e^{-\lambda_B t} \right]
$$

$$(8.9)$$

55 Setting $d^2 N/dt^2 = 0$ gives

$$
t_m = \frac{\ln \lambda_A - \ln \lambda_B}{\lambda_A - \lambda_B}
$$

$$(8.10)$$

which corresponds to the maximum values of $|dN/dt|.$

57

$$
N_2(T, \Delta t) = P_1 \lambda_1 \left[\frac{\left(1 - e^{-\lambda_1 T}\right) e^{-\lambda_1 \Delta t}}{\lambda_1 (\lambda_2 - \lambda_1)} + \frac{\left(1 - e^{-\lambda_2 T}\right) e^{-\lambda_2 \Delta t}}{\lambda_2 (\lambda_1 - \lambda_2)} \right]
$$

$$
+ \frac{P_2}{\lambda_2} \left(1 - e^{-\lambda_2 T}\right) e^{-\lambda_2 \Delta t}
$$

$$(8.11)$$

58

$$W_{0,J}^J (\theta) = \frac{(2J)! \, (2J + 1)}{(J!)^2 \, 2^{2J+1}} \sin^{2J} \theta \tag{8.12}$$

59 Use $\bar{A} = \left(A_f - \bar{v}\right) / 2$ in the five-Gaussian formula.

60 The algorithm converges to $\mu_1 = 160.264$, $\mu_2 = 79.736$, $E_1^* = 51.743 \, \text{MeV}$, $E_2^* = 104.00 \, \text{MeV}$.

61 $Z_p = 54.290$ from Gaussian fit.

62

- Using $F (A, Z) = 1$: $Z_p = 54.282$, $\sigma_Z = 0.449$.
- Using $F (A = 140, Z) = 1 + \varepsilon \cos (\pi Z)$: $Z_p = 54.524$, $\sigma_Z = 0.620$, $\varepsilon = 0.378$, which gives a better fit than a simple Gaussian.

63 $\delta = 0.237$.

64 Going from thermal to $6 \, \text{MeV}$, the symmetric region fills in, and the tails of the light and heavy mass peaks grow a bit.

65 $Y_{\max} / Y \left(\bar{A}\right) = 5.71$.

66 $d = 16.5 \, \text{fm}$.

67 $d = 20.8 \, \text{fm}$.

68

1. The total energy is

$$E_{\text{tot}} = \alpha_1 (c_1 - R_0)^2 + \alpha_2 (c_2 - R_0)^2 + \frac{e^2 Z_1 Z_2}{c_1 + c_2} \tag{8.13}$$

2. The excitation energies are

$$E_x^{(1)} = \frac{\text{TKE}^4}{4\alpha_1 e^4 Z_1^2 Z_2^2}$$

$$E_x^{(2)} = \frac{\text{TKE}^4}{4\alpha_2 e^4 Z_1^2 Z_2^2} \tag{8.14}$$

3. Using $E_{\text{tot}} = Q + E_n + S_n = 209.657 \, \text{MeV} = \text{TKE} + E_x^{(1)} + E_x^{(2)}$. and solving numerically, we find $\text{TKE} = 188.4 \, \text{MeV}$.

8.5 Solutions for Chap. 5

69 $\lambda = 7.45\,\text{fm} \Rightarrow E_n = 14.7\,\text{MeV}$.

70 $1.044\,\text{MeV} \le E_n \le 2.324\,\text{MeV}$.

71 $\int_1^\infty dv\, P(v) = 0.907$.

72 $v = 0.247\,\text{cm/ns}$.

73 $E_{\text{thr}} = 1.019\,\text{MeV}$.

74 $\bar{E} = 2T$.

75

$$N_n^{\text{lab}}(E) = \begin{cases} \dfrac{1}{4\sqrt{E_0 E_{F/N}}} & \left(\sqrt{E_0} - \sqrt{E_{F/N}}\right)^2 \le E \le \left(\sqrt{E_0} + \sqrt{E_{F/N}}\right)^2 \\ 0 & \text{otherwise} \end{cases}$$

(8.15)

76

$$\frac{v(A)}{L} = \begin{cases} \frac{A-78}{26} & 78 \le A \le 104 \\ \frac{A-132}{26} & 132 \le A \le 158 \\ 0 & \text{otherwise} \end{cases}$$

(8.16)

77 2710 decays.

78 Average, $\bar{E}_\gamma = 3T$; most probable, $E_\gamma^{\text{max}} = 2T$. If $\bar{E}_\gamma = 3\,\text{MeV}$, then $T = 1\,\text{MeV}$.

79

$$\bar{M}_\gamma = \frac{\sqrt{\pi}}{2} B \exp\left(\frac{1}{4B^2}\right) \text{erfc}\left(\frac{1}{2B}\right)$$

(8.17)

80 Centroid of $P(M_\gamma)$ shifts to higher values, and distribution broadens.

81 Assuming $\bar{E}_{\gamma,\text{tot}} = \overline{M_\gamma E_\gamma} = \bar{M}_\gamma \times \bar{E}_\gamma$, we find

$$\bar{E}_{\gamma,\text{tot}} = \frac{3}{2}\sqrt{\pi} BT \exp\left(\frac{1}{4B^2}\right) \text{erfc}\left(\frac{1}{2B}\right)$$

(8.18)

82 For approximately isotropic radiation, we need $\hbar^2 I/(\mathfrak{J}T) \ll 1$, i.e., low-spin, high moment of inertia and/or large temperature.

83 Each absorbed neutron produces on average 2.08 fission neutrons.

Table 8.2 Calculated relative populations of the ground-state members in ^{144}Ba

J_f^π	Population (arbitrary units)
0^+	620.477
2^+	2504.87
4^+	2905.7
6^+	2168.66

Table 8.3 Calculated gamma-ray intensities relative to the $2^+ \to 0^+$ transition in ^{144}Ba

Transition	E_γ (keV)	α	$I_\gamma = I_{\text{tot}}/(1+\alpha)$ (arb. units)
$6^+ \to 4^+$	432	0.01489	2136.84
$4^+ \to 2^+$	331	0.0332	4911.3
$2^+ \to 0^+$	199	0.1751	6449.86

84 The kinetic energy of the slower neutron is $E_n^{\text{slow}} = 0.349\,\text{MeV}$.

85 The average sample multiplicity is

$$\bar{\nu} = \sum_\nu \nu P(\nu) \tag{8.19}$$

and the unbiased estimate of the sample variance is given by

$$\sigma_\nu^2 = \frac{V_1}{V_1^2 - V_2} \sum_\nu P(\nu)(\nu - \bar{\nu})^2 \tag{8.20}$$

where $V_1 = \sum_\nu P(\nu) = 1$ and $V_2 = \sum_\nu P^2(\nu) \approx 0.2226$, from which we obtain $\sigma_\nu \approx 1.436$. The Gaussian distribution gives a fair approximation to the distribution.

86 Most probable, $E_\gamma^{\text{max}} = 0.3\,\text{MeV}$; average, $\bar{E}_\gamma = 0.898\,\text{MeV}$.

87 The relative populations are listed in Table 8.2:

88 The relative intensities are given in Table 8.3:

8.6   Solutions for Chap. 6

89 $E_C = 986.138\,\text{MeV}$, and $E_C/\text{BE} \approx 55\%$.

90 See Eq. (6.11).

92 Oscillatory behavior becomes more pronounced as $N \to \infty$; the first few terms give the smoothed behavior.

95 $E_{\text{GS}} = 2 \times 0 = 0\,\text{MeV}$, $E_x^{\text{max}} = 2 \times 1.5 - E_{\text{GS}} = 3\,\text{MeV}$.

96

$$S_0 = 2 \ln \left(1 + e^{\alpha}\right) + 2 \ln \left(1 + e^{\alpha - 0.5\beta}\right) + 2 \ln \left(1 + e^{\alpha - \beta}\right)$$
$$+ 2 \ln \left(1 + e^{\alpha - 1.5\beta}\right) - 2\alpha_0 + \beta_0 E$$

(8.21)

97 $E = 0.709 \, \text{MeV}$.

98 $D_0 = 0.3463 \, \text{MeV}^2$, $\rho(A, E) = 11.80 \, \text{MeV}^{-1}$

99

1. Organizing the excitations by number of particles and holes:

   (a) 1p-1h: $E_x = 2 - 1 = 1 \, \text{MeV}$, $E_x = 3 - 1 = 2 \, \text{MeV}$.
   (b) 2p-2h: $E_x = 4 - 2 = 2 \, \text{MeV}$, $E_x = 5 - 2 = 3 \, \text{MeV}$, $E_x = 6 - 2 = 4 \, \text{MeV}$.
   (c) 3p-3h: $E_x = 7 - 3 = 4 \, \text{MeV}$, $E_x = 8 - 3 = 5 \, \text{MeV}$.
   (d) 4p-4h: $E_x = 10 - 4 = 6 \, \text{MeV}$.

2. There are $3/2 = 1.5 \, \text{states/MeV}$ in the range $0.5 - -2.5 \, \text{MeV}$, $3/2 = 1.5 \, \text{states/MeV}$ in the range 2.5–$4.5 \, \text{MeV}$, and $2/2 = 1 \, \text{state/MeV}$ in the range 4.5–$6.5 \, \text{MeV}$.

100

$$Q_{20} = Q_{20}^{(L)} + Q_{20}^{(R)} + 2 \frac{A_R A_L}{A_R + A_L} d^2$$

(8.22)

101 $N = 20$.

102 $\bar{N} = 10.093$.

103

1. $G = 0.132 \, \text{MeV}$.
2. $E_{\text{GS}} = 15.049 \, \text{MeV}$.
3. $E_{\text{pc}} = 0.0486 \, \text{MeV}$.

104

1. The grand canonical partition function is

$$Z(\alpha, \beta) = \left(1 + e^{\alpha - \beta \varepsilon}\right)^{\Omega}$$

(8.23)

2. Setting $x \equiv e^{\alpha - \beta \varepsilon}$, we obtain the two equations

$$(1 + x) E - x\Omega E = 0$$
$$x\Omega - (1 + x) A = 0$$

(8.24)

from which we obtain $E = A\varepsilon$.

3. We find $D_0 = 0$. There is no mechanism to generate excited states in this model; thus the level density consists of an infinite "spike" at a single energy.

105 Remember that there are two particles occupying the state and that we must account for both. Therefore,

1. The spatial density is given by

$$\rho(x) = 2 |\psi(x)|^2 \tag{8.25}$$

2. The direct and exchange potentials are

$$\Gamma_D(x) = 2V_0 |\psi(x)|^2 \tag{8.26}$$

$$\Gamma_E(x, x') = -2V_0 \psi^*(x') \delta(x - x') \psi(x) \tag{8.27}$$

8.7 Solutions for Chap. 7

106 $\sigma = 0.8175$ mb.

107 Prob $(t) = |\Psi(0)|^2 e^{-t/\tau}$.

108

$$P(E) = \left| \frac{\Psi(0)}{\sqrt{2\pi}} \right|^2 \frac{1}{(\hbar\omega - E_r)^2 + \left(\frac{\hbar}{2\tau}\right)^2} \tag{8.28}$$

110 $\Delta t = 2\pi\hbar/D$.

111

1. $R_b \approx \frac{D}{2\pi\hbar}$
2. $\frac{1}{\tau_f} = \frac{TD}{2\pi\hbar}$
3. $T = 2\pi \frac{\Gamma}{D}$

112 Number of orientations $= 2J_C + 1$.

113

1. Orientations $= (2j_a + 1)(2J_A + 1)(2\ell + 1)$.
2. $g(J_C) = \frac{2J_C + 1}{(2j_a + 1)(2J_A + 1)(2\ell + 1)}$

114

$$\sigma_{\alpha\beta}(E, J_C) = \frac{\pi}{k^2} \frac{2J_C + 1}{(2j_a + 1)(2J_A + 1)} \frac{\Gamma_\alpha \Gamma_\beta}{(\hbar\omega - E_r)^2 + \left(\frac{\Gamma}{2}\right)^2} \tag{8.29}$$

115

$$\langle \sigma_{\alpha\beta} (E, J_C) \rangle = \frac{\pi}{k^2} \frac{2J_C + 1}{(2j_a + 1)(2J_A + 1)} \frac{2\pi}{\Delta E} \frac{\Gamma_\alpha \Gamma_\beta}{\Gamma} \tag{8.30}$$

116

$$\langle \sigma_{\alpha\beta} (E, J_C) \rangle = \frac{\pi}{k^2} \frac{2J_C + 1}{(2j_a + 1)(2J_A + 1)} \sum_n \frac{2\pi}{\Delta E} \frac{\Gamma_\alpha^{(n)} \Gamma_\beta^{(n)}}{\Gamma^{(n)}} \tag{8.31}$$

117

$$\langle \sigma_{\alpha\beta} (E, J_C) \rangle = \frac{\pi}{k^2} \frac{2J_C + 1}{(2j_a + 1)(2J_A + 1)} W \frac{T_\alpha (E) T_\beta (E)}{T (E)} \tag{8.32}$$

118 Hill-Wheeler transmission coefficient reaches its maximum value of 11 at $E_x = 6.58$ MeV.

119

$$\bar{P}_f = \frac{\bar{T}_f}{\bar{T}_f + T'} \tag{8.33}$$

Index

Printed in the United States
by Baker & Taylor Publisher Services